Objectivity

Objectivity

Lorraine Daston & Peter Galison

ZONE BOOKS · NEW YORK

2010

© 2007 Lorraine Daston and Peter Galison
ZONE BOOKS
1226 Prospect Ave.
Brooklyn, NY 11218

Printed in the United States of America.

Distributed by The MIT Press,
Cambridge, Massachusetts, and London, England

Library of Congress Cataloging-in-Publication Data

Daston, Lorraine, 1951–
    Objectivity / Lorraine Daston and Peter Galison.
        p. cm.
    Includes bibliographical references and index.
    ISBN 978-1-890951-79-5
    1. Objectivity. I. Galison, Peter. II. Title.
    BD220.D37 2007
    121′.4–dc22

                                    2007023997

Contents

To Gerald Holton, teacher and friend

Preface to the Paperback Edition

Since this book appeared in 2007, we have had the great privilege of discussing it, both viva voce and in print, with readers bringing perspectives from diverse disciplines, countries, and generations (we are especially grateful to the students who have engaged with our work so eagerly and imaginatively). Our interlocutors have corrected errors and taught us which aspects of our narrative are most surprising, most opaque, most at odds with more familiar ways of writing the history of science, the history of the self, and simply history in general. Scholars from a variety of disciplines (from philosophy, art history, and literature to Victorian Studies, history of science, and psychology) have engaged with and extended themes from the book. It has been a pleasure to follow this growing literature. We take the opportunity of the publication of the paperback edition to reflect briefly on these lessons from our readers.

Perhaps the most disorienting feature of a history of scientific objectivity was not, as we had originally assumed, the bare assertion that objectivity had a history but rather the specifics of just when and how that history began. The shock was that the emergence of objectivity in the mid-nineteenth century did not coincide with any of the conventional accounts (which differ from discipline to discipline and among national traditions) of the origins of modernity: not with the Scientific Revolution in the seventeenth century, the political revolutions of the late eighteenth century, or the industrial and technological revolutions of the turn of the twentieth century. Historians of science and philosophy in particular have imbibed a narrative, most famously expounded by Edmund Husserl in *Die Krisis der europäischen Wissenschaft* (1936) and most influentially diffused by his student Alexandre Koyré, that pinpoints the onset of modernity to the application of mathematics to nature in the works of Galileo, Descartes, and Newton. This is a powerful story about how the abstractions of magnitude and number replaced (so the claim went) the bright, raucous, pulsing world of everyday experience, and it is made all the more compelling by a plangent undertone of regret for the enchanted world that we moderns have allegedly lost sounded by cultural critics from Max Weber to Herbert Marcuse. In the past twenty years, historians of early modern science have dismantled this narrative piece by piece on the basis of deep and detailed research — but the keystone assumptions that scientific and philosophical modernity and objectivity are synonymous and that both emerged in the seventeenth century endure amidst the wreckage are still firmly in place.

Our account challenges these assumptions both chronologically (however seismic the developments of the seventeenth century were for the subsequent course of science and philosophy, objectivity wasn't one of them) and substantively (objectivity is not identical with mathematization). In fact, the history of mathematical modeling in the sciences is strewn with appeals to personal intuitions and metaphysical beliefs. Objectivity and quantification may sometimes converge, as in the case of inference statistics, but they may also diverge: for example, neither the mathematical models of planetary distances envisioned by Johannes Kepler nor those of crystal structures advanced by René-Just Haüy sought to suppress subjectivity — the *sine qua non* of objectivity. Quantification and objectivity, like certainty and precision, are distinct epistemic virtues, and each deserves its own history.

Equally perplexing to another set of readers was the claim that subjectivity also had a history: that history tracks aspects of the self, not the self *tout court*. It is one of the main messages of this book that epistemology and ethos are intertwined: mechanical objectivity, for example, is a way of being as well as a way of knowing. Specific forms of image-making sculpt and steady particular, historical forms of the scientific self. At first, this sounds paradoxical. Isn't scientific objectivity all about the escape from all aspects of the self — the details of personality, politics, religion, nationality, or even species —

4

in short, the view from nowhere? But the self that scientific objectivity seeks to transcend is of a very specific kind, one in which the
faculty of the will (as opposed to reason or judgment or imagination)
is paramount, and never more so when, as in the case of scientific
objectivity, the will turns on itself: the will to will-lessness to which
proponents of objectivity aspired. Subjectivity is as historically
located as objectivity; they emerge together as mutually defining
complements.

This convex-concave relationship of objectivity and subjectivity
forced us to rethink what historical explanation fit this case and
those like it. Scientific objectivity in its various forms is not the
result of some billiard-ball-like cause that sets a chain of events in
motion. Neither Kant nor photography nor the industrial revolution
"caused" objectivity, though all provided resources its proponents
could deploy in one setting or another. Rather, objectivity and
subjectivity emerge in tandem, and the explanation is the demarca-
tion line between them. Like the similarly complementary pair
male/female, the details of what characteristics fall on one or
another side of the boundary are less important than the extraordinarily elastic and resilient structure itself. We were driven to ask:
against what form of subjectivity was mechanical objectivity positioned? To what threat of subjectivity was structural objectivity an
appropriate response? Regardless of which traits are singled out as
objective or subjective, male or female, the work of opposing them
to their complements reinforces the fundamental division.
Explanans and explanandum lie on the surface, not in some Ur-cause
buried in the depths.

Depthlessness as a form of historical explanation troubled some
readers. Why not approach objectivity through the case study, classically construed: a laboratory studied over a few decades set within a
local context: industrial annex, for example, or aristocratic retreat?
Wouldn't such a micro-historical inquiry lead to the true, deep
cause, tying the practices within a laboratory as a surface effect
to its prime mover in, say, British industrial production circa 1870?
One problem here is that the phenomena we are looking at — transformations in the characteristics of visual-publishing practices —
were not purely local. Mechanical objectivity and its characteristic
practices emerged in mid-nineteenth century German physiology

laboratories and British photographic explorations of falling drops; it was in American radiology collections and French surveys of pathological anatomy.

Our approach demands a twin attentiveness: on one side, to the collective empiricism of these volumes of working images, watching how even within many of the atlases, images and results were drawn from a multitude of investigators; and on the other, to regulative ideals governing atlas production that can be grasped only across a broad survey of the genre. When mechanical objectivity emerged in the mid-nineteenth century, those collective genre-defining practices involved a vigilant hunt to minimize idealizing interventions in image production and reproduction, to remove long interpretive exegeses, often to publish in a triplet of languages (French, English, German), to employ larger-format, archival paper. Together these practices composed a kind of scientific self separated by an ever-brighter line from a world of working objects intended to be valid across places, times, and inquirers.

At the heart of our book are nitty-gritty visual practices, lots of them, in many disciplines — the cross-hatching of a natural history engraving, the filled-in symmetry of a snowflake, the standardized color scale of a computer-simulated wavelet. As a result, it took us an unconscionably long time to write this book, searching for annotated sketches by eighteenth-century artists and naturalists, learning the in's and out's of nineteenth-century microphotography, figuring out how solar magnetograms were "smoothed" in the mid-twentieth century, tracking new forms of "image galleries" in the early twenty-first century — and then trying to make sense of the long, convoluted history as a whole. Nonetheless, we imagined this study as a beginning rather than an end. We hoped that it would engage the curiosity of other scholars about topics that do not yet have a history, despite their leading role in creating modern science: the forms and requirements of collective empiricism, the ways in which scientific experience is molded by image-making and image-reading, the entanglement of epistemology and ethos in epistemic virtues, the mutations of the scientific self, the mesh between the most concrete image-making practices and the most abstract epistemological goals.

We hope that others will find it fruitful to use some of the tools tried out here — including historically-inflected forms of sight; the mutual formation of the scientific self and the scientific object; the usefulness of treating printed and digitally-distributed scientific images as objects in their own right (presented not just re-presented). If our conversations with readers are any indication, however, our imagination for where this kind of history might lead was far too constrained. We look forward to reading and being surprised in our turn by what others take and make out of this book, which we thank Zone Books for making available in a paperback edition.

Preface to the First Edition

We began to think, talk, and write about the history of scientific objectivity when we both had the good fortune to be fellows at the Center for Advanced Study in the Behavioral Sciences at Stanford in 1989–1990; we recall the Center's support and stimulating lunchtime discussions with gratitude undimmed by the intervening years. The article that resulted from that collaboration was published as "The Image of Objectivity."[1] Both of us then turned to other projects far removed from objectivity — or so we thought.

Yet as one of us wrote about twentieth-century physics and the other about early modern natural philosophy, we both kept watch for hints and clues concerning the prologue and aftermath of the remarkable emergence of scientific objectivity in the nineteenth century. Each of us kept files of scattered references and wrote occasional articles on the subject; we exchanged ideas whenever happy circumstances brought us together and at some point — neither of us can quite pinpoint when — decided we would broaden our article into a book. We were able to sustain the fond illusion of a simple accordion-like "expansion" until 1999, when we began to see how inextricably tied conceptions of the self were to the right depiction of nature. Slowly it dawned on us that wholesale rethinking, not just rewriting and more research, would be needed to understand the history of scientific objectivity — and its alternatives. It was then that we began to work in earnest together (in 2001–2002 in Berlin and 2002–2003 in Cambridge, Massachusetts). Chapters were plotted, researched, and written — only to be ultimately discarded. In our more despairing moments, we felt as if we had undertaken to write some Borgesian monograph on the whole of human knowledge. Objectivity seemed endless.

Gradually, very gradually, we discerned shape and contours amid the sprawl. Our topics of study — objectivity, but also the atlas of scientific images — overflowed the usual boundaries that organize the history of science, straddling periods and disciplines. The history of objectivity and its alternatives, moreover, contradicted the structure of most narratives about the development of the sciences. Ours turns out to be less a story of rupture than one of reconfiguration. We nonetheless came to believe that the history of objectivity had its own coherence and rhythm, as well as its own distinctive patterns of explanation. At its heart were ways of seeing that were at once social, epistemological, and ethical: collectively learned, they did not owe their existence to any individual, to any laboratory, or even to any discipline.

We came to understand this image history of objectivity as an account of kinds of sight. Atlases had implications for who the scientist aspired to be, for how knowledge was most securely acquired, and for what kinds of things there were in the world. To embrace objectivity — or one of its alternatives — was not only to practice a science but also to pattern a self. Objectivity came to seem at once stranger — more specific, less obvious, more recently historical — and deeper, etched into the very act of scientific seeing, than we had ever suspected.

Objectivity Shock

He lit his laboratory with a powerful millisecond flash — poring over every stage of the impact of a liquid drop, using the latent image pressed into his retina to create a freeze-frame "historical" sequence of images a few thousandths of a second apart. (See figure P.1.) Bit by bit, beginning in 1875, the British physicist Arthur Worthington succeeded in juxtaposing key moments, untangling the complex process of fluid flow into a systematic, visual classification. Sometimes the rim thrown up by the droplet would close to form a bubble; in other circumstances, the return wave would shoot a liquid jet high into the air. Ribs and arms, bubbles and spouts — Worthington's compendium of droplet images launched a branch of fluid dynamics that continued more than a century later. For Worthington himself, the subject had always been, as he endlessly repeated, a physical system marked by the beauty of its perfect symmetry.

Perfect symmetry made sense. Even if it could be trapped by the latent image left in Worthington's eye after the spark had emptied into the dark, why would one want the accidental specificity of this or that defective splash? Worthington, like so many anatomists, crystallographers, botanists, and microscopists before him, had set out to capture the world in its types and regularities — not a helter-skelter assembly of peculiarities. Thousands of times he had let splash mercury or milk droplets, some into liquid, others onto hard surfaces. In hand-drawn sketches, made immediately after the bright flash of an electric spark, he had captured an evanescent morphology of nature. Simplification through a pictorial taxonomy, explanation of the major outcomes — finally science emerged from a kind of fluid flow that had eluded experiment.

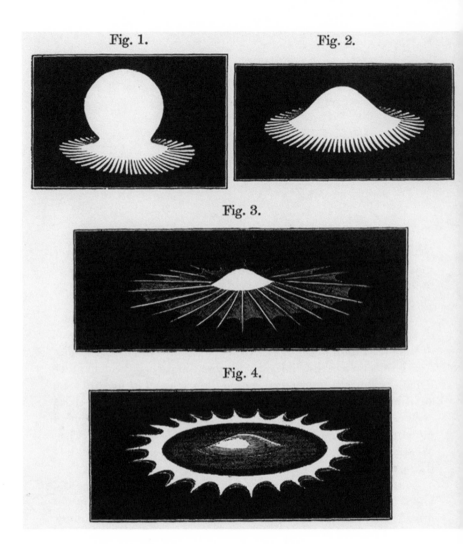

Fig. 1.   **Fig. 2.**

Fig. 3.

Fig. 4.

Fig. P.1. Symmetrical Vision. Arthur Worthington, "A Second Paper on the Forms
Assumed by Drops of Liquids Falling Vertically on a Horizontal Plate," *Proceedings of the
Royal Society* 25 (1877), p. 500, figs. 1–4. Tumbling from a height of 78 millimeters,
Worthington's mercury drops hit a clean glass plate. Just after first impact (fig. 1), "rays
too numerous to allow of an estimate of their number" race out from the contact point.
By the time of fig. 3, the "symmetrically disposed" rays coalesce "most often" into
twenty-four arms; in fig. 4, these arms, overtaken by mercury, reach maximum spread. In
addition, Worthington published numerous singular events ("variations"), but none that
violated the ideal, absolute symmetry he saw "behind" any particular defective splash.

For years, Worthington had relied on the images left on his retina by the flash. Then, in spring 1894, he finally succeeded in stopping the droplet's splash with a photograph. Symmetry shattered. Worthington said, "The first comment that any one would make is that the photographs, while they bear out the drawings in many details, show greater irregularity than the drawings would have led one to expect."[1] But if the symmetrical drawings and the irregular shadow photographs clashed, one had to go. As Worthington told his London audience, brighter lights and faster plates offered "an objective view" of the splash, which he then had drawn and etched (see figure P. 2).[2] There was a shock in this new, imperfect nature, a sudden confrontation with the broken particularity of the phenomenon he had studied since 1875. Plunged into doubt, Worthington asked how it could have been that, for so many years, he had been depicting nothing but idealized mirages, however beautifully symmetrical.

No apparatus was perfect, Worthington knew. His wasn't, and he said so. Even when everything was set to show a particular stage of the splash, there were variations from one drop to the next. Some of this visual scatter was due to the instrument, mainly when the drop adhered a bit to the watch glass from which it fell. In its subsequent oscillations the drop hit the surface already flattened or elongated. It had seemed perfectly obvious — in nearly two decades Worthington had never commented on it in print — that one always had to choose among the many images taken at any stage in order to get behind variations to the norm. Accidents happen all the time. Why publish them?

Worthington wrote, "I have to confess that in looking over my original drawings I find records of many irregular or unsymmetrical figures, yet in compiling the history it has been inevitable that these should be rejected, if only because identical irregularities never recur. Thus the mind of the observer is filled with an ideal splash — an Auto-Splash — whose perfection may never be actually realized."[3] This was not a case of bad eyes or a failed experiment — Worthington had sketched those asymmetrical drawings with his own hand, carefully, deliberately. The published, symmetrical "histories" had been successes — the triumph of probing idealization over mere mishaps: "Some judgment is required in selecting a consecutive series of drawings. The only way is to make a considerable number of drawings of each stage, and then to pick out a consecutive series. Now,

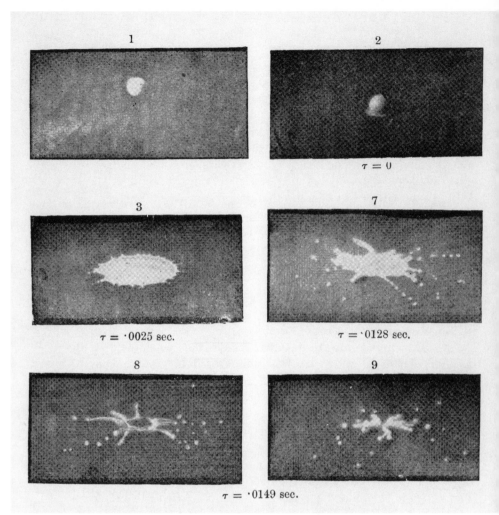

| 1 | 2 |
| | $\tau = 0$ |

| 3 | 7 |
| $\tau = \cdot 0025$ sec. | $\tau = \cdot 0128$ sec. |

| 8 | 9 |

$\tau = \cdot 0149$ sec.

Fig. P.2. Objective Splash. Engraving of "instantaneous photographs." Arthur Worthington, "The Splash of a Drop and Allied Phenomena," *Proceedings of the Royal Institution* 14 (1893–95), opp. p. 302, ser. 13. Presented at the weekly evening meeting, May 18, 1894. A milk drop splashes against a smoked glass plate, running toward the edges without adhesion just as mercury did (although without the hard-to-photograph reflectivity of the mercury surface). But now Worthington has restrained himself and is no longer struggling to see the ideal or "type" reality behind the manifest image — he called his asymmetrical images-as-they-were-recorded "objective views."

whenever judgment has to be used, there is room for error of judgment, and...it is impossible to put together the drawings so as to
tell a consecutive story, without being guided by some theory.... ⚡
You will therefore be good enough to remember that this chronicle
of the events of a tenth of a second is not a mechanical record but is
presented by a fallible human historian."[4] But now he belatedly came
to see his fallible, painstaking efforts of twenty years to impose regularity as counting for less than "a mechanical record," a kind of
blind sight that would not shun asymmetry or imperfection. Now,
unlike before, he regretted the all-too-human decisions required to
retrieve the phenomenon masked by variations. And only now did
that judgment strike him as treacherous.

For two decades, Worthington had seen the symmetrical, perfected forms of nature as an essential feature of his morphology of
drops. All those asymmetrical images had stayed in the laboratory —
not one appeared in his many scientific publications. In this choice
he was anything but alone — over the long course of making systematic study of myriad scientific domains, the choice of the perfect over
the imperfect had become profoundly entrenched. From anatomical
structures to zoophysiological crystals, idealization had long been
the governing order. Why would anyone choose as the bottom-line
image of the human thorax one including a broken left rib? Who
could want the image of record of a rhomboid crystal to contain a
chip? What long future of science would ever need a "malformed"
snowflake that violated its six-fold symmetry, a microscopic image
with an optical artifact of the lens, or a clover with an insect-torn
leaf? But after his 1894 shock, Worthington instead began to ask
himself — and again he was not alone — how he and others for so long
could have only had eyes for a perfection that wasn't there.

In the months after he first etched drawings of photographed
splashes, reeling from the impact, it may have eased the severity
of the transformation to demote the older epistemological ideal to
the merely psychological. Perhaps, he speculated in 1895, it had
been the mind's tendency to integrate variations back into regularity.
Perhaps it was an overactive attentiveness to a regular subsection
of the splash wrongly generalized to the whole. "In several cases,
I have been able to observe with the naked eye a splash that was
also photographed," he said, noting in his record book that the event

15

was "quite regular," although, on later inspection, the photograph showed the splash to be anything but symmetrical.[5] What had been a high-order scientific virtue — tracking and documenting the essential, ideal "Auto-Splash" — became a psychological fault, a defect in perception.

Now, in 1895, Worthington told his audience that the earlier images of perfect drops had to be discarded. In their place, he wanted images that depicted the physical world in its full-blown complexity, its asymmetrical individuality — in what he called, for short, "an objective view."[6] Only this would provide knowledge of what he considered "real, as opposed to imaginary fluids."[7]

Worthington's conversion to the "objective view" is emblematic of a sea change in the observational sciences. Over the course of the nineteenth century other scientists, from astronomers probing the very large to bacteriologists peering at the very small, also began questioning their own traditions of idealizing representation in the preparation of their atlases and handbooks. What had been a supremely admirable aspiration for so long, the stripping away of the accidental to find the essential, became a scientific vice.

This book is about the creation of a new epistemic virtue — scientific objectivity — that drove scientists to rewrite and re-image the guides that divide nature into its fundamental objects. It is about the search for that new form of unprejudiced, unthinking, blind sight we call scientific objectivity.

CHAPTER ONE

Epistemologies of the Eye

Blind Sight

Scientific objectivity has a history. Objectivity has not always defined science. Nor is objectivity the same as truth or certainty, and it
is younger than both. Objectivity preserves the artifact or variation
that would have been erased in the name of truth; it scruples to filter
out the noise that undermines certainty. To be objective is to aspire
to knowledge that bears no trace of the knower — knowledge un-
marked by prejudice or skill, fantasy or judgment, wishing or striving. Objectivity is blind sight, seeing without inference, interpretation,
or intelligence. Only in the mid-nineteenth century did scientists
begin to yearn for this blind sight, the "objective view" that embraces accidents and asymmetries, Arthur Worthington's shattered
splash-coronet. This book is about how and why objectivity emerged
as a new way of studying nature, and of being a scientist.

Since the nineteenth century, objectivity has had its prophets,
philosophers, and preachers. But its specificity — and its strangeness
— is most clearly seen in the everyday work of its practitioners: literally seen, in the essential practice of scientific image-making. Making pictures is not the only practice that has served scientific objectivity: an armamentarium of other techniques, including inference
statistics, double-blind clinical trials, and self-registering instruments, have been enlisted to hold subjectivity at bay.[1] But none is as
old and ubiquitous as image making. We have chosen to tell the history of scientific objectivity through pictures drawn from the long
tradition of scientific atlases, those select collections of images that
identify a discipline's most significant objects of inquiry.

Look, if you will, at these three images from scientific atlases: the first, from an eighteenth-century flora; the second, from a late nine-teenth-century catalogue of snowflakes; the third, from a mid-twen-tieth-century compendium of solar magnetograms (see figures 1.1, 1.2, and 1.3). A single glance reveals that the images were differently made: a copperplate engraving, a microphotograph, an instrument contour. The practiced eye contemporary with any one of these images made systematic sense of it. These three figures constitute a synopsis of our story. They capture more than a flower, a snowflake, a magnetic field: each encodes a technology of scientific sight impli-cating author, illustrator, production, and reader.

Each of these images is the product of a distinct code of epistemic virtue, codes that we shall call, in terms to be developed presently, truth-to-nature, mechanical objectivity, and trained judgment. As the dates of the images suggest, this is a historical series, and it will be one of the principal theses of this book that it is a series punctu-ated by novelty. There was a science of truth-to-nature before there was one of objectivity; trained judgment was, in turn, a reaction to objectivity. But this history is one of innovation and proliferation rather than monarchic succession. The emergence of objectivity as a new epistemic virtue in the mid-nineteenth century did not abolish truth-to-nature, any more than the turn to trained judgment in the early twentieth century eliminated objectivity. Instead of the anal-ogy of a succession of political regimes or scientific theories, each triumphing on the ruins of its predecessor, imagine new stars wink-ing into existence, not replacing old ones but changing the geogra-phy of the heavens.

There is a deep historical rhythm to this sequence: in some strong sense, each successive stage presupposes and builds upon, as well as reacts to, the earlier ones. Truth-to-nature was a precondition for mechanical objectivity, just as mechanical objectivity was a precon-dition for trained judgment. As the repertoire of epistemic virtues expands, each redefines the others. This is not some neat Hegelian arithmetic of thesis plus antithesis equals synthesis, but a far messier situation in which all the elements continue in play and in interaction with one another. Late twentieth-century scientists could and did still sometimes strive for truth-to-nature in their images, but they did not, could not, simply return to the ideals and practices of their eigh-teenth-century predecessors. The meaning of truth-to-nature had

18

been recast by the existence of alternatives, which in some cases figured as competitors. Judgment, for example, was understood differently before and after objectivity: what was once an act of practical reason became an intervention of subjectivity, whether defensively or defiantly exercised.

In contrast to the static tableaux of paradigms and epistemes, this is a history of dynamic fields, in which newly introduced bodies reconfigure and reshape those already present, and vice versa. The reactive logic of this sequence is productive. You can play an eighteenth-century clavichord at any time after the instrument's revival around 1900 — but you cannot hear it after two intervening centuries of the pianoforte in the way it was heard in 1700. Sequence weaves history into the warp and woof of the present: not just as a past process reaching its present state of rest — how things came to be as they are — but also as the source of tensions that keep the present in motion.

This book describes how these three epistemic virtues, truth-to-nature, objectivity, and trained judgment, infused the making of images in scientific atlases from roughly the early eighteenth to the mid-twentieth century, in Europe and North America. The purview of these virtues encompasses far more than images, and atlases by no means exhaust even the realm of scientific images.[2] We have narrowed our sights to images in scientific atlases, first, because we want to show how epistemic virtues permeate scientific practice as well as precept; second, because scientific atlases have been central to scientific practice across disciplines and periods; and third, because atlases set standards for how phenomena are to be seen and depicted. Scientific atlas images are images at work, and they have been at work for centuries in all the sciences of the eye, from anatomy to physics, from meteorology to embryology.

Collective Empiricism

All sciences must deal with the problem of selecting and constituting "working objects," as opposed to the too plentiful and too various natural objects. Working objects can be atlas images, type specimens, or laboratory processes — any manageable, communal representative of the sector of nature under investigation. No science can do without such standardized working objects, for unrefined natural objects

Fig. 1.1. Truth-to-Nature. *Campanula foliis hastatis dentatis*, Carolus Linnaeus, *Hortus Cliffortianus* (Amsterdam: n.p., 1737), table 8 (courtesy of Staats- und Universitätsbibliothek Göttingen). Drawn by Georg Dionysius Ehret, engraved by Jan Wandelaar, and based on close observation by both naturalist and artist, this illustration for a landmark botanical work (still used by taxonomists) aimed to portray the underlying type of the plant species, rather than any individual specimen. It is an image of the characteristic, the essential, the universal, the typical: truth-to-nature.

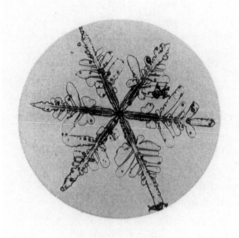

Fig. 1.2. Mechanical Objectivity. Snowflake, Gustav Hellmann, with microphotographs by Richard Neuhauss, *Schneekrystalle: Beobachtungen und Studien* (Berlin: Mückenberger, 1893), table 6, no. 10. An individual snowflake is shown with all its peculiarities and asymmetries in an attempt to capture nature with as little human intervention as possible: mechanical objectivity.

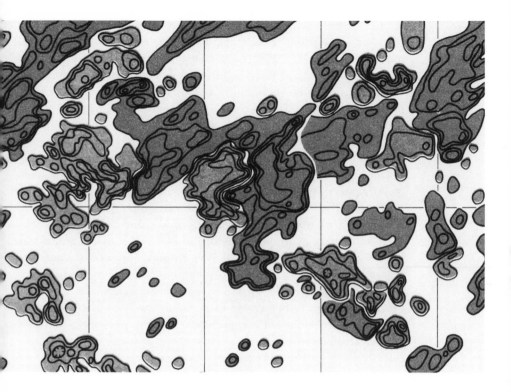

Fig. 1.3. Trained Judgment. Sun Rotation 1417, Aug.–Sept. 1959 (detail), Robert Howard, Václav Bumba, and Sara F. Smith, *Atlas of Solar Magnetic Fields*, August 1959 – June 1966 (Washington, DC: Carnegie Institute, 1967) (courtesy of the Observatories of the Carnegie Institution of Washington, DC). This image of the magnetic field of the sun mixed the output of sophisticated equipment with a "subjective" smoothing of the data — the authors deemed this intervention necessary to remove instrumental artifacts: trained judgment. (Please see Color Plates.)

are too quirkily particular to cooperate in generalizations and comparisons. Sometimes these working objects replace natural specimens: for example, a 1795 report on the collection of the vellum paintings of plants and animals at the Muséum d'Histoire Naturelle in Paris explained how such images could "reanimate, by this means, plants that blossomed ... by chance [once] in fifty or a hundred years, like the agave that flowered last year; the same goes for the animals that often pass but rarely in our climes and of which one sees sometimes only one individual in centuries."[3] Even scientists working in solitude must regularize their objects. *Collective empiricism*, involving investigators dispersed over continents and generations, imposes still more urgently the need for common objects of inquiry.

Atlases are systematic compilations of working objects. They are the dictionaries of the sciences of the eye. For initiates and neophytes alike, the atlas trains the eye to pick out certain kinds of objects as exemplary (for example, this "typical" healthy liver rather than that one with cirrhosis) and to regard them in a certain way (for example, using the Flamsteed rather than the Ptolemaic celestial projection). To acquire this expert eye is to win one's spurs in most empirical sciences. The atlases drill the eye of the beginner and refresh the eye of the old hand. In the case of atlases that present images from new instruments, such as the bacteriological atlases of the late nineteenth century and the x-ray atlases of the early twentieth century, everyone in the field addressed by the atlas must begin to learn to "see" anew. Whatever the amount and avowed function of the text in an atlas, which varies from long and essential to nonexistent or despised, the illustrations command center stage. Usually displayed in giant format, meticulously drawn and reproduced, and expensively printed, they are the *raison d'être* of the atlas. To call atlas images "illustrations" at all is to belie their primacy, for it suggests that their function is merely ancillary, to illustrate a text or theory. Some early astronomical atlases do use the figures as genuine illustrations, to explicate rival cosmologies.[4] But in most atlases from the eighteenth century on, pictures are the alpha and the omega of the genre.

Not only do images make the atlas; atlas images make the science. Atlases are the repositories of images of record for the observational sciences. The name "atlas" derives from Gerardus Mercator's world

22

map, *Atlas sive cosmographicae meditationes de fabrica mvndi et fabricati figvra* (*Atlas, or Cosmographical Meditations on the Fabric of the World*, 1595) (the title was an allusion to the titan Atlas of Greek mythology, who bore the world on his shoulders). By the late eighteenth century, the term had spread from geography to astronomy and anatomy ("maps" of the heavens or the human body), and, by the mid-nineteenth century, "atlases" had proliferated throughout the empirical sciences.[5] Even if older works did not bear the word "atlas" in their titles, they were explicitly included in the lineage that later atlas makers were obliged to trace: every new atlas must begin with an explanation of why the old ones are no longer adequate to their task, why new images of record are necessary. These genealogies define what counts as an atlas in our account. Whether atlases display crystals or cloud chamber traces, brain slices or galaxies, they still aim to "map" the territory of the sciences they serve. They are the guides all practitioners consult time and time again to find out what is worth looking at, how it looks, and, perhaps most important of all, how it should be looked at.

These reference works may be as small as a field guide that slips into a naturalist's pocket, but they tend toward the large, even the gigantic. Many are oversized volumes (an "atlas folio" is a book twenty-three to twenty-five inches tall), and some are too large and heavy to be comfortably handled by a single person. John James Audubon's *Birds of America* (1827–38) was printed as a double elephant folio (twenty-seven inches by thirty-nine inches); James Bateman's *Orchidaceae of Mexico and Guatemala* (1837–43) weighed over thirty-eight pounds. (See figures 1.4 and 1.5.) The ambitions of the authors rival the grand scale of their books. Atlas makers woo, badger, and monopolize the finest artists available. They lavish the best quality ink and paper on images displayed in grand format, sometimes life-size or larger. Atlases are expensive, even opulent works that devour time, nerves, and money, as their authors never tire of repeating. Atlas prefaces read like the trials of Job: the errors of earlier atlases that must be remedied; the long wait for just the right specimens; the courting and correcting of the artist; the pitched battle with the cheapskate publisher; the penury to which the whole endless project has reduced the indefatigable author. These pains are worth taking because an atlas is meant to be a lasting work of orientation for

23

Fig. 1.4. Double Elephant, *Stanhopea tigrina*. James Bateman, *The Orchidaceae of Mexico and Guatemala* (London: Ridgway, 1837–1843), pl. 7, drawn by Augusta Withers and lithographed by M. Gauci (Botanical Garden, Berlin). The opulently produced flora makes full use of the double elephant folio page to display the hand-colored images of the orchids but allows the accompanying text (a mere 8.5 by 11 inches) to float like an island on the facing page. The hand and surrounding normal-sized books give some idea of the scale of this expensive, enormous, and unwieldy volume, produced in a format to set off the images to maximal effect. Photograph by Kelley Wilder. (Please see Color Plates.)

Μέγα βιβλίον μέγα κακόν.

Fig. 1.5. "Big Book, Great Evil." James Bateman, *The Orchidaceae of Mexico and Guatemala* (London: Ridgway, 1837–1843), p. 8, drawn by George Cruikshank (Botanical Garden, Berlin). The Victorian cartoonist Cruikshank's vignette pokes fun at the elephantine dimensions of Bateman's atlas. A team of laborers struggles to hoist the volume with a pulley; the Greek caption is reinforced by the jeering demons looking on from the left. Since the cartoon was commissioned by Bateman himself, it probably expresses his own attitude of mingled enthusiasm and self-irony toward his magnum opus.

generations of observers. Every atlas is presented with fanfare, as if it were the atlas to end all atlases. Atlases aim to be definitive in every sense of the term: they set the standards of a science in word, image, and deed — how to describe, how to depict, how to see.

Since at least the seventeenth century, scientific atlases have served to train the eye of the novice and calibrate that of the old hand. They teach how to see the essential and overlook the incidental, which objects are typical and which are anomalous, what the range and limits of variability in nature are. Without them, every student of nature would have to start from scratch to learn to see, select, and sort. Building on the work of others would be difficult or impossible, for one could never be sure that one's predecessors and correspondents were referring to the same thing, seen in the same tutored way. Only those who had learned at the master's side would be visually coordinated. Science would be confined, as it was for many centuries, before the advent of printing made the wide dissemination of such atlases practicable, to local traditions of apprenticeship. Images like these were far from merely decorative. They made collective empiricism in the sciences possible, beyond the confines of a local school.

Making and using an atlas is one of the least individual activities in science. Atlases are intrinsically collective. They are designed for longevity: if all goes well, they should serve generations within a scientific community. Many are themselves the fruit of scientific collaborations, drawing their images from a multitude of authors or author-groups. Almost all depend on a close working relationship between scientist and illustrator. But the contributions of atlases go further: atlases make other collaborations possible, including the loose collaborations that permit dispersed observers to exchange and accumulate results. Early atlases were often written in Latin to assure maximum diffusion; after the demise of Latin as the lingua franca of the learned world, bilingual and trilingual editions were produced for the same reason. The atlas is a profoundly social undertaking, but because the term "social" carries so many and such varied connotations, it would be more precise to say that the atlas is always — and fundamentally — an exemplary form of collective empiricism: the collaboration of investigators distributed over time and space in the study of natural phenomena too vast and various to be encompassed by a solitary thinker, no matter how brilliant, erudite, and diligent.

Atlas makers create one sliver of the world anew in images — skeletons, stellar spectra, bacteria. Atlas users become the people of a book, which teaches them how to make sense of their sliver-world and how to communicate with one another about it. Certain atlas images may become badges of group identity, nowadays emblazoned on T-shirts and conference logos, in earlier decades and centuries etched in memory like icons. Dog-eared and spine-cracked with constant use, atlases enroll practitioners as well as phenomena. They simultaneously assume the existence of and call into being communities of observers who see the same things in the same ways. Without an atlas to unite them, atlas makers have long claimed, all observers are isolated observers.

In this book, we trace the emergence of epistemic virtues through atlas images — by no means the only expression of truth-to-nature or objectivity or trained judgment, but nonetheless one of the most revealing. By examining volumes of images of record (including atlases, handbooks, surveys, and expedition reports), abstractions like objectivity become concrete and visible, reflections of changing scientific ambitions for right depiction.

The history we propose raises a flock of questions: What exactly are epistemic virtues? How do lofty norms like truth, objectivity, and judgment connect with on-the-ground scientific conduct? Why try to track an entity as abstract as epistemology via the concrete details of a drawing or a photograph? And, above all, how can objectivity have a history? In the remainder of this introductory chapter, we will try to make this counterintuitive brand of history plausible, tackling the last, most burning question first.

Objectivity Is New

The history of scientific objectivity is surprisingly short. It first emerged in the mid-nineteenth century and in a matter of decades became established not only as a scientific norm but also as a set of practices, including the making of images for scientific atlases. However dominant objectivity may have become in the sciences since *circa* 1860, it never had, and still does not have, the epistemological field to itself. Before objectivity, there was truth-to-nature; after the advent of objectivity came trained judgment. The new did not always edge out the old. Some disciplines were won over quickly to the newest epistemic virtue, while others persevered in their allegiance to older ones. The relationship among epistemic virtues may be one of quiet compatibility, or it may be one of rivalry and conflict. In some cases, it is possible to pursue several simultaneously; in others, scientists must choose between truth and objectivity, or between objectivity and judgment. Contradictions arise.

This situation is familiar enough in the case of moral virtues. Different virtues — for example, justice and benevolence — come to be accepted as such in different historical periods. The claims of justice and benevolence can all too plausibly collide in cultures that honor both: for Shylock in *The Merchant of Venice*, a man's word is his bond; Portia replies that the quality of mercy is not strained. Codes of virtue, whether moral or epistemic, that evolve historically are loosely coherent, but not strictly internally consistent. Epistemic virtues are distinct as ideals and, more important for our argument, as historically specific ways of investigating and picturing nature. As ideals, they may more or less peacefully, if vaguely, coexist. But at the level of specific, workaday choices — which instrument to use, whether to retouch a photograph or disregard an outlying data point, how to train young scientists to see — conflicts can occur. It is not always possible to serve truth and objectivity at the same time, any more than justice and benevolence can always be reconciled in specific cases.

Here skeptics will break in with a chorus of objections. Isn't the claim that objectivity is a nineteenth-century innovation tantamount to the claim that science itself begins in the nineteenth century? What about Archimedes, Andreas Vesalius, Galileo, Isaac Newton, and a host of other luminaries who worked in earlier epochs? How can there be science worthy of the name without objectivity? And how can truth and objectivity be pried apart, much less opposed to each other?

All these objections stem from an identification of objectivity with science *tout court*. Given the commanding place that objectivity has come to occupy in the modern manual of epistemic virtues, this conflation is perhaps not surprising. But it is imprecise, both historically and conceptually. Historically, it ignores the evidence of usage and use: when, exactly, did scientists start to talk about objectivity, and how did they put it to work? Conceptually, it operates by synecdoche, making this or that aspect of objectivity stand for the whole, and on an *ad hoc* basis. The criterion may be emotional detachment in one case; automatic procedures for registering data in another; recourse to quantification in still another; belief in a bedrock reality independent of human observers in yet another. In this fashion, it is not difficult to tote up a long list of forerunners of objectivity — except that none of them operate with the concept in its entirety, to say nothing of the practices. The aim of a non-teleological history of scientific objectivity must be to show how all these elements came to be fused together (it is not self-evident, for example, what emotional detachment has to do with automatic data registration), designated by a single word, and translated into specific scientific techniques. Moreover, isolated instances are of little interest. We want to know when objectivity became ubiquitous and irresistible.

The evidence for the nineteenth-century novelty of scientific objectivity starts with the word itself. The word "objectivity" has a somersault history. Its cognates in European languages derive from the Latin adverbial or adjectival form *obiectivus/obiective*, introduced by fourteenth-century scholastic philosophers such as Duns Scotus and William of Ockham. (The substantive form does not emerge until much later, around the turn of the nineteenth century.) From the very beginning, it was always paired with *subiectivus/subiective*, but the terms originally meant almost precisely the opposite of what they mean today. "Objective" referred to things as they are presented to consciousness, whereas "subjective" referred to things in themselves.[6] One can still find traces of this scholastic usage in those passages of the *Meditationes de prima philosophia* (*Meditations on First Philosophy*, 1641) where René Descartes contrasts the "formal reality" of our ideas (that is, whether they correspond to anything in the external world) with their "objective reality" (that is, the degree of reality they enjoy by virtue of their clarity and distinctness, regardless of whether they exist in material form).[7] Even eighteenth-century dictionaries still preserved echoes of this medieval usage, which rings so bizarrely in modern ears: "Hence a thing is said to *exist* OBJECTIVELY, *objective*, when it exists no otherwise than in being known; or in being an Object of the Mind."[8]

The words *objective* and *subjective* fell into disuse during the seventeenth and eighteenth centuries and were invoked only occasionally, as technical terms, by metaphysicians and logicians.[9] It was Immanuel Kant who dusted off the musty scholastic terminology of "objective" and "subjective" and breathed new life and new meanings into it. But the Kantian meanings were the grandparents, not the twins, of our familiar senses of those words. Kant's "objective validity" (*objektive Gültigkeit*) referred not to external objects (*Gegenstände*) but to the "forms of sensibility" (time, space, causality) that are the preconditions of experience. And his habit of using "subjective" as a rough synonym for "merely empirical sensations" shares with later usage only the sneer with which the word is intoned. For Kant, the line between the objective and the subjective generally runs between universal and particular, not between world and mind.

Yet it was the reception of Kantian philosophy, often refracted through other traditions, that revamped terminology of the objective and subjective in the early nineteenth century. In Germany, idealist philosophers such as Johann Gottlieb Fichte and Friedrich Schelling turned Kant's distinctions to their own ends; in Britain, the poet Samuel Taylor Coleridge, who had scant German but grand ambitions, presented the new philosophy to his countrymen as a continuation of Francis Bacon; in France, the philosopher Victor Cousin grafted Kant onto Descartes.[10] The post-Kantian usage was so new that some readers thought at first it was just a mistake. Coleridge scribbled in his copy of Henrich Steffens's *Grundzüge der philosophischen Naturwissenschaft* (*Foundations of Philosophical Natural Science*, 1806): "Steffens has needlessly perplexed his reasoning by his strange use of Subjective and Objective — his S[ubjectivity] = the O[bjectivity] of former Philosophers, and his O[bjectivity] = their S[ubjectivity]."[11] But by 1817 Coleridge had made the barbarous terminology his own, interpreting it in a way that was to become standard thereafter: "Now the sum of all that is merely OBJECTIVE, we will henceforth call NATURE, confining the term to its passive and material sense, as comprising all the phaenomena by which its existence is made known to us. On the other hand the sum of all that is SUBJECTIVE, we may comprehend in the name of the SELF or INTELLIGENCE. Both conceptions are in necessary antithesis."[12]

Starting in the 1820s and 1830s, dictionary entries (first in German, then in French, and later in English) began to define the words "objectivity" and "subjectivity" in something like the (to us) familiar sense, often with a nod in the direction of Kantian philosophy. In 1820, for example, a German dictionary defined *objektiv* as a "relation to an external object" and *subjektiv* as "personal, inner, inhering in us, in opposition to objective"; as late as 1863, a French dictionary still called this the "new sense" (diametrically opposed to the old, scholastic sense) of word *objectif* and credited "the philosophy of Kant" with the novelty. When the English man of letters Thomas De Quincey published the second edition of his *Confessions of an English Opium Eater* in 1856, he could write of "objectivity": "This word, so nearly unintelligible in 1821 [the date of the first edition], so intensely scholastic, and consequently, when surrounded by familiar and vernacular words, so apparently pedantic, yet, on the other hand, so indispensable to accurate thinking, and to wide thinking, has since 1821 become too common to need any apology."[13] Sometime *circa* 1850 the modern sense of "objectivity" had arrived in the major European languages, still paired with its ancestral opposite "subjectivity." Both had turned 180 degrees in meaning.

Skeptics will perhaps be entertained but unimpressed by the curious history of the word "objectivity." Etymology is full of oddities, they will concede, but the novelty of the word does not imply the novelty of the thing. Long before there was a vocabulary that captured the distinction that by 1850 had come to be known as that between objectivity and subjectivity, wasn't it recognized and observed in fact? They may point to the annals of seventeenth-century epistemology, to Bacon and Descartes.[14] What, after all, was the distinction between primary and secondary qualities that Descartes and others made, if not a case of objectivity versus subjectivity *avant la lettre*? And what about the idols of the cave, tribe, marketplace, and theater that Bacon identified and criticized in *Novum organum* (*New Organon*, 1620): don't these constitute a veritable catalogue of subjectivity in science?

These objections and many more like them rest on the assumption that the history of epistemology and the history of objectivity coincide. But our claim is that the history of objectivity is only a subset, albeit an extremely important one, of the much longer and

31

larger history of epistemology — the philosophical examination of obstacles to knowledge. Not every philosophical diagnosis of error is an exercise in objectivity, because not all errors stem from subjectivity. There were other ways to go astray in the natural philosophy of the seventeenth century, just as there are other ways to fail in the science of the twentieth and early twenty-first centuries.

Take the case of the primary-secondary quality distinction as Descartes advanced it in the *Principia philosophiae* (*Principles of Philosophy*, 1644). Descartes privileged size, figure, duration, and other primary qualities over secondary qualities like odor, color, pain, and flavor because the former ideas are more clearly and distinctly perceived by the mind than the latter; that is, his was a distinction among purely mental entities, one kind of idea versus another — what nineteenth-century authors would (and did) label "subjective."[15] Or Bacon's idols: only one of the four categories (the idols of the cave) applied to the individual psyche and could therefore be a candidate for subjectivity in the modern sense (the others refer to errors inherent in the human species, language, and theories, respectively). Bacon's remedy for the idols of the cave had nothing to do with the suppression of the subjective self, but rather addressed the balance between opposing tendencies to excess: lumpers and splitters, traditionalists and innovators, analysts and synthesizers.[16] His epistemological advice — bend over backward to counteract one-sided tendencies and predilections — echoed the moral counsel he gave in his essay "Of Nature in Men" on how to reform natural inclinations: "Neither is the ancient rule amiss, to bend nature as a wand to a contrary extreme, whereby to set it right; understanding it where the contrary extreme is no vice."[17]

The larger point here is that the framework within which seventeenth-century epistemology was conducted was a very different one from that in which nineteenth-century scientists pursued scientific objectivity. There is a history of what one might call the nosology and etiology of error, upon which diagnosis and therapy depend. Subjectivity is not the same kind of epistemological ailment as the infirmities of the senses or the imposition of authority feared by earlier philosophers, and it demands a specialized therapy. However many twists and turns the history of the terms *objective* and *subjective* took over the course of five hundred years, they were always paired:

there is no objectivity without subjectivity to suppress, and vice versa. If subjectivity in its post-Kantian sense is historically specific, this implies that objectivity is as well. The philosophical vocabulary of mental life prior to Kant is extremely rich, but it is notably different from that of the nineteenth and twentieth centuries: "soul," "mind," "spirit," and "faculties" only begin to suggest the variety in English, with further nuances and even categories available in other vernaculars and Latin.

Post-Kantian subjectivity is as distinctive as any of these concepts. It presumes an individualized, unified self organized around the will, an entity equivalent to neither the rational soul as conceived by seventeenth-century philosophers nor the associationist mind posited by their eighteenth-century successors. Those who deployed post-Kantian notions of objectivity and subjectivity had discovered a new kind of epistemological malady and, consequently, a new remedy for it. To prescribe this post-Kantian remedy — objectivity — for a Baconian ailment — the idols of the cave — is rather like taking an antibiotic for a sprained ankle.

Although it is not the subject of this book, we recognize that our claim that objectivity is new to the nineteenth century has implications for the history of epistemology as well as the history of science. The claim by no means denies the originality of seventeenth-century epistemologists like Bacon and Descartes; on the contrary, it magnifies their originality to read them in their own terms, rather than tacitly to translate, with inevitable distortion, their unfamiliar preoccupations into our own familiar ones. Epistemology can be reconceived as ethics has been in recent philosophical work: as the repository of multiple virtues and visions of the good, not all simultaneously tenable (or at least not simultaneously maximizable), each originally the product of distinct historical circumstances, even if their moral claims have outlived the contexts that gave them birth.[18]

On this analogy, we can identify distinct epistemic virtues — not only truth and objectivity but also certainty, precision, replicability — each with its own historical trajectory and scientific practices. Historians of philosophy have pointed out that maximizing certainty can come at the expense of maximizing truth; historians of science have shown that precision and replicability can tug in opposite directions.[19] Once objectivity is thought of as one of several epistemic virtues, distinct in its origins and its implications, it becomes easier to imagine that it might have a genuine history, one that forms only part of the history of epistemology as a whole. We will return to the idea of epistemic virtues below, when we take up the ethical dimensions of scientific objectivity.

The skeptics are not finished. Even if objectivity is not coextensive with epistemology, they may rejoin, isn't it a precondition of all science worthy of the name? Why doesn't the mathematical natural philosophy of Newton or the painstaking microscopic research of Antonie van Leeuwenhoek qualify as a chapter in the history of objectivity? They will insist that scientific objectivity is a transhistoric honorific: that the history of objectivity is nothing less than the history of science itself.

Our answer here borrows a leaf from the skeptics' own book. They are right to assert a wide gap between epistemological precept and scientific practice, even if the two are correlated. Epistemology (of whatever kind) advanced in the abstract cannot be easily equated with its practices in the concrete. Figuring out how to operationalize an epistemological ideal in making an image or measurement is as challenging as figuring out how to test a theory experimentally. Epistemic virtues are various not only in the abstract but also in their concrete realization. Science dedicated above all to certainty is done differently — not worse, but differently — from science that takes truth-to-nature as its highest desideratum. But a science devoted to truth or certainty or precision is as much a part of the history of science as one that aims first and foremost at objectivity. The Newtons and the Leeuwenhoeks served other epistemic virtues, and they did so in specific and distinctive ways. It is precisely close examination of key scientific practices like atlas-making that throws the contrasts between epistemic virtues into relief. This is the strongest evidence for the novelty of scientific objectivity.

Objectivity the thing was as new as objectivity the word in the mid-nineteenth century. Starting in the mid-nineteenth century, men of science began to fret openly about a new kind of obstacle to knowledge: themselves. Their fear was that the subjective self was prone to prettify, idealize, and, in the worst case, regularize observations to fit theoretical expectations: to see what it hoped to see. Their predecessors a generation or two before had also been beset by epistemological worries, but theirs were about the variability of nature, rather than the projections of the naturalist. As atlas makers, the earlier naturalists had sworn by selection and perfection: select the most typical or even archetypical skeleton, plant, or other object under study, then perfect that exemplar so that the image can truly stand for the class, can truly represent it. By *circa* 1860, however, many atlas makers were branding these practices as scandalous, as "subjective." They insisted, instead, on the importance of effacing their own personalities and developed techniques that left as little as possible to the discretion of either artist or scientist, in order to obtain an "objective view." Whereas their predecessors had written about the duty to discipline artists, they asserted the duty to discipline themselves. Adherents to old and new schools of image making confronted one another in mutual indignation, both sides sure that the other had violated fundamental tenets of scientific competence and integrity. Objectivity was on the march, not just in the pages of dictionaries and philosophical treatises, but also in the images of scientific atlases and in the cultivation of a new scientific self.

Histories of the Scientific Self

If objectivity was so new, and its rise so sudden, how did it then become so familiar, so profoundly assumed that it by now threatens to swallow up the whole history of epistemology and of science to boot? If indeed it emerged as a scientific ideal borne out in practices only in the mid-nineteenth century, why then? What deeper historical forces — intellectual, social, political, economic, technological — created this *novum*?

These are just the sort of questions we asked ourselves when we first began to explore the history of objectivity. Certainly, great changes were under way *circa* 1800, changes so momentous that they are commonly designated as "revolutions": the French Revolution, the Industrial Revolution, the Kantian revolution, the second Scientific Revolution. We further wondered about the influence of expanding bureaucracies, with their rhetoric of mechanical rule-following, or of certain inventions, such as photography, with its aura of unselective impartiality. But after exploring these sorts of explanations, we in the end abandoned them as inadequate — not because we thought these factors were irrelevant to the advent of objectivity, but because they were only remotely relevant. What we sought was an explanation in which cause and effect meshed seamlessly, not one in which a powerful but remote force (one of those "revolutions") drove any number of the most diverse and scattered effects at a distance. We did not doubt either the existence or the efficacy of the remote forces, or even their ultimate links to our explanandum, the advent of objectivity. What we were after, however, were proximate links: an explanation on the same scale and of the same nature as the explanandum itself.

If training a telescope onto large, remote causes fails to satisfy, what about the opposite approach, scrutinizing small, local causes under an explanatory microscope? The problem here is the mismatch between the heft of explanandum and explanans, rather than the distance between them: in their rich specificity, local causes can obscure rather than clarify the kind of wide-ranging effect that is our subject here. Local circumstances that may seem to lie behind, for example, a change in surgical procedures in a late Victorian London hospital are missing in an industrial-scale, post-Second World War physics lab in Berkeley, and yet in both cases a similar phenomenon is at issue: the pitched battle over how to handle automatically produced scientific images. Looking at microcontexts tells us a great deal — but it can also occlude, like viewing an image pixel by pixel.

The very language of cause and effect dictates separate and heterogeneous terms: cause and effect must be clearly distinguished from each other, both as entities and in time. Perhaps this is why the metaphors of the telescope and microscope lie close to hand. Both are instruments for bringing the remote and inaccessible closer. But relationships of cause and effect do not exhaust explanation. Understanding can be broadened and deepened by exposing other kinds of previously unsuspected links among the phenomena in question, such as patterns that connect scattered elements into a coherent whole. What at first glance appeared to be apples and oranges turn out to grow from the same tree, different facets of the same phenomenon. This is the sort of intrinsic explanation that seems to us most illuminating in the case of objectivity.

What is the nature of objectivity? First and foremost, objectivity is the suppression of some aspect of the self, the countering of subjectivity. Objectivity and subjectivity define each other, like left and right or up and down. One cannot be understood, even conceived, without the other. If objectivity was summoned into existence to negate subjectivity, then the emergence of objectivity must tally with the emergence of a certain kind of willful self, one perceived as endangering scientific knowledge. The history of objectivity becomes, *ipso facto*, part of the history of the self.

Or, more precisely, of the scientific self: The subjectivity that nineteenth-century scientists attempted to deny was, in other contexts, cultivated and celebrated. In notable contrast to earlier views held from the Renaissance through the Enlightenment about the close analogies between artistic and scientific work, the public personas of artist and scientist polarized during this period. Artists were exhorted to express, even flaunt, their subjectivity, at the same time that scientists were admonished to restrain theirs. In order to qualify as art, paintings were required to show the visible trace of the artist's "personality"— a certain breach of faithfulness to what is simply seen. Henry James went so far as to strike the word "sincerity" from the art critic's vocabulary: praising the paintings of Alexandre-Gabriel Decamps in 1873, he observed that "he painted, not the thing regarded, but the thing remembered, imagined, desired — in some degree or other intellectualized."[20] Conversely, when James himself self-consciously tried to write with "objectivity," he described it as a "special sacrifice" of the novelist's art.[21] The scientists, for their part, returned the favor. For example, in 1866, the Paris Académie des Sciences praised the geologist Aimé Civiale's panoramic photographs of the Alps for "faithful representations of the accidents" of the earth's surface, which would be "deplorable" in art, but which "on the contrary must be [the goal] towards which the reproduction of scientific objects tends."[22] The scientific self of the mid-nineteenth century was perceived by contemporaries as diametrically opposed to the artistic self, just as scientific images were routinely contrasted to artistic ones.

Yet even though our quarry is the species, we cannot ignore the genus: however distinctive, the scientific self was nonetheless part of a larger history of the self.[23] Here we are indebted to recent work on the history of the self more generally conceived, particularly the explorations by the historian Pierre Hadot and the philosophers Michel Foucault and Arnold Davidson of the exercises that build and sustain a certain kind of self. In Greek and Roman Antiquity, for example, philosophical schools instructed their followers in the spiritual exercises of meditation, imagination of one's own death, rehearsal of the day's events before going to sleep, and descriptions of life's circumstances stripped of all judgments of good and evil.[24] Some of these techniques of the self involved only the mind; others, such as fasting or a certain habitually attentive attitude while listening, also made demands upon the body. Sometimes they were supplemented by external instruments, such as journals and other *hupomnemata* that helped disciples of this or that sage to lead the closely examined life.[25] Like gymnastics, spiritual exercises were supposed to be performed regularly and repeatedly, to prepare the self of the Epicurean or the Stoic acolyte to receive the higher wisdom of the master.

Although the scientific self of objectivity of course arose in an entirely different historical context and aimed at knowledge rather than enlightenment, it, too, was realized and reinforced by specialized techniques of the self: the keeping of a lab notebook with real-time entries, the discipline of grid-guided drawing, the artificial division of the self into active experimenter and passive observer, the introspective sorting of one's own sensations into objective and subjective by sensory physiologists, the training of voluntary attention. These techniques of the self were also practices of scientific objectivity. To constrain the drawing hand to millimeter grids or to strain the eye to observe the blood vessels of one's own retina was at once to practice objectivity and to exercise the scientific self.

Scientific practices of objectivity were not, therefore, merely illustrations or embodiments of a metaphysical idea of self. That is, our view is not that there was, before the relevant scientific work, an already-established, free-floating scientific self that simply found application in the practices of image-making. Instead, the broader notion of (for example) a will-based scientific self was articulated — built up, reinforced — through concrete acts, repeated thousands of times in a myriad of fields in which observers struggled to act, record, draw, trace, and photograph their way to minimize the impact of their will. Put another way, the broad notion of a will-centered self was, during the nineteenth century, given a specific axis: a scientific self grounded in a will to willessness at one pole, and an artistic self

38

that circulated around a will to willfulness at the other. Forms of scientific self and epistemic strategies enter together.

Epistemic Virtues

Understanding the history of scientific objectivity as part and parcel of the history of the scientific self has an unexpected payoff: what had originally struck us as an oddly moralizing tone in the scientific atlas makers' accounts of how they had met the challenge of producing the most faithful images now made sense. If knowledge were independent of the knower, then it would indeed be puzzling to encounter admonitions, reproaches, and confessions pertaining to the character of the investigator strewn among descriptions of the character of the investigation. Why does an epistemology need an ethics? But if objectivity and other epistemic virtues were intertwined with the historically conditioned person of the inquirer, shaped by scientific practices that blurred into techniques of the self, moralized epistemology was just what one would expect. Epistemic virtues would turn out to be literal, not just metaphorical, virtues.

This would take techniques of the self far beyond the ancient directive to "know thyself," which Hadot and Foucault associated with programs of spiritual exercises. Epistemic virtues in science are preached and practiced in order to know the world, not the self. One of the most deeply entrenched narratives about the Scientific Revolution and its impact describes how knower and knowledge came to be pried apart, so that, for example, the alchemist's failure to transmute base metals into gold could no longer be blamed on an impure soul.[26] Key epistemological claims concerning the character of science, which was, in principle, public and accessible to knowers everywhere and always, depend on the schism between knower and knowledge. Of course, certain personal qualifications were still deemed important to the success of the investigation: patience and attentiveness for the observer, manual dexterity for the experimenter, imagination for the theorist, tenacity for all. But these qualities have been seen in most accounts of modern science as matters of competence, not ethics.

Yet the tone of exhortation and admonition that permeates the literature of scientific instruction, biography, and autobiography from the seventeenth century to the present is hardly that of the pragmatic how-to manual. The language of these exhortations is often frankly religious, albeit in different registers — the humility of the seeker, the wonder of the psalmist who praises creation, the asceticism of the saint. Much of epistemology seems to be parasitic upon religious impulses to discipline and sacrifice, just as much of metaphysics seems to be parasitic upon theology. But even if religious overtones are absent or dismissed as so much window dressing, there remains a core of ethical imperative in the literature on how to do science and become a scientist. The mastery of scientific practices is inevitably linked to self-mastery, the assiduous cultivation of a certain kind of self. And where the self is enlisted as both sculptor and sculpture, ethos enters willy-nilly. It is useful for our purposes to distinguish between the ethical and the moral: *ethical* refers to normative codes of conduct that are bound up with a way of being in the world, an ethos in the sense of the habitual disposition of an individual or group, while *moral* refers to specific normative rules that may be upheld or transgressed and to which one may be held to account.

It is not always the same kind of ethos, or the same kind of self, that is involved: both have histories. In the period covered by this book, ethics shift from the regimens of upbringing and habit associated with the Aristotelian tradition to the stern Kantian appeal to autonomy; selves mutate from loose congeries of faculties ruled by reason to dynamic subjectivities driven by will. These changes leave their mark on the epistemologies of science and on scientific selves. It is perhaps conceivable that an epistemology without an ethos may exist, but we have yet to encounter one. As long as knowledge posits a knower, and the knower is seen as a potential help or hindrance to the acquisition of knowledge, the self of the knower will be at epistemological issue. The self, in turn, can be modified only with ethical warrant. (For this reason, even merely prudent bodily regimens of diet and exercise have, from Antiquity to the present, had a strong tendency to take on a moral tinge.) Extreme modifications of the self, through the mortification of flesh and spirit, are *prima facie* evidence of ethical virtuosity in numerous periods and cultures. Science is no exception, as the heroic literature on voyages of exploration, self-experimentation, and maniacal dedication testify.[27]

Epistemic virtues are virtues properly so-called: they are norms that are internalized and enforced by appeal to ethical values, as well as to pragmatic efficacy in securing knowledge. Within science, the specific values and related techniques of the self in question may contrast sharply with those of ancient religious and philosophical sects intent upon rites of purification and initiation preparatory to the reception of wisdom. This is why the rhetoric of the alchemists, Paracelsians, and other early modern reformers of knowledge and society rings so strangely in modern (or even eighteenth-century) ears. These visionaries sought wisdom, not just truth, and enlightenment, not just knowledge. Post-seventeenth-century epistemic virtues differ accordingly in their aims, content, and means. But they are alike in their appeals to certain tailor-made techniques of the self that were tightly interwoven with scientific practices. It is precisely this close fit between techniques and practices that supplies the rationale for the at-first-glance-roundabout strategy of studying notions as abstract as truth and objectivity through concrete ways of making images for scientific atlases. Epistemic virtues earn their right to be called virtues by molding the self, and the ways they do so parallel and overlap with the ways epistemology is translated into science.

New epistemic virtues come into being; old ones do not necessarily pass away. Science is fertile in new ways of knowing and also productive of new norms of knowledge. Just as the methods of experiment or of statistical inference, once invented and established, survive the demise of various scientific theories, so epistemic virtues, once entrenched, seem to endure — albeit to differing degrees in different disciplines. But the older ones are inevitably modified by the very existence of the newer ones, even if they are not replaced outright. Truth-to-nature after the advent of objectivity is a different entity, in both precept and practice, than before. The very multiplicity of epistemic virtues can cause confusion and even accusation, if adherents of one are judged by the standards of another. Scientific practices judged laudable by the measure of truth-to-nature — such as pruning experimental data to eliminate outliers and other dubious values — may strike proponents of objectivity as dishonest. Even without head-on collisions, the presence of alternatives, however mistily articulated, places an onus of justification on practitioners, as we shall see in the case of the atlas makers who wrestled with the merits of drawings versus photographs, idealization versus naturalism, or symbols versus images. One reason to write the history of

epistemic virtues, and to write it through a medium as specific as scientific atlas images, is that the existence and distinctness of these
virtues is clarified — as well as the possibility, even, in some cases, the
necessity of choice among them. History alone cannot make the
choice, any more than it can make the choice among competing
moral virtues. But it can show that the choice exists and what hinges
on it.

The Argument

Each chapter of this book, with a single deliberate exception, begins
with one or more images from a scientific atlas. These images lie at
the heart of our argument. We want to show, first of all, how epistemic virtues can be inscribed in images, in the ways they are made,
used, and defended against rivals. Chapters Two and Three set out a
contrast between atlas images designed to realize epistemic virtues
of truth-to-nature, on the one hand, and mechanical objectivity, on
the other. Eighteenth-century and early nineteenth-century anatomists and naturalists and their artists worked in a variety of media
(engraving, mezzotint, etching, and, later, lithography) and with a
variety of methods (from freehand sketching to superimposed grids
to the camera obscura). But almost all the atlas makers were united
in the view that what the image represented, or ought to represent,
was not the actual individual specimen before them but an idealized,
perfected, or at least characteristic exemplar of a species or other
natural kind. To this end, they carefully selected their models,
watched their artists like hawks, and smoothed out anomalies and
variations in order to produce what we shall call "reasoned images."
They defended the realism — the "truth-to-nature" — of underlying
types and regularities against the naturalism of the individual object,
with all its misleading idiosyncrasies. They were painstaking to the
point of fanaticism in the precautions they took to ensure the fidelity
of their images, but this by no means precluded intervening in every
stage of the image-making process to "correct" nature's imperfect
specimens.

   In the middle decades of the nineteenth century, at different rates
and to different degrees in various disciplines, new, self-consciously
"objective" ways of making images were adopted by scientific atlas
makers. These new methods aimed at automatism: to produce

42

images "untouched by human hands," neither the artist's nor the scientist's. Sometimes but not always, photography was the preferred medium for these "objective images." Tracing and strict measuring controls could also be enlisted to the cause of mechanical objectivity, just as photographs could conversely be used to portray types. What was key was neither the medium nor mimesis but the possibility of minimizing intervention, in hopes of achieving an image untainted by subjectivity. The truth-to-nature practices of selecting, perfecting, and idealizing were rejected as the unbridled indulgence of the subjective fancies of the atlas maker — the arc retraced by Worthington's conversion from truth-to-nature symmetry to the "objective view" described in the Prologue. These older practices did not disappear, any more than drawing did, but those who stuck to them found themselves increasingly on the defensive. Yet even the most convinced proponents of mechanical objectivity among the scientific atlas makers acknowledged the high price it commanded. Artifacts and incidental oddities cluttered the images; the objects depicted might not be typical of the class they were supposed to represent; atlas makers had to exercise great self-restraint so as not to smuggle in their own aesthetic and theoretical preferences. These features of objective atlases were experienced by authors as necessary but painful sacrifices. Mechanical objectivity was needed to protect images against subjective projections, but it threatened to undermine the primary aim of all scientific atlases, to provide the working objects of a discipline.

At this juncture, we step back from the atlas images themselves: in Chapter Four we embed the changes described in Chapters Two and Three within the history of the scientific self. We first follow the scientific reception of the post-Kantian vocabulary of objectivity and subjectivity in three different national contexts, using the German physicist and physiologist Hermann von Helmholtz, the French physiologist Claude Bernard, and the British comparative anatomist Thomas Henry Huxley as our guides. Despite wide divergences on the usage of the new terminology, these influential scientists agreed on the epistemological import of the objective-subjective distinction for their own experience of ever-accelerating scientific change. We then turn to the new kind of scientific self captured by the new terminology. The self imagined as a subjectivity is not the same as the self

43

imagined as a polity of mental faculties, as in Enlightenment associationist psychology, or as an archaeological site of conscious, subconscious, and unconscious levels, as in early twentieth-century models of the mind. The history of the scientific self was part of these broader developments, but it had its own specific character. We examine it both macroscopically, from the standpoint of the literature of scientific personas — exempla of scientific lives — and microscopically, from the standpoint of detailed activities like keeping a notebook of observations or training voluntary attention, the nodes at which scientific practices and techniques of the self intersect.

Alongside the epistemic virtues of truth-to-nature, mechanical objectivity, and trained judgment emerges a portrait gallery of scientific exempla: the sage, whose well-stocked memory synthesizes a lifetime of experience with skeletons or crystals or seashells into the type of that class of objects; the indefatigable worker, whose strong will turns inward on itself to subdue the self into a passively registering machine; the intuitive expert, who depends on unconscious judgment to organize experience into patterns in the very act of perception. These are exemplary personas, not flesh-and-blood people, and the actual biographies of the scientists who aspired to truth-to-nature, mechanical objectivity, and trained judgment diverge significantly from them. What interests us is precisely the normative force of these historically specific personas, and indeed the very distortions required to squeeze biographies into their mold, to transmute quirky individuals into exempla. These efforts are evidence of the minatory force of epistemic virtues. We are still more interested in the minutiae of the ways of seeing, writing, attending, remembering, and forgetting that concretize personas in persons and do so collectively, at least in situations in which scientific pedagogy has been institutionalized. For an account of the forging of the scientific self, pedagogy is central — as central as Plato's Academy or Aristotle's Lyceum were for the forging of the philosophical self in Antiquity.

The calibration of the eye — being taught what to see and how to see it — was a central mission of the scientific atlas. Atlases refined raw experience by weeding out atypical variations and extraneous details. Starting in the mid-nineteenth century, however, the strictures of mechanical objectivity cast doubt upon judgments of the typical and the essential as intrusions of dangerous subjectivity. Bet-

44

ter to present the object just as it was seen, to the point of leaving in scratches left by lenses or accepting distortions in perspectives introduced by the two-dimensional plane of the photograph. Some atlas makers drew the logical conclusion from these *laissez-voir* policies: readers were obliged somehow to figure out for themselves what the working objects of the discipline were; the objective atlas maker forbore to advise them. The very rationale for scientific atlases crumbled. In late nineteenth- and early twentieth-century science, this crisis provoked two diametrically opposed responses which are treated in the next two chapters. One sought to abolish images (though not diagrams) altogether, in the name of an intensified, "structural" objectivity (Chapter Five); the other abandoned objectivity in favor of trained judgment (Chapter Six).

Chapter Five alone begins without an image. Structural objectivity waged war on images in science. Its proponents, who were mostly mathematicians, physicists, and logicians, carried the selfdenial of mechanical objectivity to new extremes. Not content to censor the impulse to select and perfect images, they called for a ban on images, even on mathematical intuitions, as inherently subjective. They understood the threat of subjectivity in different terms than the advocates of mechanical objectivity had: the enemy was no longer the willful self that projected perfections and expectations onto the data; rather, it was the private self, locked in its own world of experience, which differed qualitatively from that of all other selves.

This conviction that much of mental life, especially sensations and representations, was incorrigibly private and individualized was itself the product of a highly successful late nineteenth-century scientific research program in sensory physiology and experimental psychology. Confronted with results showing considerable variability in all manner of sensory phenomena, some scientists took refuge in structures. These were, they claimed, the permanent core of science, invariant across history and cultures. Just what these structures were — differential equations, the laws of arithmetic, logical relationships — was a matter of some debate. But there was unanimity among thinkers as diverse as the logician Gottlob Frege, the mathematician Henri Poincaré, and the philosopher Rudolf Carnap that objectivity must be about what was communicable everywhere and always

45

among all human beings — indeed, all rational beings, Martians and monsters included. The price of structural objectivity was the suppression of individuality, including images of all kinds, from sensations of red to geometrical intuitions. This austere brand of objectivity is still alive and well among philosophers.[28]

But structural objectivity found little favor among the scientific atlas makers. How could they dispense with images? These scientists of the eye sought less draconian solutions to the crisis of mechanical objectivity. Chapter Six surveys these responses. Around the turn of the twentieth century, many scientists began to criticize the mechanically objective image: it was too cluttered with incidental detail, compromised by artifacts, useless for pedagogy. Instead, they proposed recourse to trained judgment, not hesitating to enhance images or instrument readings to highlight a pattern or delete an artifact. These self-confident experts were not the seasoned naturalists of the eighteenth century, those devotees of the cult of the genius of observation. It did not take extraordinary talents of attention and memory plus a lifetime's experience to discern patterns; ordinary endowments and a few years of training could make anyone an expert. Nor did the expert seek to perfect or idealize the depicted object; it was enough to separate signal from noise in order to produce the "interpreted image." Far from flexing the conscious will, the experts relied explicitly on unconscious intuition to guide them. In place of the paeans to hard work and self-sacrifice so characteristic of mechanical objectivity, practitioners of trained judgment professed themselves unable to distinguish between work and play — or, for that matter, between art and science. They pointed out the inadequacy of algorithms to distinguish pion from muon tracks in bubble-chamber photographs or the electroencephalograms of seizures caused by grand mal and petit mal epilepsy, instead surrendering themselves to the quasi-ludic promptings of well-honed intuitions.

There are novelties yet in store. We close, in Chapter Seven, with a glimpse of a new kind of atlas image — for example, one of the flow of turbulent fluids — constructed by computer simulations. These images no longer *represent* a particular fluid at a certain place and time; they are products of calculations hovering in the hybrid space between theory and experiment, science and engineering. In some of them, making and seeing are indistinguishable: the same manipulation of an atomic force microscope, for example, rolls a nanotube and projects its image. Representation of nature here gives way to presentation: of built objects, of marketable products, even of works of art. Out of the fusion of science and engineering is emerging a new ethos, one that is disturbing professional identities left and right. Once again, unease about the role and persona of the scientist is a signal that there is digging to be done — digging into the nature of the image, the dynamics of image production and use, and the status of who the scientist is or aspires to be.

Both the scope and the narrative shape of this book contrast with much of the best work in the history of science published in the past two decades, although the book is gratefully indebted to that scholarship. The lessons of these rich histories of science in context inform every page of this book. Yet we have chosen to tell this story not as a microhistory, thickly described and densely embedded in local circumstances, or even as series of such finely textured episodes. Still less is this book intended as a collection of case studies, an induction over instances in the service of a universal claim. Our study is unusually broad in geographic, chronological, and disciplinary sweep: it attempts a panoramic view of developments spread over the eighteenth through the early twentieth centuries and situated in Europe and the United States. The periodization we have adopted cuts across standard divides between the first and second Scientific Revolutions, between early modern and modern. More significantly, the momentum and contours of our periodization diverge from those of either gradual development or sharp rupture. The import and justification of these departures in scope and periodization will, we hope, be made clear by the body of the book: the proof of the writing is in the reading. But just because they are departures, it is worth being explicit at the outset about what they are.

Some significant historical phenomena are invisible at the local level, even if their manifestations must by definition be located somewhere, sometime. There are developments that unfold on a temporal and geographic scale that can only be recognized at the local level once they have been spotted from a more global perspective. Just as no localized observer alone can detect the shape of a storm front or the distribution of an organic species, so some historical phenomena can be discerned only by integrating information

47

from a spread of contexts. These phenomena will inevitably be inflected by local context, but without losing their identity. The existence, emergence, and interaction of epistemic virtues in science are phenomena on this larger scale. They are not confined to chemistry or physiology, Germany or France, a decade or even a generation. By combining broad scope with narrow focus, we aim to do justice to scale as well as texture.

The scope of this book is broad, but it is not comprehensive. It does not encompass all science, all scientists, or even all scientific images for the places and periods it treats. It is about a particular class of images in the service of a particular aspect of science: scientific atlases as an expression of historically-specific hierarchies of epistemic virtues.

Atlas images underpin other forms of scientific visualization: they define the working objects of disciplines and at the same time cultivate what might be called the *disciplinary eye*, analogous to what art historians call the *period eye*. Atlas images are therefore not just one class of images among many in science. They are the visual foundations upon which many observational disciplines rest. If atlases ground disciplines, epistemic virtues cut across them. Neither truth-to-nature nor mechanical objectivity nor trained judgment ever permeated science in its entirety, but they nonetheless overflowed the boundaries of any one discipline or even any single division of disciplines. Epistemic virtues have left their mark in the life as well as the physical sciences, in the field as well as in the laboratory. They are not ubiquitous, but in their cultivation of forms of scientific sight they are pervasive in their reach and profound in their impact.

In the first instance we base our claims about the significance of epistemic virtues on the significance of the atlas images. The atlases are not the only evidence of the existence and force of epistemic virtues such as objectivity, but as repositories of the images of record, they carry considerable weight. When similar practices justified in similar terms turn up roughly at the same time in atlases of crystallography and clinical pathology, of galaxies and grasses, these analogies give strong reasons for believing in transformations that simultaneously span many disciplines and penetrate to the roots of each. Where else might one expect such evidence? Wherever epistemological fears about this or that obstacle to knowledge are acute. As subsequent chapters will show, these fears are as various as their remedies. But in all cases, it is fear that drives epistemology, including the definition of what counts as an epistemic vice or virtue. Conversely, science pursued without acute anxiety over the bare existence of its chosen objects and effects will be correspondingly free of epistemological preoccupations. An emerging scientific-engineering ethos in the twenty-first century, for example, worries more about robustness than mirages, as we shall see in Chapter Seven. Anxiety about virtue, epistemic or otherwise, is neither omnipresent nor perpetual.

But when epistemic anxiety does break out, scientific atlases by their very nature register it early and emphatically. We therefore use atlases as a touchstone to reveal the changing norms that govern the right way to see and depict the working objects of science. These image compendiums lead us outward along various paths, sometimes to well-known scientists such as Helmholtz or Poincaré, at other times to less celebrated figures, laboratories, and representational techniques. We always return to our central question: how does the right depiction of the working objects of science join scientific sight to the scientific self?

The history of science has been imagined in both uniformitarian and catastrophist terms, that is, as either the steady, continuous growth of knowledge or the intermittent eruption of revolutionary novelty. However apt these schemata may be for one or another episode in the history of specific scientific theories or practices, they are a bad fit for the phenomena we are tracking in this book. Objectivity is neither the fruit of an incremental evolution nor a sudden explosion on the scientific scene — nor an all-at-once Gestalt switch. Scattered instances of scientific objectivity in word and deed started to appear in the 1830s and 1840s, but they did not thicken into a swarm until the 1860s and 1870s. Instead of either a smooth slope or an abrupt precipice, the emergence of scientific objectivity (and other epistemic virtues) might be imagined on the analogy of an avalanche: at first, a few tumbling rocks, falling branches, and minor snow slides amount to nothing much, but then, when conditions are ripe, individual events, even small ones, can trigger a massive, downward rush.

Of course, a great deal hinges on just how to specify "when

conditions are ripe." In the case of the avalanche, there will often be complicated combinations of slope, terrain, saturation, and snow-layer binding that set up the instability. The historical sequence of epistemic virtues also supplies something close to preconditions of instability. Even if conditions are known to be extremely dangerous, no one could say precisely when — or how — an avalanche might start. Like the formation of an avalanche, the potential for a previous epistemic virtue to be transvalued into an epistemic vice is localized in time, but not with on-the-dot punctuality. Just as in the case of the avalanche, preconditions must coincide with contingent circumstances. We can identify a rapidly proliferating and mutually conflicting set of ideals, each claiming to be the right way to depict the splash of a drop or the structure of a blood cell. We cannot say exactly when or why in a given domain scientists will begin to insist upon an "objective view." Rather than razor-sharp boundaries between periods, we should therefore expect first a sprinkling of interventions, which then briskly intensify into a movement, as fears are articulated and alternatives realized — the unleashing of an avalanche.

But the ambitious historian may persist: Isn't this problem, aren't all problems of historical timing, just due to insufficient information? If some Laplacean demon would turn its infinite industry and intelligence to a complete specification of all the circumstances at a given time and place, wouldn't it be possible to explain the emergence of objectivity — or, for that matter, the outbreak of the French Revolution, the invention of the magnetic compass, the rise of chivalry, yes, even the onset of an avalanche — with pinpoint precision? This is a persistent and revealing historical fantasy. It is fantastical to imagine that we can deterministically identify not only the "trigger" in historical processes — but also the detailed route of development. It is impossible not only because it is practically beyond our grasp, but also because it is incoherent. Just as in the case of the utterly useless Borgesian map that reproduces an empire in one-to-one facsimile, the Borgesian archive of all historical information would duplicate history, not explain it. Forget the thousands of microtriggers. Our interest here is, on the one hand, to capture the conditions of epistemic instability, and, on the other, to identify the new patterns that result — the most striking of which was objectivity.

Objectivity in Shirtsleeves

By this point, many readers will be perplexed by what is missing in this book about scientific objectivity. Some, persuaded that objectivity is a mirage, will ask: Where are the criticisms of the epistemological pretensions of objectivity? Does anyone really still believe in the possibility of the view from nowhere, a God's-eye perspective of the universe? Others, all too convinced of the existence of objectivity, will demand: What about the moral blindness of objectivity, its monstrous indifference to human values and emotions? Isn't overweening objectivity the culprit in so many techno-scientific disasters of the modern world? The one side doubts the possibility of objectivity; the other, its desirability. Both sides will protest in chorus: How can an account of the epistemological and moral aspects of objectivity decline to grapple with these questions?

Our answer is that before it can be decided whether objectivity exists, and whether it is a good or bad thing, we must first know what objectivity *is* — how it functions in the practices of science. Most accounts of objectivity — philosophical, sociological, political — address it as a concept. Whether understood as the view from nowhere or as algorithmic rule-following, whether praised as the soul of scientific integrity or blamed as soulless detachment from all that is human, objectivity is assumed to be abstract, timeless, and monolithic. But if it is a pure concept, it is a less like a bronze sculpture cast from a single mold than like some improvised contraption soldered together out of mismatched parts of bicycles, alarm clocks, and steam pipes.

Current usage allows a too easy slide among senses of objectivity that are by turns ontological, epistemological, methodological, and moral. Yet these various senses of the objective cohere neither in precept nor in practice. "Objective knowledge," understood as "a systematized theoretical account of how the world really is," comes as close to truth as today's timorous metaphysics will permit.[29] But even the most fervent advocate of "objective methods" in the sciences — be those methods statistical, mechanical, numerical, or otherwise — would hesitate to claim that they guarantee the truth of a finding.[30] Objectivity is sometimes construed as a method of understanding, as when epistemologists ponder how reliance "on the

51

specifics of the individual's makeup and position in the world, or on the character of the particular type of creature he is" might distort his view of the world.[31] And sometimes objectivity means an attitude or ethical stance, which is grounds for praise as calm neutrality or blame as icy impersonality — as proof against "blind emotional excitement... which in the end may lead to social disaster," or as an arrogant and deceitful pretense, "the God trick."[32] The debates in political, philosophical, and feminist circles now raging over the existence, desirability, or both of objectivity in science assume rather than analyze this smear of meanings, leaping from metaphysical claims of universality to moral reproaches of indifference in a single paragraph.[33] This is why conceptual analysis alone seems to be an unpromising tool for the task of understanding what objectivity is, much less how it came to be what it is.

But if actions are substituted for concepts and practices for meanings, the focus on the nebulous notion of objectivity sharpens. Scientific objectivity resolves into the gestures, techniques, habits, and temperament ingrained by training and daily repetition. It is manifest in images, jottings in lab notebooks, logical notations: objectivity in shirtsleeves, not in a marble chiton. This is a view of objectivity as constituted from the bottom up, rather than from the top down. It is by performing certain actions over and over again — not only bodily manipulations but also spiritual exercises — that objectivity comes into being. To paraphrase Aristotle on ethics, one becomes objective by performing objective acts. Instead of a pre-existing ideal being applied to the workaday world, it is the other way around: the ideal and ethos are gradually built up and bodied out by thousands of concrete actions, as a mosaic takes shape from thousands of tiny fragments of colored glass. To study objectivity in shirtsleeves is to watch objectivity in the making.

If we are right about this, then a study like this one should ultimately shed light on the grand epistemological visions and moral anxieties now associated with scientific objectivity. It should be possible to trace how specific practices came to be metaphorically extrapolated by the philosophical and cultural imagination into dreams of a view from nowhere or nightmares about heartless technocrats. It may also be possible to unravel the conceptual tangle of the current meanings of objectivity. If the concept grew historically, by gradual accretion and extension from practices, it is not so surprising that its structure is confused rather than crystalline. Chapter Seven reexamines these questions from the standpoint of the history of scientific objectivity narrated in the foregoing chapters.

More fundamentally, a historical perspective also shifts the ethical meaning of objectivity. If objectivity seems indifferent to familiar human values, this is because it is itself a code of values. The values of objectivity are admittedly specific and strange: to refrain from retouching a photograph, or removing an artifact, or completing a fragmentary specimen is not obviously an act of virtue — not even to all other scientists, much less to humanity at large. Nor will everyone acknowledge resolute passivity or willed willessness as values worth aspiring to. These are values in the service of the True, not just the Good. But they are genuine values, rooted in a carefully cultivated self that is also the product of history. The surest sign that the values of objectivity deserve to be called such is that violations ignite indignation among those who profess them. Viewed in this light, whether objectivity is a good or bad thing from a moral standpoint is no longer a question about alleged neutrality toward all values, but one about allegiance to a hard-won set of coupled values and practices that constitute a way of scientific life.

Look one last time at the three images with which we began. Each is, in its way, a faithful representation of nature. But they are not facsimiles of nature, not even the photograph; they are nature perfected, excerpted, smoothed — in short, nature known. These images substitute for things, but they are already admixed with knowledge about those things. In order for nature to be knowable, it must first be refined, partially converted into (but not contaminated by) knowledge. These images represent knowledge about nature, as well as nature itself — indeed, they represent distinct visions of what knowledge is and how it is attained: truth-to-nature, objectivity, trained judgment. Finally, they represent the knower. Behind the flower, the snowflake, the solar magnetogram stand not only the scientist who sees and the artist who depicts, but also a certain collective way of knowing. This knowing self is a precondition for knowledge, not an obstacle to it. Nature, knowledge, and knower intersect in these images, the visible traces of the world made intelligible.

53

CHAPTER TWO

Truth-to-Nature

Before Objectivity
In 1737, the young Swedish naturalist Carolus Linnaeus published a sumptuous flora of the plants cultivated in the well-stocked garden of George Clifford, an Amsterdam banker and director of the Dutch East India Company: the *Hortus Cliffortianus* (*Clifford's Garden*).[1] No expense had been spared to render the book beautiful as well as useful; Linnaeus's wealthy patron had engaged the services of the German botanical illustrator Georg Dionysius Ehret to prepare drawings of specimens, both fresh and dried, and the renowned Dutch artist Jan Wandelaar to engrave the drawings (see figure 2.1). All participants in the venture — patron, naturalist, and artists — intended it to mark an epoch in the history of botany. The book's frontispiece showed allegorical representations of the continents bearing plant offerings to an Apollo figure drawn with Linnaeus's features (see figure 2.2). Less bombastically but more influentially, working on the *Hortus Cliffortianus*, with access to Clifford's ample botanical library, as well as his garden and greenhouse, provided Linnaeus with the practical basis for his subsequent publications on botanical nomenclature, classification, description, and illustration, which have profoundly marked the development of the science of botany ever since.[2]

Yet Linnaeus's descriptions and the illustrations he commissioned and supervised closely for the *Hortus Cliffortianus* cannot be called objective. This is not just a historian's quibble about anachronism, a finicky objection to applying a term Linnaeus and his mid-eighteenth-century contemporaries would have found quaintly scholastic, if they recognized it at all.[3] Nor is it a claim that Linnaeus's work

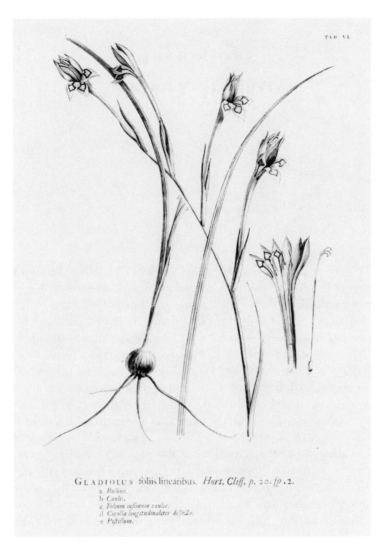

GLADIOLUS foliis linearibus. *Hort. Cliff.* p. 20. fp .2.
a *Bulbus.*
b *Caulis.*
c *Folium infimum caulis.*
d *Corolla longitudinaliter dissecta.*
e *Pistillum.*

Fig. 2.1. Species Archetype. *Gladiolus foliis linearibus*, Carolus Linnaeus, *Hortus Cliffortianus* (Amsterdam: n.p., 1737), table 6 (courtesy of Staats- und Universitäts-bibliothek Göttingen). Drawn by Georg Dionysius Ehret and engraved by Jan Wandelaar under Linnaeus's close supervision, this plate highlights the distinguishing feature of this species of gladiolus: its long, straight leaves (note the magnified leaf prominently placed in the center of the plate). Like the other figures in the *Hortus Cliffortianus*, this one aimed to convey visually the desiderata of an ideal botanical description, which according to Linnaeus should be "brief, certain, and apt" ("Lectori Botanico," *ibid.*, n.p.).

Fig. 2.2. Allegory of Botany Reformed. Frontispiece, Carolus Linnaeus, *Hortus Cliffortianus* (Amsterdam: n.p., 1737) (courtesy of Staats- und Universitätsbibliothek Göttingen). Designed and executed by Jan Wandelaar, who also wrote an accompanying explanation in verse, the allegorical engraving shows Europe being brought "the most noble plants, fruits, flowers / That ASIA, AFRICA and AMERICA can boast" ("Verklaarung van de Tytelprent," n.p.). (The gladiolus in fig. 2.1, for example, was native to Africa.) In the foreground, putti display the tools of scientific gardening: a shovel and a brazier, but also a thermometer and a geometric plan of the beds symbolic of the book's grand ambitions.

was "unscientific," flawed by prejudice, ignorance, or incompetence. The standard Linnaeus and other Enlightenment savants upheld was *truth-to-nature* rather than *objectivity*. The implications of this distinction reach far beyond the merely verbal: methods, metaphysics, and morals were all at stake. Truth-to-nature and objectivity are both estimable epistemic virtues, but they differ from each other in ways that are consequential for how science is done and what kind of person one must be to do it. Truth came before and remains distinct from objectivity, as the example of Linnaeus testifies.

Seeking truth is the ur-epistemic virtue, with its own long and variegated history, of which the quest for truth-to-nature is only one strand.[4] Among scientific atlas makers, truth-to-nature emerges as a prominent epistemic virtue in the early eighteenth century — Linnaeus is one of its earliest and most influential proponents — as a reaction to the perceived overemphasis by earlier naturalists on the variability and even monstrosity of nature, as we shall see below. Like most variants of truth, truth-to-nature had a metaphysical dimension, an aspiration to reveal a reality accessible only with difficulty. For Enlightenment naturalists like Linnaeus, this reality did not entail a commitment to Platonic forms at the expense of the evidence of the senses. On the contrary, sharp and sustained observation was a necessary prerequisite for discerning the true genera of plants and other organisms. The eyes of both body and mind converged to discover a reality otherwise hidden to each alone.

To see like a naturalist required more than just sharp senses: a capacious memory, the ability to analyze and synthesize impressions, as well as the patience and talent to extract the typical from the storehouse of natural particulars, were all key qualifications. The ideal Enlightenment naturalist, sometimes described as a "genius of observation," was endowed with an "expansive mind, master of itself, which never receives a perception without comparing it with a perception; who seeks out what diverse objects have in common and what distinguishes them from one another.... These are those men who go from observations upon observations to just consequences and who find only natural analogies."[5] Johann Wolfgang von Goethe, reflecting in 1798 on his research in morphology and optics, described the quest for the "pure phenomenon," which could be discerned only in a sequence of observations, never in an isolated instance. "To depict it, the human mind must fix the empirically variable, exclude the accidental, eliminate the impure, unravel the tangled, discover the unknown."[6] These were the concrete practices of abstract reason as understood by Enlightenment naturalists: selecting, comparing, judging, generalizing. Allegiance to truth-to-nature required that the naturalist be steeped in but not enslaved to nature as it appeared.

Linnaeus's ways of looking at, describing, depicting, and classifying plants were openly, even aggressively selective. Botanists must school themselves to concentrate on characters that are "constant, certain and organic"; they must not allow themselves to be distracted by irrelevant details of a plant's appearance and thereby unnecessarily multiply species: "93 [species] of tulips (where there is only one)." They must prevent their illustrators from rendering accidental traits, like color, as opposed to essential ones, like number, form, proportion, and position. "How many volumes have you written of specific names taken from colour? What tons of copper have you destroyed in making unnecessary plates [for engravings]?"[7]

Nor did Linnaeus strive for the self-effacement of latter-day scientists; nineteenth-century botanists would find his pronouncements too pontifical for the "self-abnegation" they demanded of themselves.[8] He, in turn, would have dismissed as irresponsible the suggestion that scientific facts should be conveyed without the mediation of the scientist and ridiculed as absurd the notion that the kind of scientific knowledge most worth seeking was that which depended least on the personal traits of the seeker. These later tenets of objectivity, as they were formulated in the mid-nineteenth century, would have contradicted Linnaeus's own sense of scientific mission. Only the keenest and most experienced observer — who had, like Linnaeus, inspected thousands of different specimens — was qualified to distinguish genuine species from mere varieties, to identify the true specific characters imprinted in the plant, and to separate accidental from essential features. Linnaeus was vehemently committed to the truth of his genera (and even to the truth of specific names), but not to objectivity, not even *avant la lettre*.

This chapter is about science before objectivity, about how the alternative epistemic way of life dedicated to "truth-to-nature" shaped the practices, personas, and, above all, the reasoned images of anatomy, botany, mineralogy, zoology, and other observational sciences from the early eighteenth through the mid-nineteenth centuries. Science pursued under the star of truth-to-nature rather than of objectivity *looked* different. To return briefly to the images of the *Hortus Cliffortianus*: the leaves of the *Gladiolus foliis linearibus*, drawn and engraved with such care by Ehret and Wandelaar, do not mimic those of any particular specimen; they do not even represent the general form of the entire species. Rather, they (like the species name Linnaeus gave the plant to signal its *differentia specifica*, "linear leaved") refer back to the essential leaf forms that, according to Linnaeus, were the underlying types of all leaves observed in individual plants. Divided into "simple," "composite," and "determinate" classes and further subdivided into subclasses ("triangular," "circular," "truncated"), these leaf schemata were presented in the book's very first figure, a visual key to the illustrations of species that followed (see figure 2.3). (The "linear"-type leaves of the *Gladiolus foliis linearibus* are number seven in the table.) A Linnaean botanical description singled out those features common to the entire species (the *descriptio*) as well as those that differentiated this species from all others in the genus (the *differentia*) but at all costs avoided features peculiar to this or that individual member of the species. The Linnaean illustration aspired to generality — a generality that transcended the species or even the genus to reflect a never seen but nonetheless real plant archetype: the reasoned image.[9] Types need not be depicted schematically, as this late eighteenth-century watercolor of leaf types by the Austrian botanical artist Franz Bauer shows (see figure 2.4). The type was truer to nature — and therefore more real — than any actual specimen.

Collectively, eighteenth-century atlas makers created a way of seeing, one that saw past the surfaces of plants, bones, or crystals to underlying forms. The choice of images that best represented "what truly is" engaged scientific atlas makers in ontological and aesthetic judgments that mechanical objectivity later forbade. Because the genre of the scientific atlas spans the mid sixteenth century to the present, it permits focused comparisons of ideals and practices associated with truth-to-nature, on the one hand, and objectivity, on the other — lofty abstractions that may otherwise dissipate into the metaphysical ether. In this chapter and the next, we will use images

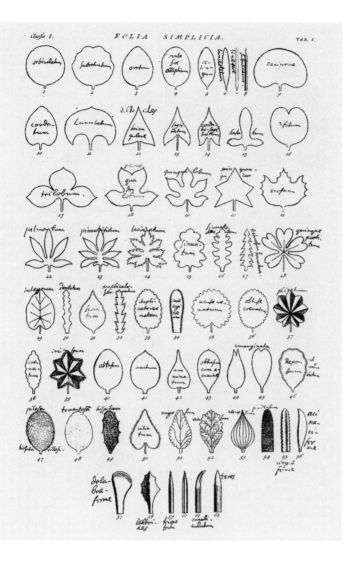

Fig. 2.3. "Types of Leaves." Carolus Linnaeus, *Hortus Cliffortianus* (Amsterdam: n.p., 1737), table 1 (courtesy of Staats- und Universitätsbibliothek Göttingen). The first plate in the volume shows simple leaf types to be used in botanical classification in deliberately schematic form, with descriptive Latin tags ("heart-shaped," "three-leaved"). The linear leaves that single out the *Gladiolus foliis linearibus* (fig. 2.1) are shown in the first row, no. 7.

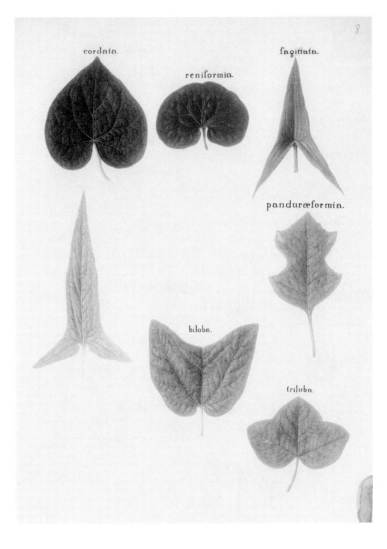

Fig. 2.4. Leaf Types Embodied. Aquarelle, Franz Bauer, Franz Bauer Nachlass, vol. 8, GR 2 COD MS. HIST. NAT. 94:VIII (courtesy of Staats- und Universitätsbibliothek Göttingen). Despite the apparent naturalism of this watercolor (probably executed *circa* 1790), the leaves depicted are the Linnaean types, labeled with the same names as the outlines in fig. 2.3: for example, "heart-shaped," "kidney-shaped," and "arrow-shaped," which correspond to nos. 9, 10, and 13 in the Linnaean schema. (Please see Color Plates.)

from scientific atlases — who made them, how, and to what end — to sharpen what may at first seem to be a paradoxical contrast between truth and objectivity, between reasoned and objective images.

Taming Nature's Variability

From the sixteenth century on, practitioners of the sciences of the eye have prepared visual surveys of their designated phenomena in the form of atlases, understood here as any compendium of images intended to be definitive for a community of practitioners. These profusely illustrated volumes depict carefully chosen observables — bodily organs, constellations, flowering plants, snowflakes — from carefully chosen points of view. As we noted in Chapter One, the purpose of these atlases was and is to standardize the observing subjects and observed objects of the discipline by eliminating idiosyncrasies — not only those of individual observers but also those of individual phenomena. Because we moderns habitually oppose the objectivity of things to the subjectivity of individuals, we fret most about idiosyncratic subjects: their "personal equations," their theoretical biases, their odd quirks. But idiosyncratic objects pose at least as great a threat to communal, cumulative science, for nature seldom repeats itself, variability and individuality being the rule rather than the exception. Even the geometric regularities of crystals are far from uniform, as the French mineralogist René-Just Haüy observed in his 1784 attempt to classify them: "Among crystals the varieties of the same kind often appear at first glance to have no relation to one another and sometimes even those [kinds] one detects become a new source of difficulties"[10] (see figure 2.5). Myriad accidents and perturbations cause deviations from mathematical perfection or organic types.

In addition to their primary function of standardizing objects in visual form, atlas pictures served other purposes in the natural sciences. They served the cause of public distribution of data for the scientific community, by preserving what is ephemeral and distributing what is rare or inaccessible to all who could purchase the volume, not just the lucky few who were in the right place at the right time with the right equipment. Seventeenth- and eighteenth-century voyages of exploration like those of Captain James Cook to the South Pacific took along not only naturalists to describe but also

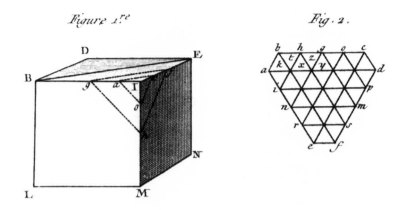

Fig. 2.5. Geometric Crystals. René Just Haüy, *Essai d'une théorie sur la structure des crystaux: Appliquée à plusieurs genres de substances crystallisées* (Paris: Chez Gogué & Née de la Rochelle, 1784), pl. 1, figs. 1–2 (courtesy of Staatsbibliothek zu Berlin Preussischer Kulturbesitz). Despite the variations and irregularities in individual crystals, Haüy maintained that all could be reduced to "a kernel of primitive form" by cutting diagonal sections parallel to the line BE. The "common fundamental forms" thereby revealed defined the various "species" of crystals that transcend the particularities of individuals (pp. 54–55).

artists to draw new flora and fauna; these images were almost always more lifelike (and intact) than the dried herbarium specimens or imperfectly preserved dead animals sent back to collections. Before early nineteenth-century improvements in taxidermy, images often supplied stay-at-home naturalists with their only exemplars of new species and genera.[11] As the Paris Académie des Sciences remarked, apropos of the 1807 publication of the discoveries of the South Seas voyages of François Péron and his artist, Charles Lesueur, the latter's drawings were decisive in combating the skepticism of European naturalists about "these extraordinary beings which seem to contradict our prior ideas" — as in the case of cassowaries, flightless birds[12] (see figure 2.6).

Pictures also served the cause of memory, for, as the atlas makers never tired of repeating, images are more vivid and indelible than

NOUVELLE-HOLLANDE : ÎLE DES KANGUROOS.

CASOAR de la N.^{le} Hollande. *(Casoarin. nova Hollandia. Lath.)*

1. *Casoar mâle.* 2. *Casoar femelle.* 3. *Jeune Casoar de 5 semaines environ. Les deux individus marqués de bandes longitudinales sont âgés de 10 à 15 jours.*

Fig. 2.6. "Cassowaries of New Zealand." François Péron, *Voyages de découvertes aux Terres australes* (Paris: Imprimerie impériale, 1807–1816), pl. 66 (courtesy of Staatsbibliothek zu Berlin Preussischer Kulturbesitz). Atlas images of animal species exotic to Europeans, such as these "cassowaries" (in fact emus), or difficult to preserve, such as jellyfish, bore witness to the existence of new species. They also served as objects of inquiry in their own right. Great care was taken in this atlas to convey colors, either by printing in multiple colors or by hand-coloring the printed illustrations. In this image, based on observations of multiple animals, the artist-naturalists have shown the adult male and female and the young of the species. The original field sketches show only the male (*left*). (Please see Color Plates.)

words. In his pioneering atlas of pathology, *Anatomie pathologique du corps humain* (*Pathological Anatomy of the Human Body*, 1829–1842), Jean Cruveilhier, the first holder of the chair of pathological anatomy in the Parisian medical faculty, underscored this point. In contrast to normal anatomy, in which there exist abundant opportunities to observe this or that organ "a second, a third, a twentieth time," the opportunities for the pathologist are rare and fleeting: "A lost occasion may perhaps never recur." Even an observer with the eyes of a lynx and the memory of an elephant cannot "fix the fugitive features, if he does not engrave them as if in bronze, so as to be able to represent them at will, to put them into relation with analogous facts."[13]

Finally, especially for early and mid-nineteenth-century authors, as we shall see in Chapters Three and Four, pictures served the cause of permanence. They would, it was hoped, endure as facts for tomorrow's researchers long after today's theories and systems had gone the way of crystalline spheres and animal spirits. The atlas distributed and preserved the working objects of science across space and time, enlarging the scope of collective empiricism.

There is no atlas in any field that does not pique itself on its fidelity to nature. But in order to decide whether an atlas picture is a faithful rendering of nature, the atlas maker must first decide what nature is. Which objects should be presented as the standard phenomena of the discipline, and from what viewpoint?

Starting in the mid-nineteenth century, as we shall see in Chapter Three, these choices triggered a crisis of anxiety and denial, for they seemed to be invitations to subjectivity. But Enlightenment atlas makers faced up to their task with considerably more confidence and candor. This is not to say that they abandoned themselves to subjectivity, in the dismissive sense of rendering specimens as their personal whims decreed. On the contrary, they were well-nigh maniacal in their precautions to ensure the fidelity of their figures, according to their own lights. However, they conceived of fidelity in terms of the exercise of informed judgment in the selection of "typical," "characteristic," "ideal," or "average" images: all these were varieties of the reasoned image. The essence of the atlas makers' task was to determine the essential. In their view, whatever merit their atlases possessed derived precisely from this discernment and from the breadth and depth of experience in their field upon which discernment rested. Later atlas makers, committed to mechanical objectivity, resisted intervention; their predecessors, committed to truth-to-nature, relished it.

Yet eighteenth-century atlas makers were not free of all epistemological anxieties. Their fears centered, rather, on the untamed variability, even monstrosity of nature. They were reacting against the preoccupation of many sixteenth- and seventeenth-century naturalists with what Francis Bacon had approvingly described in his *Novum organum* (1620) as "irregular or heteroclite" phenomena and "strange and monstrous objects, in which nature deviates and turns from her ordinary course."[14] Bacon had called for a collection of such oddities of nature, a "natural history of pretergenerations," as a corrective to the ingrained tendency of scholastic natural philosophers to generalize rashly from a handful of commonplace examples. Heeding Bacon's call, the earliest scientific societies filled their annals — the *Miscellanea curiosa* of the Schweinfurt Academia Naturae Curiosorum (established in 1652), the *Philosophical Transactions* of the Royal Society of London for Improving Natural Knowledge (established in 1660), the *Histoire* and *Mémoires* of the Paris Académie Royale des Sciences (established in 1666) — to overflowing with accounts of anomalies, singularities, and monstrosities of all kinds: strange lights in the sky, two-headed cats, luminescent shanks of veal, prodigious sleepers who slumbered for weeks on end.[15] (See figure 2.7.) These collections of anomalies and singularities, which were meant to hinder premature generalizations and promote exact observation of particulars, represent an epistemic way of life that was as opposed to that of truth-to-nature as the latter was to objectivity.

By the early eighteenth century, however, leading naturalists had begun to worry that the search for natural regularities was being overwhelmed by excessive scientific attention to nature's excesses.[16] Although anatomists might still signal anomalous conformations discovered in the course of their dissections, by the 1730s the emphasis in scientific inquiry had shifted to the quest for regularities glimpsed behind, beneath, or beyond the accidental, the variable, the aberrant in nature — the confusion of prepositions betokens metaphysical confusion about the goals of the search. Linnaeus went so far as to brand the plant varieties bred by gardeners and florists as monstrous

Fig. 2.7. Monstrous Birth. Monsieur Bayle, "A Relation of a Child which Remained Twenty Six Years in the Mothers Belly," *Philosophical Transactions* 139 (1677), pp. 979–80. The account is typical of the many reports of monsters, strange weather, and other singularities that filled the pages of the first scientific journals in the latter half of the seventeenth century. Reports like this one "took the pains to give an exact account" (*ibid.*, p. 979) of all details of an individual (and possibly unique) case, in contrast to the idealized and generalized images produced under the direction of mid-eighteenth-century naturalists such as Linnaeus .

and therefore as unworthy of scientific study: "The species of Botanists come from the All-wise hand of the Almighty, the varieties of Florists have proceeded from the Sport of Nature, especially under the auspices of the gardeners."[17]

As Linnaeus's appeal to the Almighty suggests, eighteenth-century attempts to overcome nature's profligate variability were often buttressed by an Enlightenment version of natural theology that characteristically praised the regularity of God's laws as more worthy of admiration than the exceptional marvel or miracle. Truth-to-nature, like objectivity, was historically specific. It emerged at a particular time and place and made a particular kind of science possible — a science about the rules rather than the exceptions of nature.

The Idea in the Observation

In the summer of 1794, Goethe recorded a "Fortunate Encounter" with Friedrich von Schiller. Although the two literary lions had initially regarded each other warily, they became friends through a discussion of Goethe's hypothesis concerning how all plants could be derived through metamorphosis from a single prototype, the *Urpflanze*. They famously differed on just what the *Urpflanze* was:

> Schiller: "That is not an observation from experience. That is an idea."
> Goethe: "Then I may rejoice that I have ideas without knowing it, and can even see them with my own eyes."[18]

How did ideas like the *Urpflanze* become visible on the page? What did truth-to-nature look like? Early atlas makers did not all interpret the notion of "truth-to-nature" the same way. The words *typical*, *ideal*, *characteristic*, and *average* are not synonymous, even though they all fulfilled the same standardizing purpose. These alternative ways of being true to nature suffice to show that concern for accuracy does not necessarily imply concern for objectivity. On the contrary: extracting nature's essences almost always required scientific atlas makers to mold their images in ways that their successors would reject as dangerously "subjective." Because all these methods of discovering the idea in the observation clashed with objectivity, later atlas makers tended to lump them together as regrettable meddling with the data. But in fact the practices of truth-to-nature fanned out into a spectrum of interventions.

In eighteenth-century atlases, "typical" phenomena were those that hearkened back to some underlying *Typus* or "archetype," and from which individual phenomena could be derived, at least conceptually. The typical is rarely, if ever, embodied in a single individual; nonetheless, the astute observer can intuit it from cumulative experience, as Goethe "saw" the *Urpflanze*. Goethe wrote of his archetype of the animal skeleton: "Hence, an anatomical archetype [*Typus*] will be suggested here, a general picture containing the forms of all animals as potential, one which will guide us to an orderly description of each animal.... The mere idea of an archetype in general implies that no particular animal can be used as our point of comparison; the par-

69

ticular can never serve as a pattern [*Muster*] for the whole."[19] This is not to say that the archetype wholly transcended experience, for Goethe claimed that it was derived from and tested by observation. However, observations in search of the typical must always be made in series, because single observations made by one individual can be highly misleading: "For the observer never sees the pure phenomenon [*das reine Phänomen*] with his own eyes; rather, much depends on his mood, the state of his senses, the light, air, weather, the physical object, how it is handled, and a thousand other circumstances."[20] (See figure 2.8.)

Typical images dominate the anatomical, botanical, and zoological atlases of the seventeenth through the mid-nineteenth centuries (and sometimes long thereafter), but not always in the unalloyed form celebrated by Goethe. Two important variants, which we shall call the "ideal" and the "characteristic," also appear in atlas illustrations of this period. The "ideal" image purports to render not merely the typical but the perfect, while the "characteristic" image locates the typical in an individual. Both ideal and characteristic images regularize the phenomena, and the fabricators of both insisted upon pictorial accuracy. But the ontology and aesthetics underlying each contrasted sharply with one another, as the following examples show.

With the collaboration of Wandelaar, the Dutch artist and engraver enlisted by Linnaeus,[21] Bernhard Siegfried Albinus, the professor of anatomy at Leiden, produced several of the most influential eighteenth-century anatomical atlases of the idealized sort, including the *Tabulae sceleti et musculorum corporis humani* (*Tables of the Skeleton and Muscles of the Human Body*, 1747). In the preface to this work, Albinus described his goals and working methods in considerable detail, in terms that would seem self-contradictory by later standards of mechanical objectivity. He was committed at once to upholding the most exacting standards of visual fidelity in depicting his specimens and to creating images of "the best pattern of nature." (See figure 2.9.)

To the former end, he went to lengths until then unheard of among anatomists meticulously cleaning, reassembling, and propping up the skeleton, checking the exact positions of the hipbones, thorax, clavicle, and so on, by comparison with a very skinny man made to stand naked alongside the prepared skeleton. (This test cost

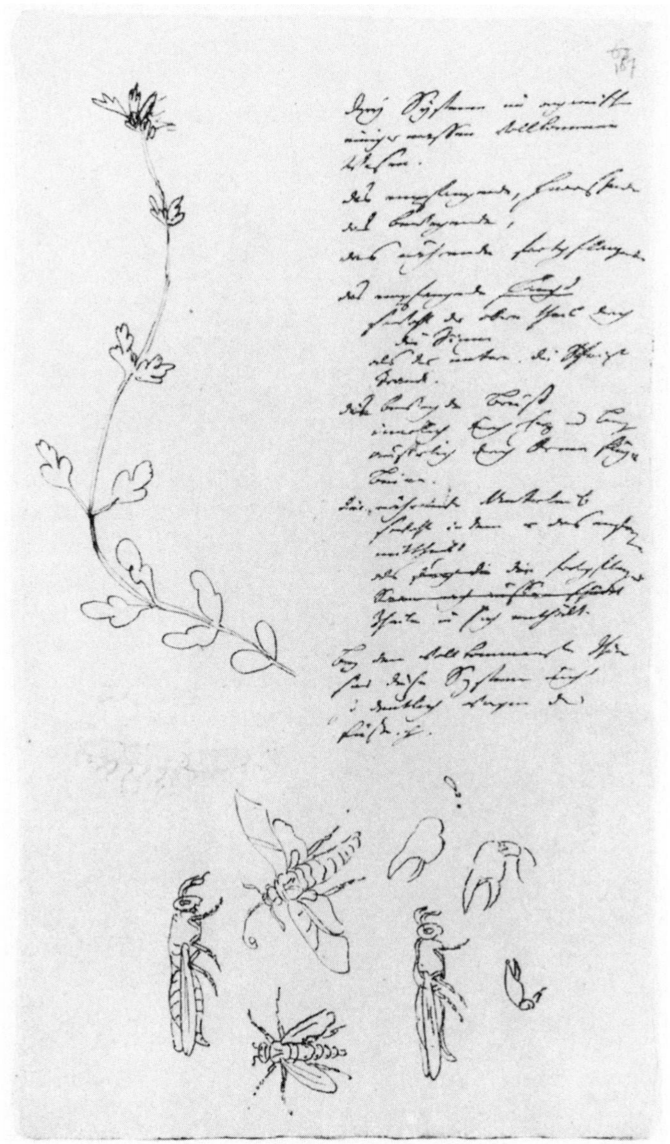

Fig. 2.8. "*Typus* **of Higher Plant and Insect.**" Johann Wolfgang von Goethe, *Die Schriften zur Naturwissenschaft*, vol. 9A, *Zur Morphologie*, ed. Dorothea Kuhn (Weimar: Böhlau, 1977), table 9 and pp. 239–40. Goethe's pencil-and-ink sketch from the early 1790s is surrounded by his notes on the three "organic systems" (the sensitive, the mobile, and the nutritive) and their essential characteristics. Goethe detected the *Typus* of the *Urpflanze* throughout the plant kingdom: "I grow ever more certain that the general formula that I have discovered is applicable to all plants. With it I can already explain the most idiosyncratic forms, for example passion flower, arum [lily], and place them in parallel to one another." Goethe to Karl Ludwig von Knebel, Oct. 3, 1787, *ibid.*, p. 373.

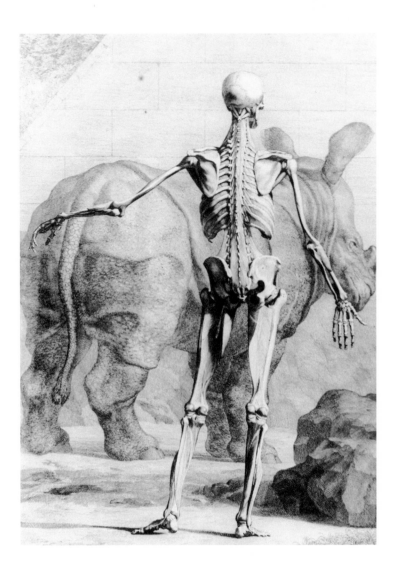

Fig. 2.9. Idealized Skeleton with Rhinoceros. Bernhard Siegfried Albinus, *Tabulae sceleti et musculorum corporis humani* (Leyden: J. & H. Verbeek, 1747), table 8 (courtesy of Staats- und Universitätsbibliothek Göttingen). Although Albinus monopolized the skills of the draftsman and engraver Jan Wandelaar for some ten years and corrected both the drawings and the engravings, he permitted the artist to add "ornaments" to the backgrounds of the tables to enhance the beauty of the plates. The rhinoceros shown in this plate was included for its agreeable rarity; the copy of the *Tabulae sceleti* belonging to the library of the University of Göttingen reports in a handwritten annotation that the animal "was shown for money in France, Holland, [and] Germany" in the 1740s — so it is probably the animal depicted in the Venetian artist Pietro Longhi's *Exhibition of a Rhinoceros at Venice* (*circa* 1751).

Albinus some anxiety as well as time and trouble, for the naked man demanded a fire to ward off the winter chill, greatly accelerating the decay of the skeleton.) Still worried lest the artist err in the proportions, Albinus erected an elaborate double grid, one mesh at four Rhenish feet from the skeleton and the other at forty, then positioned the artist at precisely the point where the struts of the grids coincided to the eye, drawing the specimen square by square, onto a plate Albinus had ruled with a matching pattern of "cross and streight [sic] lines." This procedure, suggested by Albinus's Leiden colleague, the natural philosopher Willem 'sGravesande, is strongly reminiscent of the Renaissance artist Leon Battista Alberti's instructions for drawing in perspective, and amounts to a kind of remote tracing of the object. The fixed viewpoint of the artist and the mapping of visual field onto plane of representation by means of the grids subject the artist to an exacting discipline of square-to-square correspondence in the name of naturalism. Albinus, like the Renaissance practitioners of perspective, also prescribed how the finished engravings should be viewed, as well as drawn.[22]

Yet these remarkable figures, which occasioned three months of "an incredible deal of trouble to the ingraver," were not actually of the particular skeleton Albinus so painstakingly prepared. Like Goethe, like Linnaeus, he was after truth-to-nature, the idea in the observation, not the raw observation itself. Having thus taken every ordinary and several extraordinary measures to ensure the integrity of object and subject, Albinus's pronouncements about just what the finished pictures are pictures *of* comes as a distinct shock to the modern reader. They were pictures of an ideal skeleton, which may or may not be realized in nature and of which this particular skeleton is at best an approximation. Albinus was all too aware of the atlas maker's plight: nature is full of diversity, but science cannot be. He must choose his images, and Albinus's principle of choice was frankly normative:

> And as skeletons differ from one another, not only as to the age, sex, stature and perfection of the bones, but likewise in the marks of strength, beauty and make of the whole; I made choice of one that might discover signs of both strength and agility; the whole of it elegant, and at the same time not too delicate; so as neither to shew a juvenile or feminine roundness and slenderness, nor on the contrary an unpolished

73

roughness and clumsiness; in short, all of the parts of it beautiful and pleasing to the eye. For as I wanted to shew an example of nature [*naturae exemplum*], I chused to take it from the best pattern of nature.[23]

Accordingly, Albinus selected a skeleton "of the male sex, of a middle stature, and very well proportioned; of the most perfect kind, without any blemish or deformity." (For Albinus it went without saying that a perfect skeleton was perforce male; in 1797, the German anatomist Samuel von Soemmerring constructed an "ideal" — and ideology-laden — female skeleton.)[24] But still the skeleton was not perfect enough, and Albinus did not scruple to improve nature by art: "Yet however it was not altogether so perfect, but something occurred in it less compleat than one could wish. As therefore painters, when they draw a handsome face, if there happens to be any blemish in it mend it in the picture, thereby to render the likeness the more beautiful; so those things which were less perfect, were mended in the figure, and were done in such a manner as to exhibit more perfect patterns; care being taken at the same time that they should be altogether just [*adhibita cura, ne quid a vero discederetur*]."[25]

"Perfect" and "just [*vero*]" (that is, true, exact): these were Albinus's polestar and compass, and he saw no contradiction between the two. Albinus could hold both aims simultaneously because of a metaphysics and an attitude toward judgment and interpretation that contrasted sharply with those of the later nineteenth century, as we shall see in Chapter Three. In effect, Albinus believed that universals such as his perfect skeleton had equivalent (or superior) ontological warrant to particulars; the universal might be represented in a particular picture, the reasoned image, if not actually embodied in a particular skeleton. The universal, like Goethe's "pure phenomenon," could only be known through minute acquaintance with the particular in all its details, but no image of a mere particular, no matter how precise, could capture the ideal. Only the observer with the experience and perspicacity of the sage could see it.

Nor was anatomy anomalous in its idealizing tendencies. Until well into the nineteenth century, paleontologists reconstructed and "perfected their fossil specimens," a practice sharply criticized by their successors, who prided themselves on "represent[ing] actual specimens with all their imperfections, as they are, not what they

74

may have been."[26] Mid-nineteenth-century anatomists and paleontologists believed that only particulars were real; to stray from particulars was to open a door to distortions in the service of dubious
theories or systems. In contrast, Albinus and other idealizing atlas
makers did not hesitate to offer pictures of objects they had never
laid eyes upon, like Goethe's *Urpflanze* — but in the service of truth-
to-nature rather than in violation of it.

Idealizers of Albinus's stamp were not unaware of the "naturalistic" alternative — that is, the attempt to portray *this* particular object
just as it appeared, to the limits of mimetic art.[27] There were eighteenth-century representatives of the naturalistic alternative in
anatomical illustration, but it was considerations as much of aesthetics as of accuracy that determined their quite explicit choice. The
British anatomist William Hunter's *Anatomy of the Human Gravid
Uterus* (1774), for example, opted for "the simple portrait, in which
the object is represented exactly as it was seen," as opposed to "the
representation of the object under such circumstances as were not
actually seen, but conceived in the imagination," on grounds of "the
elegance and harmony of the natural object" (see figure 2.10).

Hunter used thirteen different subjects in his atlas, at various
stages in pregnancy from three weeks to nine months. Each of his
thirty-four large (twenty-seven-inch) plates depicts an individual
corpse, often dissected and drawn over the course of months. Although Hunter emphasized the corpses' portrayal as individual
objects, he clearly intended them to be characteristic of the anatomy
of pregnant women in general. He asserted that a "simple portrait"
bore "the mark of truth, and becomes almost as infallible as the
object itself," but acknowledged that "being finished from a view of
one subject, [it] will often be somewhat indistinct or defective in
some parts," whereas the figure "made up perhaps from a variety of
studies after NATURE, may exhibit in one view, what could only be
seen in several objects; and it admits of a better arrangement, of
abridgement, and of greater precision."

Hunter's preference for the portrait of the individual object was
not unqualified, for he admitted that considerations of precision
might favor the composite or typical alternative. Nor did he regard
aesthetic considerations with suspicion, as being at odds with scientific accuracy. On the contrary, Hunter, like Albinus, considered the

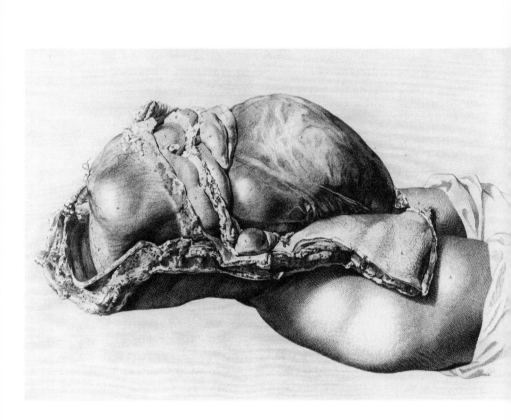

Fig. 2.10. Dissected Womb. William Hunter, *The Anatomy of the Human Gravid Uterus, Exhibited in Figures* (Birmingham: Baskerville, 1774), pl. 2, drawn by Jan van Rymsdyk and engraved by Gérard Scotin (courtesy of Staatsbibliothek zu Berlin Preussischer Kulturbesitz). In the legend to this figure (ZZb) of the anatomy of a woman who died in the ninth month of pregnancy, Hunter remarks on the accidental circumstances that altered the appearance of the veins (which had been injected with wax), details faithfully recorded in the image: "But when this drawing was made, the part, having been some-time in the air, had become a little dry, and the veins projected, as they appear in the figure" (n.p.). He chose the luxury printer Baskerville "principally for the advantage of his paper and ink," to ensure the work's durability (preface, *ibid.*, n.p.).

beauty of the depiction part and parcel of achieving that accuracy, not a seduction to betray it. Hence he defended the extra expense of large, "highly and delicately" finished engravings because they revealed small details of organs "new, or only imperfectly known" to the anatomist, whereas more well-known or repetitious parts were reduced to "bare outlines."[28]

It would be a mistake, however, to take Hunter entirely at his word—to believe that his figures did indeed represent the object "exactly as it was seen." Like the photographs of the nineteenth century, Hunter's figures carry the stamp of the real only to eyes that have been taught the conventions (for example, sharp outlines versus the soft edges actually perceived) of that brand of realism.[29] Moreover, Hunter's specimens, like all anatomical "preparations," were injected with wax or dyes to keep vessels dilated and "natural"-looking even after death—making them already objects of art, even before they were drawn.[30] Although Hunter claimed to have moved "not so much as one joint of a finger" of his specimens, he considered it part of truth-to-nature to inject the womb with "some spirits to raise it up, as nearly as I could guess, to the figure it had when the abdomen was first opened."[31] Hunter's atlas is instructive for our purposes because it shows, first, that scientific naturalism and the cult of individuating detail long antedated the technology of the photograph and, second, that naturalism in scientific atlases need not go hand in hand with fear of distortion or distrust of aesthetics.[32]

Even the naturalism of the camera obscura (a dark chamber into which light enters through a pinhole fitted with a lens, projecting an inverted image of external objects onto a screen) did not obviate the need for intervention and extended commentary on the part of the atlas maker. The English anatomist William Cheselden persuaded his two Dutch artists, Gerard van der Gucht and Shinevoet, to use "a convenient camera obscura to draw in" so that they could accomplish their figures for his *Osteographia* (1733) "with more accuracy and less labour." (See figure 2.11.) Yet the mechanical precision of the camera obscura was no substitute for the learned anatomist, who chose his specimens with discernment, carefully posed them in dramatic stances (for instance, an arched cat skeleton facing off against a crouching dog skeleton), and vouched for every drawn line as well as every printed word: "The actions of all the skeletons both human

Fig. 2.11. Skeleton Drawn with Camera Obscura. Title-page illustration, William Cheselden, *Osteographia, or, The Anatomy of Bones* (London: Bowyer, 1733). Cheselden persuaded his two artists to use the camera obscura device depicted here in order to "overcome the difficulties of representing irregular lines, perspective, and proportion" ("To the Reader," *ibid.*, n.p.). The half skeleton is suspended upside down because camera obscura images are inverted. But the traced camera obscura image was the beginning, not the end, of the image-making process, as Cheselden's emendations testify. He further specified that some parts of the figures be etched rather than engraved, the better to express certain bone textures, thus asserting his control over every aspect of the plates as well as the text.

and comparative, as well as the attitudes of every bone, were my own choice: and where particular parts needed to be more distinctly expressed on account of the anatomy, there I always directed; sometimes in the drawings with the pencil, and often with the needle upon the copperplate, and where the anatomist does not take this care, he will scarce have this work well performed."[33] The camera obscura — like photography, which largely took its place in the nineteenth century — helped illustrators render a wealth of detail with comparatively little effort, but eighteenth-century atlases demanded more than mere accuracy of detail. What was portrayed was as important as how it was portrayed, and atlas makers were expected to exercise judgment in both cases, even as they tried to eliminate the wayward judgments of their artists with grids, measurements, or the camera obscura.

Art and science converged in intertwined judgments of truth and beauty. Eighteenth-century scientific atlas makers referred explicitly and repeatedly to coeval art genres and criticism. Like Hunter, the English naturalist and artist George Edwards, the Library Keeper to the Royal College of Physicians of London, promised readers of his *Natural History of Uncommon Birds* (1743–1751) drawings "after LIFE," of "a most religious and scrupulous strictness," in contrast to the liberties taken by painters of historical scenes, in which the artist "has liberty to carry to what degree of Perfection or Imperfection he can conceive, provided alway [sic] he doth not contradict the Letter of his Historian." Yet Edwards, again like Hunter, thought nothing of coloring his birds (some of which were dried or preserved in spirits) and posing them in "as many different Turns and Attitudes as I could invent."[34] It is a sign of how dramatically scientific attitudes toward such artfulness had changed by the mid-nineteenth century that while Edwards's invented poses won him the Royal Society of London's Copley Medal in 1750, John James Audubon's elegantly symmetrical and sometimes anthropomorphized compositions of birds in his *Birds of America* (1827–1838) were sharply criticized by some contemporary naturalists as falsifications of nature.[35] (See figure 2.12.)

Not only the atlas makers themselves but also their artists were supposed to be familiar with a broad range of exemplars, so that each image would be the distillation of not one but many individuals carefully observed — Goethe's idea *in* the observation. The ways

79

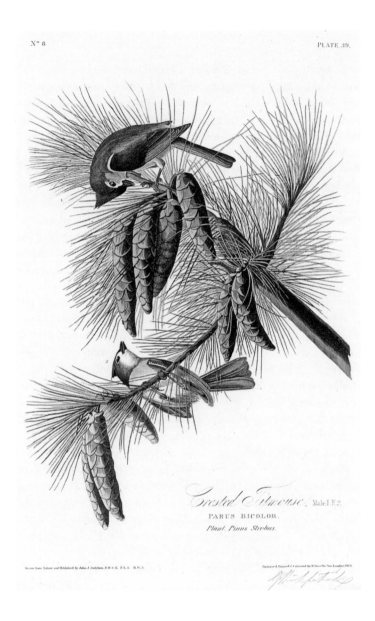

Crested Titmouse, Male 1. F 2.

PARUS BICOLOR.

Plant, Pinus Strobus.

Drawn from Nature and Published by John J. Audubon, F.R.S.E. F.L.S. M.W.S.   Engraved, Printed & Coloured by R Havell Sen London 1828

Fig. 2.12. Posed Tufted Titmouse. *Parus bicolor Linnaeus*, John James Audubon, *The Birds of America* (London: Published by the author, 1827–1838), pl. 39. Engraved and hand-colored by a team of London artists, Audubon's bird drawings were printed on double elephant folio paper in order to approximate life size as closely as possible. Yet Audubon's insistence that birds be depicted in natural habitats and poses, observed first-hand by the artist-naturalist, did not preclude mannered compositions like this one or anthropomorphic stances and descriptions. (Please see Color Plates.)

naturalists and artists achieved such distillations were conceived along similar lines and in both cases touted as a title to genius, a faculty of synthetic perception that elevated the master above the mere amateur or artisan. David Hume, for example, contended that all perceptions, whether epistemological, moral, or aesthetic, came to be infused with judgment through reflection on accumulated experience, just as post-Cartesian optics showed "how we transfer the judgments and conclusions of the understanding to the senses."[36] Anatomists from Andreas Vesalius in the mid-sixteenth century to Soemmerring in the early nineteenth century prided themselves on representations of a "canonical" body, a term that can be traced back to Galen, who in turn drew it from the classical sculptor Polykleitos.[37]

Sometimes the complexity of the phenomena overwhelmed synthetic perception. The Göttingen anatomist Albrecht von Haller complained of the "infinite labor" required to trace the labyrinthine variety of the arteries, which even numerous dissections had failed to coalesce into a clear pattern. He counseled the reader of this part of his *Icones anatomicae (Anatomical Images*, 1752) to heed the text more than the images, since the latter might not correspond to the typical case.[38] Haller is reputed to have prepared specimens of some anatomical regions as many as fifty times to make sure that the artist had a representative rather than anomalous model, displayed in characteristic circumstances.[39]

The more successful synthetic image was described by the artist Sir Joshua Reynolds in his 1769 *Discourses Delivered to the Students of the Royal Academy*. Through long observation of the individuals in a class, Reynolds claimed the artist "acquires a just idea of the beautiful form; he corrects Nature by herself, her imperfect state by her more perfect." Naturalist and painter alike sought the "invariable general form," incorporating the beautiful and the true: "Thus amongst the blades of grass or leaves of the same tree, though no two can be found exactly alike, the general form is invariable: a Naturalist, before he chose one as a sample, would examine many; since if he took the first that occurred, it might have been an accident or otherwise such a form as that it would scarce be known to belong to that species; he selects as a Painter does the most beautiful, that is the most general form of nature."[40] The French *philosophe* Louis de Jaucourt, writing on "beautiful nature [*la belle nature*]" in the *Ency-*

clopédie of Denis Diderot and Jean d'Alembert, had endorsed similar neoclassical aesthetic views: "[The ancient Greeks] understood clearly that it was not enough to imitate things, that it was moreover necessary to select them."[41] Nature was the model, the final court of appeal, for all art and science — but nature refined, selected, and synthesized. This convergence of artistic and scientific visions arose from a shared understanding of mission: many observations, carefully sifted and compared, were a more trustworthy guide to the truths of nature than any one observation.

Atlases of "characteristic" images can be seen as a hybrid of the idealizing and naturalizing modes: although an individual object (rather than an imagined composite or corrected ideal) is depicted, it is made to stand for a whole class of similar objects. It is no accident that pathological atlases were among the first to use characteristic images, for neither the *Typus* of the "pure phenomenon" nor the ideal, with its venerable associations with health and normality, could properly encompass the diseased organ. Cruveilhier's exquisitely colored and mostly lithographed plates, drawn by André Cazal and lithographed by Benard and Langlumé, testify to the necessity of new dimensions of representation, as well as of greater specificity, in depicting the pathological.[42] (See figure 2.13.) Even the practice of averaging, with its emphasis on the precise measurement of individual objects, could be made to serve the ends of essentialism.[43]

The characteristic atlases of the early and mid-nineteenth century mark a transition between the atlases that had sought truth-to-nature in the unabashed depiction of the typical — be it the reasoned image of the *Typus*, ideal, characteristic exemplar, or average — and those later atlases that strove for mechanical objectivity, as we shall see in Chapter Three. Like the latter, the characteristic atlases presented figures of actual individuals, not of types or ideals that could not be observed in a single instance. But like the former, these individuals simultaneously embodied types of whose reality the atlas maker was firmly convinced.

To learn to see the typical was the achievement of a lifetime, what the atlas maker aspired to and what the atlas was supposed to teach its readers. Yet it was not enough for the naturalist to see; an atlas was also supposed to depict. In order to convey the idea in the observation by an image, atlas makers had to impose their specialized

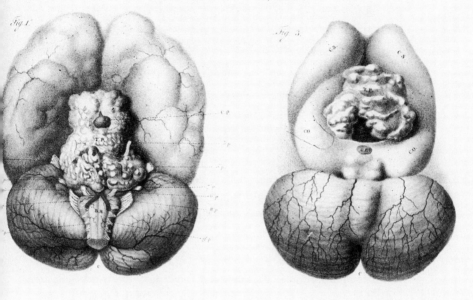

Fig. 2.13. Pathology in Color. "Diseases of the brain," Jean Cruveilhier, *Anatomie pathologique du corps humain* (Paris: Baillère, 1829–1842), vol. 1, pl. 6, drawn by André Chazal, lithographed by Langlumé, and hand-colored. These two figures depict a brain tumor found in an eighteen-year-old girl who died two hours after being brought to the Hôpital de la Charité in Paris. The individualization of such cases was characteristic of Cruveilhier's atlas, which was intended to acquaint physicians with rare maladies that they might encounter only once in a lifetime of practice. Numerous trials were required to achieve coloring "more natural and more true than that previously employed" (*ibid.*, p. vii). (Please see Color Plates.)

vision on their artists: they had to practice four-eyed sight.

Four-Eyed Sight

When René-Antoine Ferchault de Réaumur, Sieur de La Rochelle and a renowned French naturalist, died on October 17, 1757, his last will and testament left everything legally possible to the illustrator of many of his works, Hélène Dumoustier de Marsilly. No doubt anticipating some raised eyebrows, Réaumur justified his choice of heir at length:

> I would like to be able to show all the gratitude that I owe for the use she granted to me, with such patience and constancy, of her talent for drawing. It is she who made my Mémoires sur l'histoire des insectes and subsequent works presentable to the public. Whatever taste I might have had for this work, I would have despaired of finishing it and would have abandoned it, in consideration of the time I would have lost had I been obliged to continue to supervise ordinary draughtsmen with my [own] eyes...the taste and intelligence of Mademoiselle du Moutier [sic] equaling her talents, I could rely almost entirely on her. That which she drew under my eyes was not more correct than that which she drew in my absence. Not only did she know how to enter into my views, she knew and knows how to divine them, since she knows how to recognize that which is most remarkable in an insect and the position in which it should be represented.[44]

Here was the dream of the Enlightenment naturalist: the artist who understood the views of the naturalist so thoroughly that she divined them without being told, whose skilled hand was guided by them even without supervision, who saw with his eyes. (See figure 2.14.)

It was a dream rarely realized, as Réaumur knew all too well. He had worked with other artists, only to throw up his hands in frustration, as he hinted in his will. He had even gone to the length of lodging "chès [sic] moi" a young man who showed some aptitude for drawing, in order to train and monopolize him specially for the task of illustrating the six-volume Mémoires pour servir à l'histoire des insectes (Natural History of Insects, 1734– 1742) — only to have him die, thereby (as Réaumur remarked with some exasperation) delaying the publication of the monumental work still further. Like

84

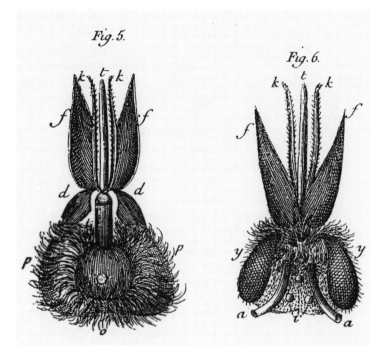

Fig. 2.14. Geometrized Bee. Head and proboscis of wood-boring bee, René-Antoine Ferchault de Réaumur, *Mémoires pour servir à l'histoire des insectes* (Paris: Imprimerie royale, 1734–1742), vol. 6, pl. 5, figs. 5–6. Although these magnified views were drawn by Hélène Dumoustier de Marsilly, they are signed only by the engraver, Philippe Simonneau. The symmetrical arrangement of the letters keyed to the textual description of the anatomical parts emphasizes the strict symmetry of the image itself. The geometric rendering of the parts as cylinders and spheres echoes Réaumur's description of how the bee uses its long trunk to pierce an "approximately cylindrical" piece of wood with strokes "parallel to the axis" (*ibid.*, p. 42): the idea in the observation.

countless other early modern naturalists (and many modern ones), Réaumur insisted that even skilled and intelligent artists had to be closely supervised, no matter how much time this took, for "it is impossible for him [the draftsman] to enter into the views of an author, if the author does not guide, so to speak, his brush."[45] Otherwise, artists were prone to be struck by certain irrelevant parts of the object, to choose an unrevealing perspective or position, to render all too exactly individual peculiarities of the specimen, or, worst of all, to depict exactly what they saw, hence obscuring the type of a skeleton or plant or insect. In the visual tug-of-war between Enlightenment naturalist and artist, the naturalist fought for the realism of types against the artist, who clung to the naturalism of appearances. Because the reasoned image could be seen only by the mind's eye, the social and cognitive aspects of the relationship between naturalist and artist blurred.

The obvious solution to Réaumur's dilemma, as he himself admitted, would have been to learn to draw himself. Some early modern naturalists — Konrad Gesner, Jan Swammerdam, and Charles Plumier, for example — do seem to have mastered the necessary drawing skills, though not many did so before the latter half of the eighteenth century.[46] And still fewer knew how to engrave or make woodcuts, the necessary preconditions for reproducing a drawing in a publication. But even for gentlemanly naturalists who could sketch, it was considered a liberal skill, not to be confused with the mechanical skills of the paid illustrator. Still less was a sketch to be confused with engraving. Eighteenth-century draftsmen at least contracted individually with their employers, albeit from a position of social inferiority. Engravers had, with the exception of some virtuosos, been commercialized and subjected to a shop-floor division of labor that lowered both their wages and their status *vis-à-vis* other artists.[47]

The distinction between liberal and mechanical drawing, depending on the identity of the draftsman, left visual traces in the drawings themselves. The drawings of the naturalists were thickly surrounded by handwritten text: scribbled annotations, measurements, ruminations. Their sketches were deliberately integrated into the processes of observation and reflection: they were tools to think with rather than illustrations to market. In the opinion of the natu-

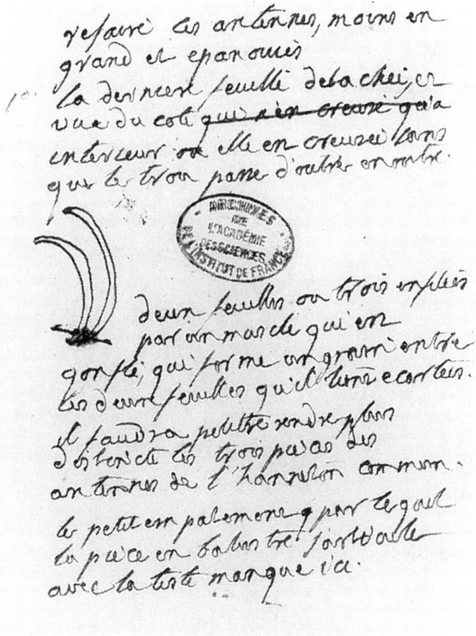

Fig. 2.15. Correcting the Artist. Insect antennae, René-Antoine Ferchault de Réaumur, Dossier Réaumur, Archives de l'Académie des Sciences, Paris (courtesy of the Archives de l'Académie des Sciences). Réaumur here corrects drawings intended for his treatise on insects: "Redo these antennae, not so large and spread out." His own attempt at a sketch in the margins suggests how urgently he required the services of a trained artist.

ralists, these handwritten borders converted craft into intelligence, handiwork into headwork.[48] As for Réaumur, he was perhaps correct in his assessment of his own meager gifts in this line, as his attempts to correct a drawing of insect antennae suggest (see figure 2.15). Faced with a similar situation with regard to the illustrations for his *Météores* (1637), Descartes had written, in a letter to Constantijn Huygens, that he could no sooner learn to draw than a deaf-mute from birth could learn to speak.[49]

Most naturalists who published illustrated works found themselves at the mercy of a draftsman, and almost all required the services of an engraver. By the early eighteenth century, it was a settled matter that works of natural history, anatomy, and other observational sciences required illustrations, despite sixteenth- and seventeenth-century controversies on this score.[50] Indeed, in some fields, such as botany and anatomy, the illustrations bid fair to become the chief justification for the publication, even in the view of an author who supplied only the text. But the objects depicted in these works were emphatically not given by nature alone. To find the idea in the observation beneath the swarm of variations that this or that individual specimen of orchid or skeleton presented to the eye required a special talent, perhaps even genius. This is why the eighteenth-century naturalists tried to guide the pencils, brushes, and burins of their artists. Ideally, as in the case of Réaumur and Dumoustier de Marsilly, the visions of the naturalist and the artist fused in something like four-eyed sight.

In practice, the collaborations of Enlightenment naturalists and artists to produce working objects for the sciences of the eye were taut with tensions: social, intellectual, and perceptual. Battles of wills, eyes, and status were joined when the naturalist peered over the shoulder of the artist, correcting every pen stroke. Naturalists and artists were necessary to one another, a fact appreciated by both, but in terms of authorship, the naturalists had the upper hand. In all but a few exceptional cases, it was the naturalist's name that appeared on the title page, while the names of the artist and the engraver huddled in small, faint print at the bottom of the plates: *Del.*[*ineavit*] ("drawn by") X; *Sculp.* [*sit*] ("engraved by") Y, conventions established in the seventeenth century.[51] But the title to that title-page top billing was wobbly, unless the naturalists could claim to have somehow authored the images as well as the texts. Naturalists longed for knowledgeable artists, and it was, in fact, far more frequent for an artist to become a proficient naturalist, as did Linnaeus's artist Ehret, than the reverse. By Réaumur's own admission, Dumoustier de Marsilly became a highly competent observer of insects, but Réaumur never learned to draw. Paradoxically, the more scientifically knowledgeable the artist, the more uneasy the naturalist became about who exactly was the author, as artists sometimes discovered.

From the standpoint of savants like Réaumur, these collaborations aimed at a fusion of the head of the naturalist with the hand of the artist, in which the artist surrendered himself (or, often, herself) entirely to the will and judgment of the naturalist. This relationship of subordination to the point of possession or thought transference frequently exploited other forms of social subordination in order to render the artist as pliant as possible: the subordination of servant to master, of child to adult, of woman to man. Some naturalists went so far as to train their own artists while they were still children, as in Réaumur's ill-fated experiment, in order to form their style completely.

Such relationships of near-total dependence fell into the category of domestic servitude. More ambiguous was the feminization of scientific, especially botanical, illustration already under way in the eighteenth century. On the one hand, there were the many wives, daughters, and sisters of naturalists who drew specimens for their menfolk: Sophie Cuvier sketched birds for her father, the French naturalist Georges Cuvier; Joseph Dalton Hooker's daughter Harriet painted plants for the journal edited by her father, as did the womenfolk of many other British botanists. These were genteel pastimes and familial favors, part of the semivisible network of women helpmeets — wives, daughters, sisters — who translated science into a private idiom.[52] On the other hand, there were the women artists who earned their keep from their work: Madeleine Basseporte, Barbara Regina and Margaretha-Barbara Dietzsch, Emilie Bounieu, Marie-Thérèse Vien — and Réaumur's artist, Dumoustier de Marsilly. No doubt external pressures played a role here: barred from the more prestigious genres of historical and religious painting, these eighteenth-century women artists often specialized in still lifes and natural history illustration. Freed from these constraints during the French Revolution, Bounieu, for example, abandoned natural history for the more lucrative commissions offered by history painting and portraits.[53] It is more speculative but still plausible to suggest that naturalists encouraged women artists because the double inferiority of their status as artisans and as women promoted the visual and intellectual receptivity that made the illustrator, as Albinus had put it, "a tool in my hand."

Conflicts flared up when the artist refused to accept the inferior role assigned by the naturalist. In a contretemps over payment and the ownership of some drawings, Réaumur, an aristocrat and member of the Paris Académie Royale des Sciences, haughtily described the artist Louis Simonneau as a mere "worker from whom one orders various products." Simonneau, himself a member of the Académie Royale de Peinture et de Sculpture, reacted with indignation. He protested Réaumur's condescending tone, "setting himself up as superior and making a comparison with products ordered by a master by a worker, [though] M. Simonneau is not in the least his inferior, being in his field an academician like him [Réaumur]."[54] (See figure 2.16.)

When political upheavals loosened the social hierarchies that had kept man under master, the relationships between naturalist and artist were also reordered, a sign of how one set of roles was closely patterned on the other. When, for example, in 1793 the Muséum d'Histoire Naturelle was created out of the former Jardin du Roi as the flagship scientific institution of the French revolutionary republic, the resident illustrator Gérard van Spaendonck campaigned for and won a chair in "natural iconography." This promotion put him at least nominally on equal terms with the professors of anatomy, chemistry, botany, and zoology — apparently above their protests.[55]

However subordinate, illustrators were seldom invisible. They signed their plates, were acknowledged and praised in prefaces, and were sought after, even monopolized, for their skills.[56] The labor of the illustrator, in contrast to that of laboratory assistants, was conspicuous and esteemed.[57] Some succeeded in gaining the upper hand over the naturalists, especially if they found a powerful and wealthy patron, as in the case of the early nineteenth-century French artist Pierre-Joseph Redouté, Spaendonck's successor as botanical illustrator at the Muséum d'Histoire Naturelle. (See figure 2.17.) The fact that Redouté's *Liliacées* (1802–1816) was published with his own name featured first and large on the title page, and with his own preface, instead of one by the botanists (including the Swiss botanist Augustin Pyrame de Candolle) who provided the plant descriptions, was a notable anomaly.[58] (See figure 2.18.) Only Redouté's fame as "the Raphael of flowers," the patronage of the empress Joséphine, and the considerable wealth he had amassed as a result permitted him to upstage botanists like Candolle.[59] Yet the sales of natural-history works languished or prospered according to the quality and quantity of illustrations, as naturalists themselves were acutely aware, even as they vied with their artists for credit.

Such was the strained relationship between James Sowerby, a portrait painter who became first a scientific illustrator and then a self-taught botanist, and his sometime employer and patron Sir James Edward Smith, the president of the Linnean Society of Lon-

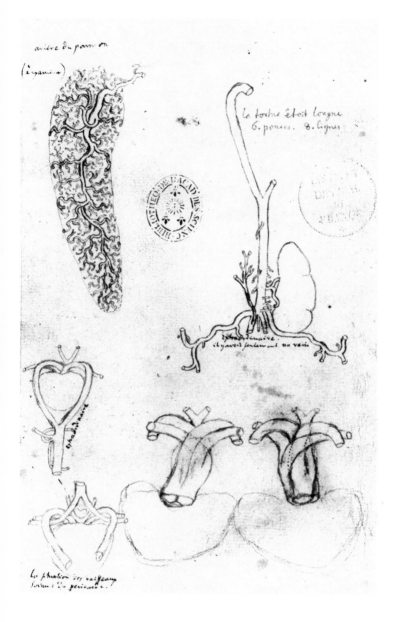

Fig. 2.16. Catching Anomalies. Tortoise lungs and heart, "Manuscrits non-datés: Dessins et textes non-datés pour Histoire Nat. des Animaux par Perrault," Archives de l'Académie des Sciences, Paris (courtesy of the Archives de l'Académie des Sciences). These sketches, probably drawn by Sébastien Leclerc and annotated by Claude Perrault, were made in conjunction with the comparative anatomy of animals undertaken by the Académie Royale des Sciences, the results of which were published in Claude Perrault, *Mémoires pour servir à l'histoire naturelle des animaux* (Paris: Imprimerie royale, 1671). Here the anatomist marks a part of the tortoise heart as "extraordinary"—that is, anomalous and therefore not characteristic of the organ.

Strelitzia Reginæ

Strelitzia de la Reine.

Fig. 2.17. Flowers for the Queen. *Strelizia Reginae*, Pierre-Joseph Redouté, *Les liliacées* (Paris: Didot Jeune, 1802–1816), vol. 2, p. 78. Redouté was among the few scientific illustrators able to publish works as chief or even sole author. These celebrity artists profited from patronage in high places, following the example set by the botanists themselves. When Sir Joseph Banks, the honorary director of the Royal Botanic Gardens at Kew, in England, from 1772 to 1820, received the first specimen of this bird of paradise flower from South Africa in 1773, he named it after Princess Charlotte Sophia of Mecklenburg-Strelitz, the wife of King George III of England. (Please see Color Plates.)

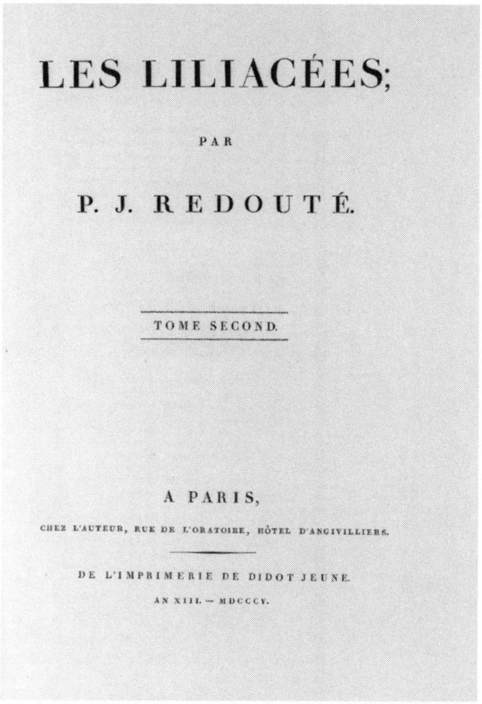

LES LILIACÉES;

PAR

P. J. REDOUTÉ.

TOME SECOND.

A PARIS,

CHEZ L'AUTEUR, RUE DE L'ORATOIRE, HÔTEL D'ANGIVILLIERS.

DE L'IMPRIMERIE DE DIDOT JEUNE.

AN XIII. — MDCCCV.

Fig. 2.18. Authorial Status. Pierre-Joseph Redouté, *Les liliacées* (Paris: Didot Jeune, 1802–1816), vol. 2, title page. Here Redouté's name stands big and bold as the sole author. Such top billing was a rare privilege for scientific illustrators, who were seldom acknowledged as authors and whose names generally appeared in small print below those of the scientists, if at all, on the title pages of atlases. As in the case of Audubon, highly placed patrons and luxury editions helped boost Redouté's standing.

don. Sowerby had illustrated Smith's *Exotic Botany* (1804–1805), and Smith had been warm in his praise for his artist in the preface: "I enter on a new work, assisted by his pencil, with the most perfect confidence."[60] Perfect confidence did not, however, preclude the usual close monitoring of each and every drawing, as this sketch by Sowerby annotated by Smith makes clear (see figure 2.19). Smith has penciled peremptory corrections: "This is not a very happy sketch, for this species is much larger in the flower & every part than either of the others the leaves broader, and not revolute. Pray alter it. The leaves too seem lighter and yellower."[61] Sowerby himself qualified as an artist-naturalist, having published his own *Coloured Figures of English Fungi or Mushrooms* (1797–1815) and supplied many illustrations for William Curtis's *Botanical Magazine* (established in 1787). His eye for plant structures was therefore a practiced one. Yet Smith's vigilance over the drawings was constant and unbending, despite the

93

Fig. 2.19. **Visual Tug-of-War.** *Tetratheca thymifolia*, James Sowerby, Sowerby Collection, box 35, folder B63, sheet 22 (© The Natural History Museum, London). Annotated sketches like this one (see also fig. 2.13) are among the few surviving traces of the close — and usually hierarchical — relationship between artist and naturalist, who often worked side by side rather than communicating in writing. Naturalists staked their claim to authorship of images as well as texts in atlases by closely monitoring shapes and shades at every stage of production, from rough sketch to engraving. The sketches also served as tools for the artists themselves: under Smith's criticisms, Sowerby has added "anthers mag[nified]d too much." (Please see Color Plates.)

occasional penciled demur in Sowerby's hand in reply to a crisp command to widen a petal or apply another shade of yellow (Sowerby: "anthers mag[nifie]d too much." Smith: "I think not"). Smith's condescension turned to pique when Sowerby was cited as the principal author of their *English Botany* (1790–1814), for which Sowerby supplied the figures and Smith the descriptions. "The flippancy," complained Smith, "with which every body quotes 'Sowerby,' whom they know merely as the delineator of these plates, without adverting to the information of the work, or the name of its author, leads on to the mortifying conclusion, that all I have done is of little avail, except to the penetrating eyes of the scientific few, who stand less in need of such assistance."[62]

To be made into another's tool, as Albinus put it, had epistemological and ethical as well as social dimensions. In sharp contrast to the mid-nineteenth-century rhetoric of scientific objectivity we shall encounter in Chapter Three, it was the artist who was here enjoined to submit passively to the will of the naturalist, not the naturalist who was supposed passively to register data from nature. The naturalist who pursued truth-to-nature was, on the contrary, exhorted to be active: observing and interpreting nature, monitoring and correcting the artist. The conflicts between Réaumur and Simonneau or between Smith and Sowerby were about more than social status and authorial vanity. They were also about sympathy (as Réaumur chose to interpret his relationship with Dumoustier de Marsilly) and servility (as Simonneau refused to interpret his relationship to Réaumur) and about seeing *as* versus seeing *that*. The reasoned image was authored: synthesized, typified, idealized by the intellect of the naturalist. In order to transfer that reasoned image to the page, the artist had to become something like a medium, not merely a subordinate.[63]

By the mid-nineteenth century, scientists themselves aspired to waxlike receptivity. They admonished one another to listen attentively to nature, and "never to answer for her nor hear her answers only in part," as the French physiologist Claude Bernard advised fellow experimenters in 1865.[64] The fantasy of the perfect scientific servant persisted among proponents of objectivity — but this servant was no longer imagined as the compliant draftsman who drew what the naturalist knew rather than what the artist saw. Instead, the ideal scientific domestic became an uneducated blank slate who could see without prejudice what his or her too-well-in-formed master might not.

Bernard's own example of the assistant "who had not a single scientific idea" was revealingly distorted. According to Bernard, the servant François Burnens "represented the passive senses" for his blind master, the late eighteenth-century Swiss naturalist François Huber. In fact, Burnens was Huber's reader and hence learned natural history alongside his master; he was, moreover, by Huber's own admission, a gifted naturalist who understood their joint investigations of bees "as well as I did." Only once Huber had satisfied himself of Burnens's skill and sagacity by having him repeat observations and experiments by Réaumur did Huber award Burnens "my complete confidence, perfectly assured of seeing well in seeing with his eyes."[65] Far from enlisting the "passive senses" of an ignorant servant, Huber trusted Burnens's eyes because his domestic had been trained as an active observer in the truth-to-nature style. Bernard's utter misunderstanding of Burnens's role measures the distance between divergent ideals of scientific passivity and its optimal distribution.

Metaphors of passive receptivity — minds as mirrors, soft wax, and, eventually, photographic plates — have permeated scientific epistemology since at least the seventeenth century, but they have been applied to different actors and to different ends. When Enlightenment savants dreamed of knowledge without mediation, they usually meant dispensing with their illustrators, or at least their engravers, not with their own senses and discernment. In contrast, mid-nineteenth-century men of science like Bernard hoped to eliminate themselves from observation — either by delegating the task to a scientifically untutored assistant or by reining in their own tendencies to intervene actively.

The inherent difficulties of imposing the naturalist's will and vision upon the artist, especially an artist knowledgeable about the subject matter, were exacerbated by a new ideology of drawing that took root in France, Britain, and the German lands in the latter half of the eighteenth century. Since the late seventeenth century, mercantilist monarchies had encouraged the reform of artisanal education in an effort to weaken the guilds domestically and to quicken trade internationally. During the mid-eighteenth century, this state program to renovate the arts and trades received a new impetus from the Encyclopedists' attack on the regime of blind habit and instinct enforced by backward guilds.[66] One goal of the *Encyclopédie*'s editors, Diderot and d'Alembert, was to intellectualize handiwork, and many people believed drawing instruction to be the best means to do so. Drawing would provide the mute craftsman with a language in which to express the ideas and designs that underlay skill, cultivating reflection, taste, and ingenuity.[67] In a trend that began in the 1740s and continued unabated into the nineteenth century, numerous schools offering free drawing instruction to children of the industrious poor opened in Paris, Vienna, Leipzig, Lyon, Glasgow, and Dresden — often in connection with local manufacturing interests, on the model of the school established at the Manufacture des Gobelins in 1667 to train children in drawing and design. In 1771, there were over three thousand students, most between the ages of eight and sixteen, receiving free drawing instruction in Paris alone.[68]

These schools were billed as a way of improving both craft and craftsmen by instilling discipline, technique, and a self-conscious, systematic way of working. The symbol and substance of forethought and reflection in handiwork was the sketch that guided the weaving of tapestry, the printing of textiles, the cutting of stone, or the painting of porcelain. Scientific illustrators were seldom members of academies, since the artistic genres they worked in (mostly still life and the decorative arts) were rated low in the hierarchy topped by history paintings. Yet *disegno* had, since Giorgio Vasari, been regarded by art critics as the intellectual heart of great painting, revealing the spiritual principle of nature.[69] All drawing, however humble, basked in the reflected glory of *disegno* and its intellectual ambitions.

Although public drawing schools were never meant to perturb the social order, they beckoned ambitious artisans as routes to upward social mobility. Ehret, Sowerby, and the Bauer brothers, Franz and Ferdinand, were among those who eventually attained the status of naturalists through drawing. Scientific illustration was among the few careers that placed men and women on more or less equal footing: Basseporte, who succeeded Claude Aubriet and was herself succeeded by Spaendonck at the Jardin du Roi, not only was paid (eventually) for her work; she also carried the same official title as her male colleagues.[70] The autonomy won by these artists was

97

social and intellectual, as well as financial. Smith noted of Sowerby that "had he not prefered [sic] the independence of profits arising from his own publications he would have become Draughtsman to his Majesty."[71] Ehret made no secret of his lowly beginnings as a gardener's apprentice to his uncle near Darmstadt, but he insisted that he was under no one's tutelage, not even that of Linnaeus himself: "I profited nothing from him in the dissection of the plants; for all the plants in the 'Hortus Cliffortianus' are my own undertaking, and nothing was done by him in the way of placing all the parts before me as they are figured."[72] In the latter part of the eighteenth century, drawing took on associations diametrically opposed to the submissive pliability expected by the naturalists.

In four-eyed sight, epistemology and ethos merged along with the vision of naturalist and artist. For naturalists who sought truth-to-nature, a faithful image was emphatically not one that depicted exactly what was seen. Rather, it was a reasoned image, achieved by the imposition of reason upon sensation and imagination *and* by the imposition of the naturalist's will upon the eyes and hands of the artist. The exercise of will and reason in tandem forged an active scientific self, which we will explore in more detail in Chapter Four. The question as to whether the receptivity of the artist should be celebrated or scorned paralleled the debate over dominant values in eighteenth-century moral philosophy: the faculty of sympathy enshrined by David Hume and Adam Smith versus the absolute autonomy expounded by Immanuel Kant. But the artists had no need of learned treatises to make sense of their own lived experience. By the late eighteenth century, the four-eyed sight that transferred the naturalist's idea via the artist's hand to the atlas page came to look less like sympathy and more like servility.

Drawing from Nature

If artists balked at subservience to naturalists, did they nonetheless bow to nature? Didn't the artistic traditions of mirroring nature with mimetic accuracy contradict the intellectualized true-to-nature images? The words "drawn from nature," half boast, half warranty, recur in the prefaces of illustrated scientific works of the eighteenth and early nineteenth centuries. Yet their meaning was not obvious. The qualifications "after life" (*ad vivum*) or "drawn from nature," invoked by artists from at least the sixteenth century on, must themselves be qualified.[73] It was standard practice for botanical drawings to represent the fruit and flower of a plant in the same drawing, as never occurred at the same time in nature; many of the most opulent flower paintings were drawn from desiccated herbarium specimens.[74] Illustrators often worked at top speed, especially under the adverse conditions of expeditions, using rough sketches as aide-mémoires to complete their drawings upon returning home. For example, Aubriet, the illustrator who accompanied the botanist Joseph Louis Pitton de Tournefort on a voyage to the Levant in 1700–1702, would trace the outlines of a plant while Tournefort dictated color annotations for later reference — both of them as often as not seated on balky mules in the pouring rain.[75]

The contrast conjured up by the phrase "drawn from nature" was not only between reality and fantasy but also between drawing from a model or, often, models (even if these were dried, flattened herbarium specimens or bloated anatomical preparations pickled in alcohol) and copying another drawing — since copywork was how almost every eighteenth-century artist and illustrator had been taught to draw. At least three sets of practices shaped the meanings of "drawn from nature" for illustrators of scientific atlases during this period: first, the pedagogy of drawing, especially the extensive use of models and copybooks; second, the ornamental and artistic deployment of certain images, especially those of flowers and the human body; and third, the characteristics and conventions of the various media (for example, watercolor, gouache, and pastels) and reproductive techniques (such as engraving, etching, and lithography). Built into the very practices of eighteenth-century drawing were norms and standards that countered extreme mimesis in the depiction of individual naturalia.

The *Encyclopédie* article "Drawing" laid out the standard steps by which students were taught to draw throughout the eighteenth century. It was best to start young, at "the age at which the docile hand lends itself most easily to the flexibility required by this kind of work." After learning to handle the pencil or red chalk by drawing parallel lines in all directions, the student would be given drawings by "clever masters" to copy. Only after long practice in imitating the drawings of others would the student be allowed to graduate to sketching from a three-dimensional object — in the case of the human body, a nude model, known as an "academic" study in honor of the Académie Royale de Peinture et de Sculpture, which had introduced such exercises in France in imitation of the Roman Academy of Saint Luke. Even then the student did not draw the whole object but built up to it, part by part.[76] "Drawing from nature" was the final stage of a long, regimented process that, in the free drawing schools for working-class children, submitted pupils to a discipline of time, vision, and motion that became paradigmatic for most later forms of technical education.[77] Starting in the late seventeenth century, numerous copybooks were published to provide aspiring draftsmen with patterns to copy (see figure 2.20). By the early nineteenth century, the most popular copybook series in French, German, and English ran to scores of volumes each.[78] By the time drawing students were admitted to "academic" exercises or even to sketching plants, they had already calibrated eye and hand by copying hundreds of model drawings.

A minor printing industry sprang up to supply these models. Already in the seventeenth century, copybooks specializing in floral patterns were much in demand for draftsmen and other artisans employed in the luxury trades: embroidery, miniature painting, porcelain painting, silk weaving. In 1666, the artist Nicolas Robert was appointed by Louis XIV as "peintre ordinaire du Roi pour la miniature" and painted 727 vellum (*vélin*) flower portraits, most of them edged in gilt. Subsequent illustrators employed by the naturalists at the Jardin du Roi added steadily to the collection of *vélins*, as the paintings came to be called; as director of the Muséum d'Histoire Naturelle, Cuvier was still contracting for additions to the collection of drawings in the early nineteenth century.[79] These paintings were as influential for the decorative arts as for natural history, and most of the artists who supplemented the collection after Robert — Basseporte, Spaendonck, Redouté — were employed to ornament *objets de luxe*, such as porcelain and embroidered garments, as well as to illustrate scientific works.[80] (See figure 2.21.) The movement to establish free drawing schools in the latter half of the eighteenth century further tightened the connections between botanical illustration and ornament.[81]

Whereas flowers were aestheticized in the context of the deco-

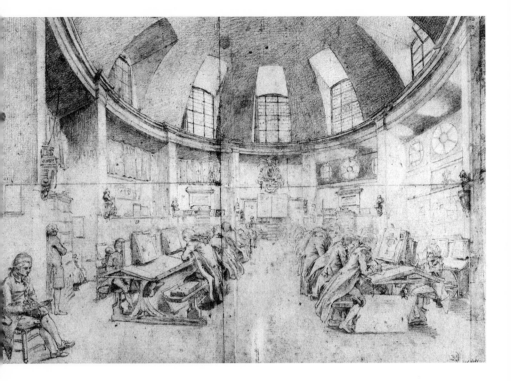

Fig. 2.20. Drawing by the Book. Ecole Gratuite de Dessin, Paris, 1780 (courtesy of Musée Carnavalet, Paris). Students practice drawing from copybooks propped in front of them. Only after years of copying sketches from these models, after their eyes and hands had been drilled and their penstrokes standardized, were advanced students allowed to draw from nature or given "academic" training in life studies.

rative arts, the human body occupied a more elevated place in the hierarchy of artistic genres. As the object of portraiture and history paintings, it was embedded within the more prestigious (and better-paid) fine arts. A painter of flowers, insects, shells, and other natu-ralia might occasionally win entry to the annual Paris salon displays with a still life or a landscape, but these were lowly genres.[82] The elite among eighteenth-century artists graduated from the drawing schools to the academies of fine arts set up in various European cap-itals.[83] Renowned anatomists wrote textbooks for this audience.[84]

Neither artists nor anatomists sensed any tension between the demands of truth and those of beauty; on the contrary, an ugly draw-ing was more than likely a false one.[85] Like the discipline taught by the drawing schools, the halo of aesthetic appreciation surround-ing the subject matter of botany and anatomy licensed naturalists and their illustrators to standardize and idealize objects drawn from nature. Soemmerring, for example, was quite aware of his debt to the copybook: "Since the anatomic description of any part, generally speaking, is just as idealistic as the representation and description of that same organ in a sketchbook, so one should follow the same prin-ciple in describing it.... Everything that the dissector depicts with anatomical correctness as a normal structure [Normalbau] must be exceptionally beautiful."[86] The perceived beauty of flowers or the human body need not have necessarily led naturalists and illustrators in the idealizing, classicist direction followed by Albinus and Soem-merring; more individualizing, naturalistic aesthetics were possible, as Hunter's case shows. But it would have hardly been possible to purge these charged objects of all aesthetic aura, given their promi-nence in both the decorative and the fine arts.

The techniques of reproduction — engraving, mezzotint, lithog-raphy — also imposed a grid of artifice upon drawings from nature.[87] In the case of engraving, the grid was literal: the art historian William Ivins has written forcefully of the engraver's cross-hatching as a "net of rationality."[88] (See figures 2.22, 2.23, and 2.24.) The vir-tuoso engravers (who might qualify for admission as artists in an academy) concentrated on making highly finished, large-scale, expensive copies of portraits and paintings for well-heeled cus-tomers; in contrast, the majority of engravers worked anonymously for printers at much lower wages.[89] Scientific works were usually

Fig. 2.21. Luxury Botanicals. Flora Danica serving platter, *Menyanthes trifoliata*, Winfried Baer, *Das Flora Danica-Service 1790–1802: Höhepunkt der Botanischen Porzellanmalerei* (Copenhagen: Kongelinge Udstillingsfond Køpenhavn und Autoren, 1999), p. 97 (courtesy of Prussian Palaces and Gardens Foundation Berlin-Brandenburg). The opulent table service "Flora Danica" was originally commissioned by the Danish court in the 1790s, probably as a diplomatic offering to the Empress Catherine the Great of Russia, a passionate collector of porcelain (which was so precious it was known as "white gold"). The paintings of plants carefully copied the figures of the monumental botanical atlas *Flora Danica* (1761–1888), begun by botanist Georg Christian Oeder with the patronage of the Danish monarchy. This platter was in all likelihood painted from sketches by the Nuremberg artist Johann Christoph Bayer, who worked as an illustrator for the *Flora Danica* as well as for the Royal Porcelain Factory, Copenhagen. Note the Linnaean botanical analyses of the flowers, upper right. (Please see Color Plates.)

handed over to an engraving shop, unless the naturalist went to the extra expense of seeking out his own engraver or securing, as Albinus did in Wandelaar, a draftsman who could and would also engrave.[90] Redouté experimented with new stipple techniques in order to give his engravings a softer texture better suited to coloring than the network of lozenges typical of the engraved image.[91]

Other techniques, such as etching and mezzotint, demanded different but equally distinctive conventions of visual representation. Neither medium was suited to the cheap printing of a normal run of an illustrated book. This may be why engraving was the preferred reproduction method for illustrated scientific works until the invention of lithography in 1798 by Alois Senefelder in Munich and the improvement of lithographic printing methods by Godefroy Engelmann in Paris during the 1820s. The great appeal of the lithograph, both artistic and economic, lay in its immediacy: the image could be printed directly from a drawing made in some greasy medium (chalk, ink, wash) on a dampened stone, eliminating the engraver.[92] Moreover, limestone was cheaper than the copper plates used in engraving. Cruveilhier's atlas of pathological anatomy was among the first to use the technique, on grounds of cost and because it rendered "the touch of the painter" better than engraving.[93] (See figure 2.25.)

Given these layers of art and artifice, convention and conception surrounding the image "drawn from nature," one may be tempted to dismiss the very notion as an illusion or a fraud. The naturalists and illustrators of the eighteenth and early nineteenth centuries were not, however, self-deceived or hypocritical, preaching fidelity to nature while practicing manipulation in the service of preconceived notions. They deemed the crafting — they would have called it "perfecting" — of images to be their scientific duty rather than a guilty distortion, and they practiced it openly. The nature they sought to portray was not always visible to the eye, and almost never to be discovered in the individual specimen. In their opinion, only lax naturalists permitted their artists to draw exactly what they saw. Seeing was an act as much of integrative memory and discernment as of immediate perception; an image was as much an emblem of a whole class of objects as a portrait of any one of them. Seeing — and, above all, drawing — was simultaneously an act of aesthetic appreciation, selection, and accentuation. These images were made to serve the ideal of truth — and often beauty along with truth — not that of ob-jectivity, which did not yet exist.

Truth-to-Nature after Objectivity

In Chapter Three, we shall examine the rise of mechanical objectiv-ity and how it changed the ways scientific atlas images were made and understood. From the perspective of atlas makers committed to objectivity, selection, synthesis, and idealization all looked like sub-jective distortions. These atlas makers sought images untouched by human hands, "objective" images. Mechanical objectivity did not, however, extinguish truth-to-nature. At times coexisting, at times colliding with the precepts and practices of mechanical objectivity, truth-to-nature continued to command the loyalty of some scientists and even whole disciplines throughout the nineteenth and twentieth centuries.

Botany was one discipline in which truth-to-nature persisted as a viable standard in the realm of images. Some botanists, to be sure, followed the beckoning mirage of an image made by nature itself, seemingly without human intervention. Authors of treatises on the application of photography to the sciences urged botanists and other naturalists to use the camera in order to capture "the thousands of details of the veining of leaves" and to achieve "a rigorous exacti-tude, an exactitude which they have so much difficulty in obtaining from artists, always too prone to correct nature." But even boosters admitted that photography would never replace drawings in botany and that floras illustrated with photographs, for example, of trees, would not release the botanist from the responsibility of choosing models that each "well represented all the characters of the species to which it belonged and whose form presents no abnormal peculi-arity, be it natural or artificial."[94] Experts in scientific photography warned botanists that when some feature was to be highlighted amid a welter of detail, drawing pencil and brush bested the camera. Moreover, photographs were not immune to subjectivity: "Nature photos are also subject to subjective influences; no two photogra-phers, no two different cameras, portray the objects in the same way."[95] This was photography pressed into service for truth-to-nature, not objectivity.

In general, however, late nineteenth-century botanists dis-

De Seve del.   Eli. Haussard. Sc.

LE CORLIEU *ou* PETIT COURLIS.

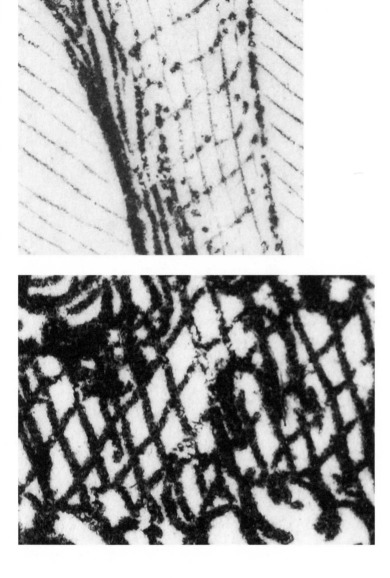

Figs. 2.22, 2.23, 2.24. Standardized Burin-Strokes. Curlew, Georges Louis Leclerc, Comte de Buffon, *Histoire naturelle, générale et particulière* (Paris: Imprimerie royale, 1770–1790), vol. 23, pl. 3, p. 28. The engravings for this edition of Buffon's enormously popular survey of natural history were executed by many hands, using standard techniques of cross-hatching, regardless of the object (fig. 2.22) to be rendered — whether (as in this case) ocean waves (fig. 2.23) or speckled feathers (fig. 2.24).

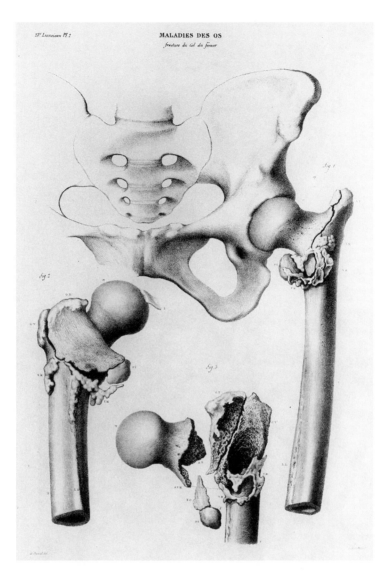

Fig. 2.25. Lithographed Textures. Bone diseases, Jean Cruveilhier, *Anatomie pathologique du corps humain* (Paris: Baillère, 1829–1842), vol. 2, fasc. 23, pl. 2. Cruveilhier distinguished two kinds of organic lesions, those of form and those of texture: "Nothing is easier than to render the first, nothing more difficult than to render the second" (*ibid.*, vol. 1, p. vii). Cruveilhier's artist, André Chazal, exploited the textural possibilities of both lithography and, for some plates, color (see fig. 2.13) to meet this challenge — possibilities of verisimilar representation that photography was long unable to rival.

dained photography and other mechanical means of making images of plants, such as the *Naturselbstdruck* (autoprint, literally "nature prints itself") technique (see figure 2.26). Few floras used either. Taking stock of available methods of botanical illustration in his *Phytographie* (1880), the Swiss botanist Alphonse de Candolle (son of Augustin Pyrame de Candolle, the botanist who had collaborated with Redouté) complained about both, regarding lithographs and woodcuts as more promising for botanical illustrations. No illustration, including a photograph, could in his mind compete with the authenticity of a herbarium specimen, however flat and faded.[96] Ludolph Treviranus, a professor of botany in Bonn and the author of an 1855 treatise on the use of woodcuts to picture plants, had earlier argued that woodcuts highlighted characteristic plant form and habitus in ways that other media could not. Above all, concluded Treviranus, plant illustrations in all media must preserve the botanist's discretion in choosing the right specimen and in "the constant monitoring of the draftsman's work, so that he expresses exactly the characteristic parts."[97] A century later, the standard twentieth-century work on botanical illustration echoed Candolle's and Treviranus's warnings against "crassly verisimilar" renderings of plants. Artists ought not to render blossoms "all too accurately," especially in the case of highly variable plants like orchids, lest they inadvertently occasion "the creation of a new species or variety."[98] As long as botanists insisted on figures that represented the characteristic form of a species or even genus, photographs and other mechanical images of individual plants in all their particularity would have little appeal. Truth-to-nature spoke louder in this case than mechanical objectivity.

This is not to say that botanists in the late twentieth or even the late nineteenth century pursued their science with more or less the same epistemic virtues as those espoused by Linnaeus. Objectivity did make inroads into other areas of botanical practice, such as the introduction of the "type method" in the late nineteenth and early twentieth centuries in order to stabilize nomenclature. At the level of species, the type method fixed the name to an individual specimen, called the "holotype," usually the first found by the discoverer or "author" of the new species. This specimen need not be (and often is not) typical of the species it represents, but it is the court of

Naturselbstdruck.

Aus der k. k. Hof- und Staatsdruckerei zu Wien. 1853.

Fig. 2.26. Nature Prints Itself. Autoprint of leaf, Alois Auer, "Die Entdeckung des Naturselbstdruckes," *Denkschriften der Kaiserlichen Akademie der Wissenschaften, Mathematisch-Naturwissenschaftliche Classe* (Vienna: Kaiserlich-Königliche Hof- und Staatsdruckerei, 1853), vol. 5, pt. 1, pp. 107–10, table 4. This nonphotographic method of mechanical self-registration pressed the object to be represented between copper and lead plates until it left an imprint in the soft lead, which could then be printed off like a copper plate. Auer, the inventor of the process, boasted that it marked the third great moment in the cultural history of humanity, after the inventions of writing and Gutenberg's movable type: it was "the discovery of how nature prints itself" (*ibid.*, p. 107). (Please see Color Plates.)

last appeal for all future questions about the definition of the species, as its official name-bearer. Holotypes are preserved with great care, specially labeled and stored at the major herbaria of the world, to which botanists seeking to clarify taxonomic questions must travel to inspect the specimen firsthand. Each one is as unique as a Vermeer or a Cézanne, and, at least to botanists, almost as valuable. Even fragments that break off the brittle, flattened, desiccated specimen are swept up and reverently preserved in an envelope with the holotype itself. (See figure 2.27.)

Botanists long accustomed to using the word "type" (recall Goethe's *Typus*) to refer to the ideal or typical found these new practices confusing. In 1880, Alphonse de Candolle tried to sort out this newly emerged ambiguity in natural history between the "authentic specimen [*échantillon authentique*]," which was an individual plant, and the "typical specimen [*échantillon typique*]," an individual that embodied "the true ideal type" of a species.[99] This revealing conflation of type specimen and typical specimen was to exercise naturalists for some fifty years in the protracted late nineteenth- and early twentieth-century debate over the definition and use of type specimens in botany and zoology. Both opponents and proponents of the method of type specimens conceived the battle as one between the personal discretion of a few elite botanists, mostly located at powerful institutions in European capitals, and mechanical rules applicable to all cases by all botanists, everywhere and always. Depending on which side one was on, type specimens promised to eliminate the "purely personal and arbitrary," the "personal equations" of botanists, in favor of a "fixed rule" — or they threatened to rigidly restrict "freedom to use personal judgment."[100]

Once these rules were accepted by the 1910 International Botanical Congress, in Brussels, and eventually incorporated into the *International Code of Botanical Nomenclature* (and the equivalent zoological code), they came to be seen as a triumph of objectivity in taxonomy: "It is obvious that a secure standard of reference is needed to tie taxonomic names unequivocally to definite, objectively recognizable taxa."[101] It is no surprise that the one place where photography gained a firm foothold in botanical illustration was in the representation of type specimens, in all their individuality and militant objectivity.[102]

111

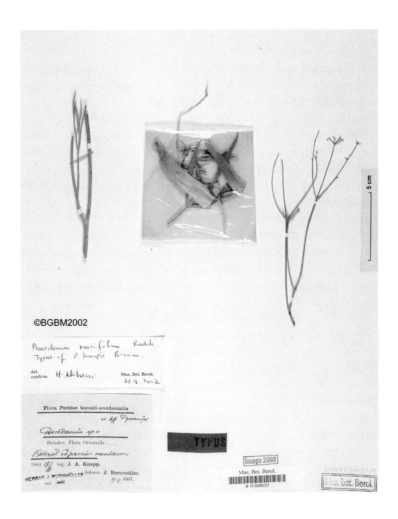

Fig. 2.27. Holotype. *Peucedanum paucifolium*, B 100086233, Botanisches Museum, Berlin (courtesy of Botanischer Garten und Botanisches Museum Berlin-Dahlem). This herbarium specimen is labeled in red as a type specimen ("Typus"), and its fragments are carefully preserved in a cellophane packet for possible future consultation by botanists. Layers of inscription (handwritten identification, red holotype label, bar code) bear witness to taxonomic shifts over time. Despite the echo to Goethe's "Typus" (see fig. 2.6), modern type specimens broke with the metaphysics and practices that underpinned the *Urpflanze*. Although botanists have preserved and consulted herbarium specimens since the sixteenth century, only in the late nineteenth and early twentieth centuries did a single individual plant, one not necessarily characteristic of the species, come to be designated the official name-bearer of the species (a practice made official by the Brussels International Botanical Congress in 1910). (Please see Color Plates.)

As this example shows, mechanical objectivity did not drive out truth-to-nature, but nor did it leave truth-to-nature unchanged. Epistemic virtues do not replace one another like a succession of kings. Rather, they accumulate into a repertoire of possible forms of knowing. Within this slowly expanding repertoire, each element modifies the others: mechanical objectivity defined itself in contradistinction to truth-to-nature; truth-to-nature in the age of mechanical objectivity was articulated defensively, with reference to alternatives and to critics. Epistemic virtues emerge and evolve in specific historical contexts, but they do not necessarily become extinct under new conditions, as long as each continues to address some urgent challenge to acquiring and securing knowledge.

The problem of variability in right depiction stretches from the beginning to the end of the period we have treated here. It haunted scientific atlas makers who pursued truth-to-nature as much as it did their successors dedicated to objectivity. But different epistemic ways of life made for different diagnoses of the sources of variability. Eighteenth-century savants tended to locate variability in the objects themselves — in the accidental, the singular, the monstrous. By the mid-nineteenth century, the chief source of variability had shifted inward, to the multiple subjective viewpoints that shattered a single object into a kaleidoscope of images. The earlier naturalists had attempted actively to select and to shape both their objects and their illustrators, whereas later naturalists aspired to hands-off passivity. The meaning of the images changed accordingly. Instead of portraying the idea in the observation, atlas makers invited nature to paint its own self-portrait — the "objective view."

CHAPTER THREE

Mechanical Objectivity

Seeing Clear

In 1906, two histologists, the Spaniard Santiago Ramón y Cajal and the Italian Camillo Golgi, shared the Nobel Prize for Physiology or Medicine. For both men that put one too many neuroscientists in Stockholm. Golgi reckoned that Ramón y Cajal's starting point had been in Golgi's own development of the "black reaction" to make visible through staining the delicate nerve cells in the brain. (The idea was to treat the tissue first with potassium dichromate for variable amounts of time, then with silver nitrate — the resulting black silver chromate salts revealed the shape of neurons in stunning detail.) In any case, the scientific orientation (the neuron doctrine) central to all that Cajal had achieved was (according to Golgi) on the way out. Indeed, there was not a single part of Cajal's program — the claim that each neuron was functionally, developmentally, and structurally independent — that Golgi accepted. In the first instance, as Golgi openly argued in his Nobel Prize acceptance speech, neurons could not be isolated from one another because the finest branches of their axons intermingled, giving rise to an inextricable network or net. Even if no actual continuity of the fibrils originating from different nerve cells could be seen, why (he asked) should one assume that such continuity did not exist? For decades, Golgi had defended his holistic view of the brain — that its elements formed a "diffuse nervous network." Surveying the field from embryology to anatomy to physiology, Golgi found not a shred of support for his rival's doctrine of the neuron: "However opposed it may seem to the popular tendency [that is, that of Cajal and his allies] to individualize the elements, I cannot abandon the idea of a unitary action of the nervous system, without bothering if, by that, I approach old conceptions."[1]

One of the elements that makes this episode so compelling is that there is no reason at all to think that either Golgi or Cajal was acting in bad faith. Both were passionately committed to depicting rightly the cells they were studying. Both had in their hands a method, invented by Golgi, that opened up for visual inspection aspects of the nervous system that had never before been seen in such extraordinary detail.

Cajal, for his part, later recalled listening in horror at the prize ceremony as Golgi relaunched the theory of interstitial nerve nets, a doctrine Cajal thought he had long since killed, replacing it with the idea of autonomous neurons that were "polarized," receiving signals through dendrites and sending them through axons. Neurons connected to one another only across gaps, according to Cajal and by 1906 many others, by "induction." He was "trembling with impatience as I saw that the most elementary respect for the conventions prevented me from offering a suitable and clear correction of so many odious errors and so many deliberate omissions."[2]

Images were central to the Cajal-Golgi battle. Cajal found Golgi's drawings and descriptions of the cerebrum, cerebellum, spinal cord, and hippocampus to have utterly failed to articulate properly the arrangements that Cajal had so painfully elicited from the silver chromate. Golgi himself had proclaimed in his atlas of 1885 that his pictures were "exactly prepared according to nature" (meaning, as we saw in Chapter Two, drawn as he was examining the microscopic specimen)—but then went on to modify the figures so they were, as in figure 3.1, "less complicated than in nature."[3] Between these two scientists lay the charge that objectivity had been violated: the one defended his undistorted sight (Cajal) and charged the other (Golgi) with having intervened, deliberately, and in the process having bent depiction to conform to his theoretical predilections.

Golgi's interventions to support his views were anathema to Cajal, and never more so than that day, December 11, 1906, in Stockholm: "When [Golgi] showed a glimpse of one [of his figures], it was artificially distorted and falsified in order to adopt it, *nolens volens*, to his capricious ideas." Golgi rose to give the first of the two acceptance speeches. Immediately, he put on the screen two images that

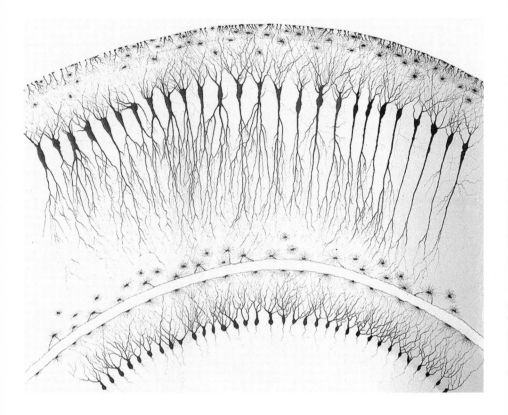

Fig. 3.1. Simpler than Nature. Camillo Golgi, *Untersuchungen über den feineren Bau des centralen und peripherischen Nervensystems*, trans. R. Teuscher (Jena: Fischer, 1894), fig. 25; translation of Golgi's *Sulla fina anatomia degli organi centrali del sistema nervoso* (Milan: Hoepli, 1886); original figure is table 21. Golgi was often adamant about drawing "after life" or "exactly prepared according to nature" — which meant that he had the histological sample before him as he drew. In this 1886 atlas, he made it clear that he had simplified some figures: "It is superfluous to say that the fibers of the Alveus invade continuously the grey layer, and thus between these two layers, instead of the clear limit which it is possible to see in the drawing [this figure], there is a gradual transition of the one into the other." Also: "Of the neuroglia elements which are diffusely distributed, only a few were drawn in the Table." Facing complex objects fraught with difficulties of preparation and observation, Golgi considered it a virtue — not a vice — to have his figures represent a reality "less complicated than in Nature." (Please see Color Plates.)

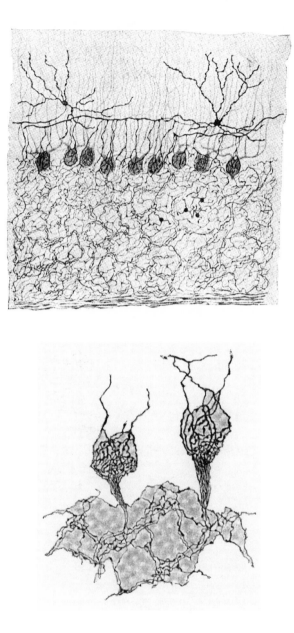

Figs. 3.2, 3.3. Golgi's Nobel Net. Camillo Golgi, "The Neuron Doctrine — Theory and Facts," Nobel Lecture, Dec. 11, 1906, repr. in Nobel Foundation, *Physiology or Medicine, 1901–1921* (Amsterdam: Elsevier, 1967), pp. 191 and 192 (© 1906, The Nobel Foundation). For Golgi, the fibers coming from the molecular layer passed by the Purkinje cells (the large oblong shapes) and continued down into the granular layer below. This was precisely what Ramón y Cajal had insisted for years was *not* the case.

must have figured among the most provocative to Cajal (figures 3.2 and 3.3). Based on a close comparison, the first of these figures is apparently a hand-drawn (and modified) version of an earlier image that Golgi reported to have been drawn "from life," probably using a camera lucida. Both Nobel images showed fibers from the "molecular layer" (above the large Purkinje cells), crossing the Purkinje cell layer, and joining the diffuse neural net of the lower ("granular") layer. It was these direct cross-links, the very existence of which Cajal categorically denied, that stood at the heart of the battle. Were they there, they would support Golgi's idea of a diffuse network and directly oppose Cajal's neuron doctrine.[4] "I have verified," Golgi insisted to the Nobel audience, "that the fibres coming from the nerve process of the cells of the molecular layer only pass near the cells of Purkinje to proceed into the rich and characteristic network existing in the granular layer."[5] These were fighting words — and fighting images. For Cajal, the descending branches of the axons of the cells in the molecular layer (dubbed "stellate" and "basket" because of their appearance) wrapped around and met the cell body and the initial segment of the axon of Purkinje cells. Each neuron stood by itself.

Here was a fiercely consuming debate between the two competitors, fought to a large extent over the objectivity of images — an all-out image war. Both scientists brought numerous figures to their presentations. Furious at what he considered Golgi's visual manipulations, Cajal accusingly wrote of his rival's "strange mental constitutio[n]," one "hermetically sealed" against criticism by its "egocentricity." Golgi was closed to the evidence (according to Cajal), and his inability to register faithfully the outside world of nature had plunged him into an "absurd position" for which one could only appeal to psychiatry for adequate terms. To Cajal, their joint presence in Stockholm was a grotesque injustice: "What a cruel irony of fate to pair, like Siamese twins united by the shoulders, scientific adversaries of such contrasting character!"[6] True, Cajal is generally seen as having won this debate, but it is also true that Cajal's theoretical stance (endorsing the neuron doctrine) shaped some of his own depictions. Our interest, however, here and throughout, is not so much in awarding victory or credit, but in tracking the struggle over images — along with their ethical and epistemological stakes.

All his life, Cajal wrote of his struggle to find a way to "see clearly" — a theme that saturated his scientific writings, his laboratory work, his autobiographical reflections, and even (as we will see in the next chapter) his fiction. It is perhaps fitting that, in 1933, when Cajal was eighty, just a year before he died, he titled his last work, his synthetic polemic, *Neuron Theory or Reticular Theory? Objective Evidence of the Anatomical Unity of Nerve Cells.*[7] Seeing clearly, seeing honestly (finding "*las pruebas objetivas*") was, for Cajal, absolutely necessary for the epistemic virtue of objectivity. Objectivity was at once the guiding and the unifying theme for his self-representation as a moral figure of science, for his insistence on rigorously faithful pictures of the nerve cells, and, most specifically, for his career-spanning defense of the neuron doctrine. The confrontation between Golgi and Cajal was emblematic of that between competing epistemic ideals, which had played out over the question of objectivity in the latter half of the nineteenth century. We will return to the dueling neuroanatomists several times as we map the new configuration of epistemological convictions, image-making practices, and moral comportment that aimed to quiet the observer so nature could be heard: *mechanical objectivity*.

"Let nature speak for itself" became the watchword of the new scientific objectivity. It provoked an inversion of values in scientific image-making. Where idealizing intervention had been upheld as a virtue by earlier scientific atlas makers, it became a vice in the eyes of many of their successors: witness Cajal's anger at Golgi's simplifications. (There was also an issue of technique: Golgi and his students accused Cajal of not being able to reveal the complexity of the nervous system because of their ineptitude in carrying out the silver impregnation.) At issue was not only objectivity but also ethics: all-too-human scientists now had to learn, as a matter of duty, to restrain themselves from imposing the projections (which Cajal called Golgi's "capricious ideas") of their own unchecked will onto nature. To be resisted were the temptations of aesthetics, the lure of seductive theories, the desire to schematize, beautify, simplify — in short, the very ideals that had guided the creation of true-to-nature images. Wary of human mediation between nature and representation, researchers now turned to mechanically produced images. Where human self-discipline flagged, machines or humans acting as

will-less machines would take over. Scientists enlisted self-registering instruments, cameras, wax molds, and a host of other devices
in a near-fanatical effort to create images for atlases documenting
birds, fossils, snowflakes, bacteria, human bodies, crystals, and flowers — with the aim of freeing images from human interference. Not
only would all schematization be avoided, one turn-of-the-century
atlas author assured his readers, but the object of inquiry would also
"stand truly before us; no human hand having touched it."[8]

This chapter is an account of the ethical-epistemic project of producing a visually grounded mechanical objectivity in the late nineteenth and early twentieth centuries. By *mechanical objectivity* we
mean the insistent drive to repress the willful intervention of the
artist-author, and to put in its stead a set of procedures that would, as
it were, move nature to the page through a strict protocol, if not
automatically. This meant sometimes using an actual machine, sometimes a person's mechanized action, such as tracing. However accomplished, the orientation away from the interpretive, intervening
author-artist of the eighteenth century tended (though not invariably) to shift attention to the reproduction of individual items —
rather than types or ideals. The working objects would be gathered
into systematic visual compendiums that were supposed to preserve
form from the world onto the page, not to part the curtains of experience to reveal an ur-form. Depicting individual objects "objectively" required a specific, *procedural* use of image technologies —
some as old as the lithograph or camera lucida, others as freshly late-
nineteenth-century as photomicrography. These protocols aimed to
let the specimen appear without that distortion characteristic of
the observer's personal tastes, commitments, or ambitions. Technology and its accompanying procedures, however, were not enough.
Mechanical objectivity required a certain kind of scientist — long on
diligence and self-restraint, scant on genial interpretation.

Was mechanical objectivity ever completely realized? Of course
not, and its advocates knew they faced a regulative ideal. That is,
they saw objective depiction in their sciences as a guide point. If they
could replace speculation with close observation of an individual,
that was good. If they could find a procedure that would hem in freehand drawing, even better. And if they found a way to minimize
interpretation in the process of image reproduction — better still. It

is easy to assume that objective depiction was *either* an ideal *or* consequential — but it was, in fact, both. Analogously, fairness in the organization of a game may never be complete, but it can nonetheless shape the procedures that its participants adopt.

We do not here — any more than in Chapter Two — intend anything approaching a comprehensive, encyclopedic survey of the genre and history of the scientific atlas. In this central period of the scientific atlas (roughly 1830 to 1930), there are approximately two thousand distinct (nongeographical) atlas titles, alongside hundreds of other forms of systematic assemblages of images — their number begins relatively modestly and then accelerates significantly after 1860 or so. Associated with atlases are natural historical expedition reports, handbooks, and atlas-type compendiums issued under other names — purveying images of everything from spectra to embryos.[9] Adding to (rather than displacing) the continuing genre of idealizing atlases, our focus in this chapter will be on the new kind of scientific atlas that arose in the nineteenth century, one that explicitly militated for a newly disciplined scientific self bound to a highly restrained way of seeing.

The product of this double reformation of self and sight came to be known as scientific objectivity. Like almost all forms of moral virtuosity, nineteenth-century objectivity preached asceticism, albeit of a highly trained and specialized sort. Its temptations and frailties had less to do with envy, lust, gluttony, and other familiar vices than with witting and unwitting tampering with the visual "facts." The relation of this particular form of disciplining the self and the kind of image desired was close: just insofar as one could restrain the impulse to intervene or perfect, one could allow objects — from crystals to chrysanthemums — to print themselves to the page. Put conversely: Seductive as it might be to "see as" this or that ideal, the premium for objective sight was on "seeing that," full stop. But in the view of late nineteenth-century scientists, these professional sins were almost as difficult to combat as the seven deadly ones, and they required a scientific self equipped with a stern and vigilant conscience, in need not just of external training but also of a fierce self-regulation.

Mechanized or highly proceduralized science initially seems incompatible with moralized science, but in fact the two were closely related. While much is and has been made of those distinctive traits — emotional, intellectual, and moral — that distinguish humans from machines, it was a nineteenth-century commonplace that machines were paragons of certain human virtues. Chief among these were those associated with work: patient, indefatigable, ever-alert machines would relieve human workers whose attention wandered, whose pace slackened, whose hand trembled. Where intervening genius once reigned, there, the nineteenth-century scientists proclaimed ever more loudly, hard, self-disciplined and self-restrained work would carry the day.

In addition to the sheer industriousness of machines, there was more: levers and gears did not succumb to temptation. Of course, strictly speaking, no merit attached to these mechanical virtues, for their exercise involved neither free will nor self-command. But the fact that the machines had no choice but to be virtuous struck scientists distrustful of their own powers of self-discipline as a distinct advantage. Instead of freedom of will, machines offered freedom from will — from the willful interventions that had come to be seen as the most dangerous aspects of subjectivity. Machines were ignorant of theory and incapable of speculation: so much the better. Such excursions were the first steps down the slippery slope toward intervention. Even in their failings, machines embodied the negative ideal of noninterventionist objectivity.

Machines did not run themselves, of course. All through the nineteenth century, scientists worked with experts on microscopic photography, engraving, or botanical and anatomical illustration. But whereas eighteenth-century savants had sought to impose their will and way of seeing on such helper-collaborators to achieve four-eyed sight, by the mid-nineteenth century this relationship was undergoing dramatic change. On the one side, the nineteenth-century author spoke incessantly of "policing" the illustrator. On the other, the scientist relied on the illustrator to *check* the author's flights of fancy or speculation. Many forms of restraint were needed to prevent the work's breaking loose from its visual moorings. To capture an unvarnished, objective photomicrograph or drawing of a snowflake, bacillus, or hemoglobin crystal was — and was often recognized as — an operation of consummate skill. Whatever their views on the proper division of credit, scientific atlas makers very frequently commented on the skills of their illustrators, even if these were skills hemmed, even policed, by the supervising scientist. Alfred Donné, a Parisian professor of medicine, not only praised the daguerreotypes made by Léon Foucault for Donné's 1844–45 microscopy atlas of bodily fluids but also listed Foucault as his coauthor on the title page. An effective illustrator came to embody an essential component of a composite scientific self — that part of the self capable of amplifying the moral "no" that nature whispered against the scientist's much-loved hypothesis. Increasingly in search of mechanical objectivity, scientists demanded images, machines, and illustrators that would not budge even to obey the scientist's own misdirected will.

This form of image-based scientific objectivity emerged only in the mid-nineteenth century. It appeared piecemeal, haltingly at first and then more intensively, positioned against idealizing, truth-to-nature images that themselves never died out completely. Like the spring melt of an ice-bound northern river, the change begins with a crack here and there; later come the explosive shears that throw off sheets of ice, echoing through the woods like shotgun blasts, followed eventually by a powerful rush of water that should not, for all its drama, obscure the myriad local changes that preceded it. Objectivity entered the practical domain of scientific atlas making slowly, throughout the 1840s, then gained momentum, until it could be found almost everywhere in the rush of the 1880s and 1890s.

Mechanical objectivity is strikingly distinct from earlier attempts to depict nature rightly in its methods (mechanical), ethics (restrained), and metaphysics (individualized). Although mechanical objectivity can be found in other scientific endeavors of the period, for the same reasons given earlier, we largely restrict our attention to atlases (along with various kinds of scientific handbooks). Here we see images of record designed to last for generations, concrete visual practices rather than oratory alone, and a long historical baseline that offers a window onto the joined shift of scientific ethos and epistemic virtues. Atlases in the age of objectivity taught simultaneously what there was and how scientists must restrain themselves in order to know. Although the ambition of objective vision never entirely replaced seeing as truth-to-nature, the atlases of mechanical objectivity stood for a new and powerful alternative form of scientific vision — blind sight.

By the late nineteenth century, mechanical objectivity was firmly installed as a guiding if not *the* guiding ideal of scientific representation across a wide range of disciplines. Ethics and epistemology fused as atlas makers strove to supervise not only their artists but also themselves. The image, the standard bearer for objectivity as it had been for truth-to-nature, marched before a relentless army attempting to replace willful depiction with mechanical reproduction. This mechanizing impulse was present at once in scientific technique and as moral vision; indeed, the two were inseparable. Nothing in the works of William Cheselden, Bernhard Siegfried Albinus, or Carolus Linnaeus quite prepares us for the fervor of self-denying ethics that animated the late nineteenth-century brief for mechanized representation. Image, author, and technique joined to create a new form of scientific sight.

Before proceeding to the broader category of automatic image production, we must address the form of automatic reproduction of the mid-nineteenth century that looms so large in retrospect: photography. Was the rush for objectivity simply due to a fascination with the new medium? Tempting as this simple explanation may be, the evidence militates against it. Far from being the unmoved prime mover in the history of objectivity, the photographic image did not fall whole into the status of objective sight; on the contrary, the photograph was also criticized, transformed, cut, pasted, touched up, and enhanced. From the very first, the relationship of scientific objectivity to photography was anything but simple determinism. Not all objective images were photographs; nor were all photographs considered *ipso facto* objective.

Photography as Science and Art

Photography was not one but several inventions. Developed in the 1820s and 1830s using different media and different methods, this family of techniques produced strikingly different visual results. Louis-Jacques-Mandé Daguerre, who had previously earned his living in Paris by painting illusionist panoramas, produced a method of chemically fixing an image from a camera obscura on a highly polished silver plate (or a copper plate coated in silver); the resulting image was a unique object, remarkable for its sharp-edged rendering of minute detail.[10] Working independently of Daguerre, the British polymath William Henry Fox Talbot experimented with paper treated with salt and silver nitrate, against which he pressed various flat objects, such as leaves and lace (and, later, camera obscura projections) to obtain a negative reminiscent of a watercolor or silhouette. Talbot initially called his invention "photogenic drawing"; he hoped it would replace the camera lucida for maladroit draftsmen like himself and perhaps also provide a way of reproducing paintings more cheaply and faithfully than engraving.[11] Talbot's countryman and friend the astronomer and physicist Sir John Herschel also saw the potential of photography as a means of making and copying pictures, but his chief interest in the process, to which he contributed numerous chemical improvements in correspondence with Talbot, was its potential to create a scientific instrument for the investigation of the properties of light, such as the detection of ultraviolet light (which was invisible to the naked eye). From the outset, scientific photography partook of this variety of means and ends.[12] (See figures 3.4 and 3.5.)

But scientific photography was only one species of nineteenth-century photography, and objective photography was in turn only one variety of scientific photography.[13] Starting with Herschel's experiments on ultraviolet light, photography was ingeniously deployed to make visible phenomena otherwise invisible to the human eye: light polarization, bullets streaking through the air, birds in flight.[14] In these cases, photographers used their images as instruments of scientific discovery. Photography could also be used to reproduce known phenomena, especially in the field of natural history, with an extraordinary density of detail, extending the precision of lithography.[15] The Swiss-born American naturalist Alexander Agassiz hoped photography would "give figures with an amount of detail which the great expense of engraving or lithographing would usually make impossible, even were it mechanically practicable."[16] (See figure 3.6.) In the service of discovery or detail, scientific photography need not lay claim to mechanical objectivity; sometimes quite the contrary. Our focus here is on that subspecies of scientific photography that did make such claims.

Both artists and scientists were quick to appreciate that photography could be used for registering details, but they split over its usefulness for promoting mechanical objectivity. In his sensational

Fig. 3.4. Arrangement of Fossil Shells. Louis-Jacques-Mandé Daguerre, 1837–1839, Conservatoire National des Arts et Métiers, Paris, daguerreotype (© Musée des arts et métiers-CNAM, Paris/Photo Studio, CNAM). The daguerreotype method exposed a polished silver plate coated with a layer of silver iodide to light in a camera, producing a latent image on the plate itself that became visible when the plate was fumed with mercury. This image, one of Daguerre's earliest, displays the remarkable finish and detail that fascinated contemporaries. But it could not be reproduced, except by engraving the daguerreotype itself, thereby destroying it. (Please see Color Plates.)

Fig. 3.5. Three Leaves. Photogram, William Henry Fox Talbot, 1839, Fox Talbot Museum, Lacock, England (courtesy of the British Library). This photogenic drawing negative resulted from the exposure to sunlight of paper impregnated with light-sensitive silver chloride. It is, in fact, a photogram: the leaves have been pressed directly against the paper under glass and exposed for about a quarter of an hour, turning the silver chloride into metallic silver. The resulting image could then be used, by repeating the process, to create a "positive" in which light and dark areas were reversed. Because of the long exposure times required by the process, images were often indistinct.

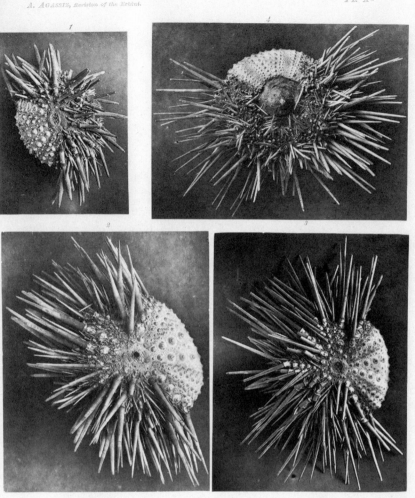

A. Sonrel, Photo.   American Photo-Relief Co., Printers, Phila.

Fig. 3.6. Echinoderms in Detail. *Echinometra viridis* (fig. 1, *upper left*) and *Echinometra subangularis* (figs. 2–4), Woodburytypes, Alexander Agassiz, *Revision of the Echini* (Cambridge: Cambridge University Press, 1872–1874), pl. 10 (Museum of Comparative Zoology, Harvard University. Photograph © President and Fellows of Harvard College). Agassiz was among the first to use novel techniques like the Woodburytype and Albertype to mechanically reproduce photographic images for scientific publications. His survey of echini specimens held in collections throughout the world was illustrated with both lithographs and photographs, the latter made by Auguste Sonrel, who had also been Louis Agassiz's (Alexander's father) scientific illustrator and lithographer. The polished style that made Sonrel's natural-history lithographs famous was continued in the new medium. Here, scientific photography aimed at the near-effortless registration of detail, not at objectivity.

public presentation of Daguerre's invention to a joint public session of the Académie des Sciences and the Académie des Beaux-Arts in Paris on August 19, 1839, the French astronomer and physicist François Arago exclaimed over the possibilities the new medium offered as a scientific recording device and light detector; quoting the painter Paul Delaroche, he also envisioned photographs as a means of perfecting "certain conditions of art, so that they become for painters, even the most clever, a subject of observation and studies." Though scientists might want photography to provide them with a hands-off epistemology, and artists might be after photography's soft light, chiaroscuro, and richness of tone, there were those on both sides of the divide who admired the photograph's ability to render each and every tiny detail effortlessly. Arago imagined how useful the new invention would have been to the Napoleonic expedition to Egypt in order to record "the millions and millions of hieroglyphics" covering temples; Delaroche marveled at the "unimaginably exquisite finish" of daguerreotypes.[17]

Because photography was at first conceived as a substitute for drawing and engraving, it was imagined as a marvel of saved artistic labor. "It is so natural," remarked Talbot, apropos of his "photogenic drawings," "to associate the idea of *labour* with great complexity and elaborate detail of execution, that one is more struck at seeing the thousand florets of an *Agrostis* [blossom] depicted ... than one is by the picture of a large and simple leaf of an oak ... but in truth the difficulty is the same."[18] Reviewers of Talbot's *Pencil of Nature* (1844–1846) compared one of the calotype images (images made on photosensitized high-quality writing paper) favorably to a seventeenth-century Dutch painting of a domestic scene. Apparently to allay skepticism, Talbot inserted slips in some copies of his book: "The plates of the present work are impressed by the agency of Light alone, without any aid whatever from the artist's pencil. They are the sun-pictures themselves, and not, as some persons have imagined, engravings in imitation."[19] The capacity to freeze detail with negligible labor remained a lauded feature of nineteenth-century photography for scientific illustration — and of photography as a new, better way to reproduce artwork.[20]

Very soon, however, another argument was advanced in favor of photography as a distinctly *scientific* medium. The automatism of the photographic process promised images free of human interpretation — *objective* images, as they came to be called.[21] The multiple inventors of photography had all emphasized the wondrous spontaneity of the images, "impressed by nature's hand," in Talbot's phrase.[22] Automatism and objectivity converged in one of the earliest scientific atlases to boast of its use of photographic images, Donné's *Cours de microscopie complémentaire des études médicales* (*Course in Microscopy to Complement Medical Studies*, 1844–1845). Alongside drawings of microscopic views of blood, milk, semen, and other bodily fluids, Donné included photographs "exactly representing the objects as they appear, and independently of all interpretation; to achieve this result, I did not want to trust either my own hand or even that of a draftsman, always more or less influenced by the theoretical ideas of the author; profiting from the marvelous invention of the daguerreotype, the objects are reproduced with rigorous fidelity, unknown until now, by means of photographic processes." Donné hoped his images would extinguish the oft-repeated objection of his medical colleagues that the microscope showed only "illusions." Who could resist this wonder? An object that "painted itself, fixed itself upon the plate without the help of art, without the least contribution of the hand of man, by the sole effect of light, and always identical in the least details."[23] (See figures 3.7 and 3.8.)

In contrast to the argument from detail, the argument from objectivity undercut the artistic claims of photography. The Salon of 1859, the first official Parisian art exhibition to include photographs, divided critics. Charles Baudelaire railed against slavishly naturalistic landscapes and the still more slavish artistic photography, deploring an art so lacking in self-respect as to "prostrate itself before external reality." To "copy nature" was to forsake not only the imagination but also the individuality Baudelaire and other Romantic critics believed essential to great art: "The artist, the true artist, must never paint except according to what he sees or feels. He must be *really* faithful to his own nature." Photography might be admirable in the hands of the naturalist or the astronomer, but the "absolute material exactitude" sought by science was inimical to art.[24] Reviewing the same exhibition, Louis Figuier (a professor at the Ecole de Pharmacie in Montpellier and science journalist and popularizer) defended photography as art, citing the photographer's individual style and "sentiment." No

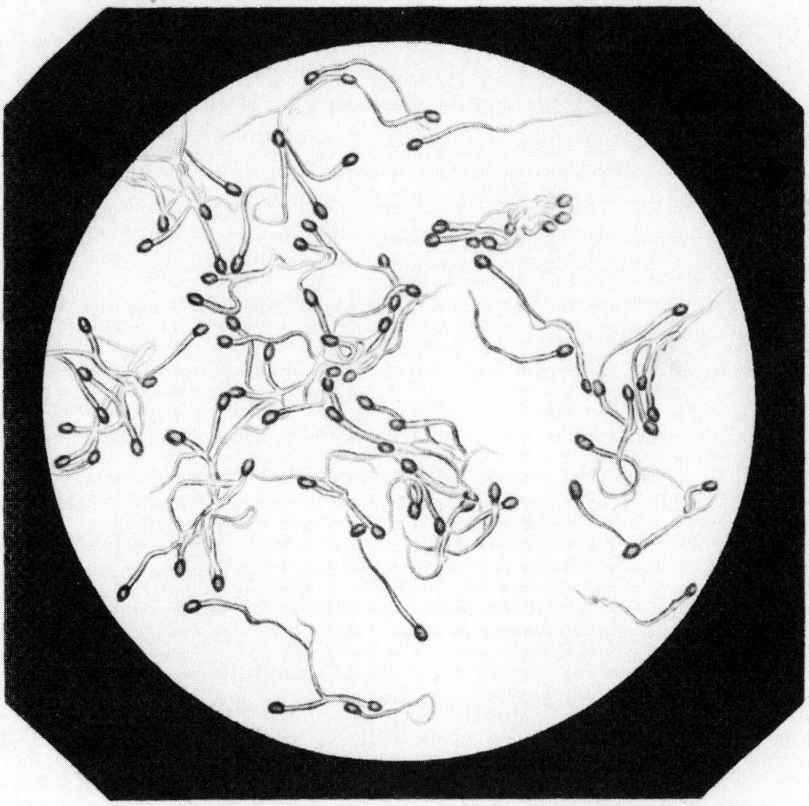

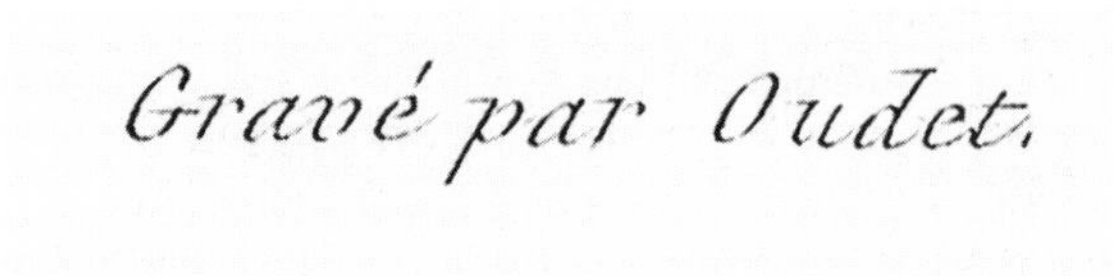

Gravé par Oudet.

Figs. 3.7, 3.8. Mechanical Objectivity Before Mechanical Reproduction. Bat spermata, Alfred Donné and Léon Foucault, *Cours de microscopie complémentaire des études médicales: Anatomie microscopique et physiologie des fluides de l'économie* (Paris: Baillère, 1844–1845), atlas, pl. 15, fig. 62 (*top*), magnified detail (*bottom*). This figure is labeled as "taken by means of a microscope daguerreotype by L. Foucault," but it is, in fact, a lithograph based on the daguerreotype, since the latter could not be mechanically reproduced. The magnified detail shows the signature of the lithographer Oudet. Until the 1880s, however, lithographs or wood engravings (see fig. 3.12) copied from photographs were often assumed to carry the latter's imprimatur of objectivity.

one, Figuier was certain, could mistake the full-blooded work of a French photographer for the wan images of the English. How could such originality be reduced to a "simple mechanism"?[25]

Opposed as Baudelaire and Figuier were on whether photography qualified as art, they agreed entirely on the criterion for defining art. Genuine art must bear the stamp of the maker's individuality and imaginative interpretation; no "mechanical" copy of nature could qualify. This was the same criterion that scientists invoked to distinguish artistic from scientific images, albeit with reversed valuation. By the 1860s, the term "mechanical photography" was being used in opposition to aesthetic photography (for example, in portraiture).[26] It was a sign of the new opposition of science and art that the mixing of genres of objective (scientific) and subjective (artistic) photography could provoke scandal, as when it was revealed that the California photographer Eadweard Muybridge, who as a commercial photographer would have routinely retouched his landscapes, had done the same for his famous photographs of a galloping horse, touted as a scientific rebuttal to artistic misconceptions.[27] Whereas photography trade journals and handbooks were full of advice on how to retouch photos and the best way to secure copyright protection for artistic property (see figure 3.9), self-consciously "mechanical" photography eschewed all such aesthetic interventions.[28] The mechanical, objective photograph had allegedly been traced by "nature's pencil" alone, and nature was entirely artless.

Were such claims anything more than rhetorical flourishes? Historians of photography point out the considerable skill and judgment required to make a photograph; nature emphatically does not paint itself by itself.[29] Historians of art call attention to the aesthetic context that shaped the making and seeing of photographs, even scientific and medical ones.[30] Historians of science note that nineteenth-century photographers and scientists and their audiences were perfectly aware that photographs could be faked, retouched, or otherwise manipulated.[31] (See figures 3.10 and 3.11.) Almost any article of the period on how to make a photograph for scientific purposes gives pages of detailed, difficult instructions; it required effort and artifice to persuade nature to imprint its image. In what sense, then, could these images be described by atlas makers as objective and mechanical?

133

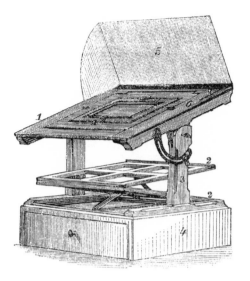

Figs. 3.9, 3.10, 3.11. Artful Photography.
Retouching apparatus, Alois von Reiter,
"Retouchirpult für Negativs," *Photographische Correspondenz* 2 (1865),
pp. 17–18 (*top*); girl in winter clothes
and retouched snowfall, H. Collischon,
1897, from Timm Starl, *Im Prisma des
Fortschritts: Zur Fotografie des 19.
Jahrhunderts* (Marburg, Germany: Jonas
Verlag, 1991), p. 52 (*bottom left and
right*). Professional journals for photographers often featured devices like this one
for retouching negatives — in this case, by
directing light to the desired spot on the
negative by means of a mirror. Portraits
could be "improved" and special effects
added, as in the case of the artificial
snowfall shown here. Late nineteenth-
century commercial photographers — and
their customers — were fully aware that
photographs could be manipulated.

When nineteenth-century scientists called for objective photographs to supplement, correct, or replace subjective drawings, they did not, in the first instance, fear imposture, except perhaps in cases such as inquiries into spiritualism.[32] Rather, they worried about a far more subtle source of error, one more authentically subjective and specifically scientific: the projection of their own preconceptions and theories onto data and images. Therefore, the fact that photographs may require filters, sophisticated lenses, special preparation of the object, long exposure time, or darkroom manipulation was irrelevant to the issue of objective or indexical depiction, so long as none of these operations colluded in the scientist's wishful thinking. Often, a division of labor in which technicians supposedly ignorant of the theoretical stakes made and developed the photographs was proposed as a precaution. Even in the late nineteenth century, after photogravure techniques made it possible to reproduce photographs cheaply and accurately, scientific drawings still survived. Photographs were preferred for subject matter that might arouse skepticism — because it was rare or spectacular or controversial. Manuals on scientific photography recommended that ethnographers use photographs rather than drawings, because European artistic conventions might otherwise distort non-European bodies: "The draftsman, whatever might otherwise have been his talent, did not know how to see and always drew people of the white race whom he later colored in black or red."[33] (See figure 3.12.) Similarly, the persistent visual ambiguities of microscopy demanded photographic illustration, to forestall the observer's tendency "to insert involuntarily his hypothetical explanation into the depiction."[34] A photograph was deemed scientifically objective because it countered a specific kind of scientific subjectivity: intervention to aestheticize or theorize the seen.

The term "mechanical" must also be understood in context, a task made more difficult by the pervasive conflation of two conceptually and historically distinct processes via the single phrase "mechanical reproduction."[35] In one sense, the phrase refers to the automatic production of an image without the interventions of an artist. In another sense, it refers to the "automatic" multiplication of images (which could be lithographs or engravings as well as photographs) so that they could be accurately, widely, and inexpensively

Fig. 3.12. "Polynesian Types." Wood engraving, E. Hamy, "Polynésiens et leur extinction," *La nature* 3 (1875), pp. 161–63. Highly illustrated popular science journals like *La nature* used a range of reproductive media, including lithographs, engravings, and — in order to reproduce photographs for mass print runs — wood engravings like this one, done after a photograph by Commander Miot. *La nature* typically turned to wood-engraved photographs (as opposed to lithographed drawings) when the object was exotic (as in this case), singular (for example, conjoined twins), or spectacular (for example, a solar eclipse).

disseminated. Although photographs became prototypical of the first sense of mechanical, they did not fall under the second until the 1880s, when new techniques, such as the Woodburytype and half-tone photolithography, made mass printings of photographs practicable.[36] Earlier published photographs had to be either printed by hand from the negative or reproduced through woodcut, engraving, or lithography. Look closely at the "microphotograph" printed in Donné's 1845 atlas (figure 3.7): it is, in fact, an engraving, signed by the engraver Oudet. Indeed, "photographs" in the scientific and the popular press were often wood engravings from photographs (as in figure 3.12), carrying the assurances that they had not been re-touched.[37] As the science popularizer Gaston Tissandier wrote in 1874, only with the means to insure "the inalterability and the indefinite multiplication" of photographs would Daguerre's mechanical art be complete.[38]

When the term "mechanical" was applied to photographs prior to *circa* 1880, it referred to the process by which light imprinted an image on specially prepared metal, paper, or glass. Because the image was likened to a drawing or engraving, the absent human hand implied by the word "mechanical" was that of the artist, not the photographer. Fixated upon the delineation of the image itself, early photographers and their audiences compared photography to drawing. Even if aided by a camera obscura or a camera lucida, the draftsman must still trace the projected image onto paper — no easy task, as Talbot had discovered to his chagrin. However arduous preparing the apparatus, composing the picture, operating the camera, and developing the image were, the process was (in the particular cultural context of the time) perceived as requiring negligible labor compared to the task of putting pencil to paper. This was why the image counted as "mechanical."

"Mechanical" had long referred to an inferior brand of human labor executed with the hands, not the head (Shakespeare's "rude mechanicals"). As the Industrial Revolution transformed work in nineteenth century, "mechanical" retained its pejorative, manual associations, but now referred dismissively to actual machines and the workers who tended them, suggesting they were repetitive, mindless, automatic.[39] Eighteenth-century scientific atlas makers had longed for artists talented enough to render kangaroos and crystals truthfully and elegantly but pliant enough to bow to the naturalist's judgment: the clever but docile servant. Nineteenth-century atlas makers derived their image-making ideals from the factory rather than the atelier. As the British mathematician and political economist Charles Babbage put it apropos of the calculation of logarithms, what was wanted was a mechanical "substitute for one of the lowest operations of the human intellect."[40] It was, he thought, but a short step from unlettered drudges to unthinking machines.[41] Haunted by anxieties about their own subjective representations, scientists discovered the ethical-epistemic consolations of the mechanical image, in which, by a supreme act of self-effacing will — or by deploying procedures and machines that bypassed the will — they could ensure that no intelligence would disturb the image.

Automatic Images and Blind Sight

Scientific photography held out a promise of automaticity, although it clearly could not do without real human hands and heads. Conversely, there were numerous forms of procedural, mechanical reproduction (such as tracing or even highly supervised wood-engraving) that were not photographic. Most important, however, the ethical-epistemic stance that scientists began to take after the 1830s increasingly insisted on a ferocious devotion to depicting what was seen on the surface, not what was deduced or interpreted. This emphasis was not simply the reflection of this or that bit of the history of photography. In short, the photographic and the mechanical were *not* coextensive, and the shift from depiction that celebrated intervention to one that disdained it did not come about *because* of photography.

For the scientific atlas makers of the late nineteenth century, the machine was both a literal and a guiding ideal. Machines assisted where the will failed, where the will threatened to take over, or where the will pulled in contradictory directions. Machine-regulated image making was a powerful and polyvalent symbol, fundamental to the new scientific goal of objectivity.

First, the machine's ability to turn out thousands of identical objects linked it with the standardizing mission of the atlas. The machine provided a new model for the perfection toward which working objects of science might strive. Echoes of the popular fascination with the ubiquity and standardized identity of manufactured goods crop up throughout nineteenth-century scientific literature. Following Herschel, James Clerk Maxwell even used the mass production of identical bullets as a metaphor for atoms too similar to be distinguished.[42] The identical form of bullets suggested a maker — and for Maxwell the identical form of atoms pointed to a Maker. Though often lost on moderns who fetishize the handmade, there was, in the nineteenth century, an aesthetic pleasure in identical objects.

Second, as it took the form of new scientific instruments, the machine embodied a positive ideal of the observer, but one that contrasted sharply with the eighteenth-century genius of observation. The machine was patient, indefatigable, ever alert, probing beyond the limits of the human senses. Once again, scientists took their cue from popular rhetoric on the wonder-working machine. Babbage rhapsodized about the advantages of mechanical labor for tasks that required endless repetition, great force, or exquisite delicacy. He was especially enthusiastic about the possibilities of using machines to observe, measure, and record, for they counteracted all-too-human weaknesses: "One great advantage which we may derive from machinery is from the check which it affords against the inattention, the idleness, or the dishonesty of human agents."[43] Just as manufacturers admonished their workers with the example of the more productive, more careful, more skilled machine, scientists admonished *themselves* with the more attentive, more hard-working, more honest instrument.

Third, and most significant for our purposes, the machine seemed to offer images uncontaminated by interpretation. This promise was never actually fulfilled — neither the camera obscura nor smoked-glass tracings nor the photograph could altogether rid the atlases of interpretation. Nonetheless, the scientists' continuing claim to such judgment-free representation is testimony to the intensity of their longing for the perfect, "pure" image. In this context, the machine stood for authenticity: it was at once observer and artist, free from the inner temptation to theorize, anthropomorphize, beautify, or interpret nature. What the human observer could achieve only by iron self-discipline, the machine effortlessly accomplished — such, at least, was the hope. Here the machine's constitutive and symbolic functions blur, for the machine seemed at once a means to and a symbol of mechanical objectivity.

The observer now aimed to be a machine — to see as if his inner eye of reasoned sight were deliberately blinded. By the middle of the nineteenth century, Otto Funke, a physiological chemist at the University of Leipzig, was doing everything in his power to transform himself into such a recording device. Not for him were wild flights of fancy, interpretive schemes, or even pretensions of wide knowledge — anything that might reshape the image as seen through his microscope. Among their other aims, physiological chemists such as Funke sought to sort out the chemical constituents of bodily fluids. Funke himself had been the first to crystallize hemoglobin in 1851, a crucial step in unraveling its function as a transporter of oxygen. Two years later, in his *Atlas of Physiological Chemistry*, he insisted: "I have attempted to reproduce the natural object in its minutest details, and even with pedantic accuracy, as far as pencil and graver would permit; above all things prohibiting the slightest idealization, either by myself or the lithographer." Quick to acknowledge that this absolute fastidiousness was a bold project, impossible to carry out completely, he nonetheless took it as his "imperative duty" to try. Not a single drawing was borrowed from predecessors, Funke claimed. Indeed, he could "conscientiously affirm" that the drawings were from actual microscopic objects, every single crystal or cell, "exactly as they appear under the microscope; not according to ideal models."[44]

For Funke, it was obvious that it was as important for someone entering a zoochemical laboratory to "learn to '*see*'" as it was to know chemical analysis. Use a microscope, of course, Funke admonished. But learning the proper mode of graphical representation was just as important as controlling the instrument. Images, he insisted, would serve the neophyte "as a grammar of the language of the microscope." Learning this plain, blind sight was no mean feat; while its necessity was granted by others, he saw his predecessors as having failed by presenting diagrams or drawings "too much idealized." More specifically, some atlases (Funke named Donné's *Atlas*) failed due to their limited scope. Others stumbled because of "false idealization" wrongly based on (perfect) crystallographic diagrams: "There are indeed hundreds of instances in which it is not the crystalline form which characterises bodies, but precisely the deviations from the perfect figure."[45] Those idealizations were such that it "might reasonably be doubted whether any impartial observer could tell what they were intended to represent. I could point out cholesterin plates, with angles of 50°; urate of soda... in the form of a spider."[46] (See figures 3.13 and 3.14.) For Funke, such claims to see beyond the plainly visible risked tumbling the observer into a chaos of conflicting, unconfirmed images.

By contrast, Funke aimed in his *Atlas* to achieve pure receptivity. He sought to discard nothing on the basis of ancillary observations, theories, or interpretations. Where others might depict an object in isolation, Funke demanded "natural mutual relations," down to the right grouping, quantitative proportions, "in short, true reflections of the microscopic field of vision," no matter what should fall in that domain. In a move that would have seemed unimaginable among the idealizers he was criticizing, Funke went so far as to record artifacts: "I have... copied even the optical deceptions which are owing to the different refractive powers of crystalline substances, as for instance, the apparent displacement of the under planes and edges of a crystal when seen through its substance. I have faithfully copied the shadows produced by the illumination of microscopic objects from beneath or from the side, and have represented the various aspects of certain objects dependent upon the focal adjustment of the lenses." Yet even Funke did not withdraw from the visual field entirely. He was willing to join objects from various preparations and from different sectors of the microscopic view, combining all he had seen into one dense drawing. After all, he remarked almost apologetically, it very rarely happens that all forms and modes of grouping are combined in one view.

Funke's drive to reproduce the scene in the microscope's eye-piece on the page extended to the minute details of the image production. "I have in all cases delivered the drawings to the lithographer in a perfectly finished state, and have not let him add a single line to them." Unlike the four-eyed sight of the eighteenth century, the illustrator's contribution was not, according to Funke, artistic skill, and indeed on the title page the lithographer's name is nowhere to be found. Indeed, for Funke, the lithographer's virtue was precisely in his capacity to *re*produce Funke's own faithful rendition of what Funke's eye saw through the lens: "I cannot sufficiently acknowledge the extraordinary fidelity and care with which Herr Wilhelmi has copied my drawings,... point for point, and the trouble

Figs. 3.13, 3.14. Spiders and Crystals. Golding Bird, *Urinary Deposits: Their Diagnosis, Pathology, and Therapeutical Indications*, 2nd ed. (London: Churchill, 1846), p. 92, fig. 9 (*top*), p. 100, fig. 20 (*bottom*). "At the risk of exposing myself to the charge of self-laudation," Otto Funke remarked, "I must confess that in most of the zoo-chemical figures with which I am acquainted, both draughtsman and lithographer . . . disguise the natural object in such a manner as to render its recognition impossible." Among his primary targets was Golding Bird, whose unblemished crystals (fig. 3.13) offended him, as did the "urate of soda (sic,) in the form of a spider" (no doubt Funke is targeting Bird's fig. 3.14; "sic" is in the original). Instead, Funke wanted an atlas with "pedantic accuracy": objectivity without a whiff of idealization.

which he has taken to adapt certain technical modes of operation to the representation of pencil work." *Everything* the lithographer did aimed to efface itself, down to the quality of Funke's pencil. The force of the striving for an ideal of self-effacement is clear not only positively but also negatively — in the *failure* to reproduce. True, the lithographic process sometimes exhibited "deficiency" in its inability to surmount the difficulty of depicting those "delicate and uniform shadow tints" that pencil and stump captured so easily — even when rendered upon stone with the finest diamond shading. Outlines, especially faint ones, inexorably appeared "somewhat more harsh and distinct upon the stone." Color was even more elusive, as it was "to some extent dependent upon subjective conditions."[47] (See figures 3.15 and 3.16.) Objectivity was the goal.

Funke argued that even the words used — the captions — should be hemmed into the briefest of expressions dictated by two rules. First, give the object's source, name, and mode of preparation. Second, describe only the *optical part of the subject*. Anything exceeding "what the plates themselves" afforded, was, for the author, beyond his remit.

Objectivity was a desire, a passionate commitment to suppress the will, a drive to let the visible world emerge on the page without intervention. When Funke could restrain his own selective, idealizing, interpreting impulses, when he could confine "his" lithographer, Herr Wilhelmi, to pure reproduction — he was proud of these accomplishments. Conversely, when the physiologist failed to live up to the demands of his self-restriction to the purely optical — when the image failed with a too-harsh outline or a subjective tint — he apologized. Objectivity was an ideal, true, but it was a regulative one: an ideal never perfectly attained but consequential all the way down to the finest moves of the scientist's pencil and the lithographer's limestone.

William Anderson captured the will to objectivity in his 1885 introductory address to the Medical and Physical Society of St. Thomas's Hospital. Anderson had studied at the Lambeth School of Art and then advanced through the medical ranks to become a lecturer in anatomy at St. Thomas's (where, to his students' admiration, he composed medical figures on the blackboard using both hands simultaneously). His address sketched the history of the relation of

Plate X.

Fig. 1.

Fig. 2.

Fig. 3.

Fig. 4.

Fig. 5.

Fig. 6.

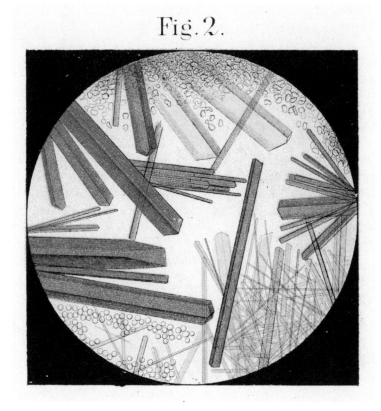

Fig. 2.

Figs. 3.15, 3.16 (detail). Blood Crystals. Otto Funke, *Atlas of Physiological Chemistry* (London: Cavendish Society, 1853), pl. 10. Fig. 3.15, within which subfig. 2, for example, shows detached crystals in a yellow "mother liquor," with much smaller blood crystals that were pale yellowish red, spotted, "mixed with irregular, scaly incipient crystals" (see detail, *lower left*) — all this "irregularity," Funke warned the reader, was due to exposure to atmospheric air. Other crystals hid one another or, in an optical illusion, refracted angles. In fig. 3.16, the crystals appeared "full of cavities" and, often, "broken." Seeing an *im*perfect world took strict discipline — and self-discipline. (Please see Color Plates.)

art to medical science, and his message was clear: the medicine of the late nineteenth century no longer could employ the great artists of the age, as Andreas Vesalius had done in the Renaissance. This loss was, however, not necessarily a bad thing. Scientific understanding had not only made artistic insight supererogatory; it had also shown that the artist could prove to be a liability. The seventeenth-century Amsterdam anatomist Govard Bidloo, for example, struck Anderson in 1885 as "too naturalistic both for art and science, but the man who was usually almost Zolaesque in his superfluous realism could not always resist the temptation to pictorial allegory."[48] If even Bidloo had fallen prey to the temptation to transgress a flat objectivity, how greatly needed was a machine that would automatically and forcibly resist temptation and exclude imposed meaning. John Bell, to whose 1810 *Engravings of the Bones, Muscles, and Joints* Anderson granted artistic merit, was saved because "he was above all a man of science, and as he did not care to risk any sacrifice of accuracy by trusting the unaided eye of the draughtsman, he had each specimen drawn under the camera obscura."[49]

The secret to surpassing the titanic artists of yore, according to Anderson, lay in the control of the representational process by automatic means. Only in this way could "temptation" be avoided, whether it proceeded from artistry (as in Bidloo's case) or from systems of thought. In the age of science, mechanization trumped art: "We have no Lionardo de Vinci [sic], Calcar, Fialetti, or Berrettini, but the modern draughtsman makes up in comprehension of the needs of science all that he lacks in artistic genius. We can boast no engravings as effective as those of the broadsheets of Vesal, or even of the plates of Bidloo and Cheselden, but we are able to employ new processes that reproduce the drawings of the original object without error of interpretation, and others that give us very useful effects of colour at small expense."[50] Such a "mechanical" elimination of the engraver cut one (too-active) handworker out of the cycle of image reproduction and therefore, Anderson believed, contributed to the eradication of interpretation. The virtue of four-eyed sight had become, for Anderson, the vice of unharnessed artistry.

Artists, even militantly realistic ones, agreed that their very presence meant that images were mediated. Champfleury, the novelist ally of Gustave Courbet and spokesman for the realist movement in

France, insisted that "the reproduction of nature by man will never be a reproduction and imitation, but always an interpretation ... since man is not a machine and is incapable of rendering objects mechanically."[51] Courbet even included the figure of Champfleury in his painting *The Painter's Studio; A Real Allegory* — the title suggests that the real and the allegorical could and should enter together. Of course, Champfleury was lauding interpretive intervention on the part of the artist, while Anderson lambasted it from the point of view of the scientist. But *both* scientific objectivity *and* artistic sub-jectivity turned on the valuation of the active, interpreting will.

Policing of subjectivity by the partial application of photographic technology was widespread in the last decades of the nineteenth cen-tury, even where the actual use of photographs in an atlas was im-practical — too expensive, too detailed, or even insufficiently detailed. For example, a quite common use of the photograph was to interpose it in the drawing stage of representation. Typical of such a strategy was the careful selection of photographs by the authors of the 1885 *Johnston's Students' Atlas of Bones and Ligaments*. Only after making such a selection did they turn the image over to an artist, who traced the photograph as the basis for the final drawing.[52] Similarly, when the pathologist Emil Ponfick (who had been Rudolf Virchow's first assistant and studied with some of Germany's leading mid-nine-teenth-century anatomists and surgeons) turned to atlas making, he too demanded control over artistry. In his 1901 magnum opus, an atlas of medical surgical diagnostics, Ponfick reassured the reader that his strict rules had limited the artist's actions. He had recorded outlines of organs on a plate of milk glass mounted over the body, then transferred the image from glass to transparent paper; from the transparent paper, he had inscribed the image onto paper destined for the full watercolor painting. While this series of putatively homo-morphic actions is by no means fully mechanical (hands-free), at every stage possible the pathologist sought all the automatism that he could implement. "As I [Ponfick] observed the work of the artist con-stantly and carefully, re-measuring the distances and comparing the colours of the copy with those of the original section, I can justly vouch for the correctness of every line."[53]

Eighteenth-century observers had also employed devices like the camera obscura — but they prided themselves on their *correction* of the resulting images (think of Cheselden). For Ponfick, on the contrary, the purpose of the apparatus was, at each step, precisely to extirpate interpretation and idealization — to remeasure, to check and compare. Instead, Ponfick's obsessive concern with the "correctness of every line" was key for the establishment of mechanical objectivity. In the precision of their depiction, objects became specific, individual, no longer representative of a type but instead the end product of a series of certifiably "automatic" copies.

But concern for the particularity of the object was neither restricted to the medical nor peculiar to the photographic. Take snowflakes — about as far from the lymph system or dissected brain as one could get. Their history tracks our larger ethico-epistemic history of scientific depiction in a particularly striking way. For hundreds of years, naturalists and scientists had attempted to characterize the delicate structure of these crystalline forms. Robert Hooke had tried drawing them in his *Micrographia* (1665), as had a myriad of authors throughout the eighteenth and early nineteenth centuries.[54] John Nettis, the eighteenth-century "doctor of physic, and oculist to the Republic of Middleburg," sketched the perfect symmetry. He depicted stars of six-plane rhomboid particles, sometimes plane hexangular particles of equal sides or oblong hexangulars. Some had hexangular lamellae of equal sides, and others were "ornamented" with six rays to which were fixed "the most slender lamellae," also hexangular. He found and drew quite stunning plates of this beautiful symmetry in 1755–56. At the very end of his article, Nettis added, as if in apology, "N.B. Number 57 and 84, are anomalous figures of snow; of which there is an infinite variety, that may be observed." Asymmetry and irregularity were footnotes to right depiction — even when their number was infinite.[55] (See figures 3.17 and 3.18.)

Nettis was just one in a long line of systematic snowflake hunters. The explorer Sir Edward Belcher came to appreciate flakes as they landed on his sextant and perused their shape under the instrument's microscope. For years, Belcher had been navigating through Arctic straits, dodging ice floes, and preserving his fleet through the harsh winter. Snowflakes were one more natural sign to be read. Stars and garters ("from their resemblance to the order of knighthood and perfection of crystal") were there, as was the frozen analogue of

J. Mynde sc.

84

Figs. 3.17, 3.18. Nota Bene: Anomalies. John Nettis, "An Account of a Method of Observing the Wonderful Configurations of the Smallest Shining Particles of Snow, with Several Figures of Them," *Philosophical Transactions* 49 (1755), table 21, p. 647. John Nettis used a compound microscope to study snow crystals. In a great harvest of flakes during the "intense cold" of January and February 1740, he landed nearly eighty different types. Nettis found that his catch followed the strict geometric patterns of "parallelograms, or oblong, strait, or oblique quadrangles, rhombs, rhomboids, trapezia, or of hexagonal forms of equal or unequal sides, whole angles are sixty degrees." Some crystals reminded him of city fortifications; all were "beautiful." Orphaned on the last page of the article was a single sentence telling the reader to note well that nos. 57 and 84 were "anomalous figures of snow." Within that *post scriptum*, Nettis remarked there was an "infinite variety" of such sports. Yet mere infinity could not shake symmetry from observation: geometric perfection ruled over mere sight.

light rain. Heavy, flocculent snow corresponded to rain, "warning
the intelligent officer that he had better pitch his tent," while fine,
spicular snow was "bad omened." At root, he believed that storms
(and meteorology more generally) had a scientific regularity, a pre-
dictability that could be mastered. Like Nettis, Belcher insisted that
snow was, in its originary form, perfect; deformities were mere late
additions. As Belcher wrote in 1855, "I detected the perfect hexagonal prismatic formation of every ray, and that the additional rays
disposed themselves invariably at angles of 60° and 120° to the
primitive six-rayed crystal, followed in succession by others...pro-
ducing eventually the most complicated and beautiful star."[56]

That same year, James Glaisher, a meteorologist, balloonist, and
the superintendent of the department of meteorology and magnet-
ism at the Royal Greenwich Observatory from 1838 to 1874, assembled a great collection of snow figures. Like Nettis, Belcher, and
William Scoresby before him, Glaisher believed in the perfection of
the snow crystals and incorporated that faith in the very fabrication
of the images. In four intense weeks of observations, he sketched
some 150 ephemeral snow figures, which his wife then carefully
redrew and completed according to the principle of symmetry, since
he had been able to sketch only a fragment of each original form.[57]
Idealization in Glaisher's figures was not an incidental supplement
but implicated in the very procedure of their fabrication. Here was
built-in truth-to-nature.

In the late 1880s, the Berlin meteorologist Gustav Hellmann joined
the illustrious lineage of snow men — but was bound and determined
to serve with mechanical objectivity. Hellmann explained that he,
too, had spent years racing to draw the fragile forms, extending by
symmetry what he had succeeded in depicting before the snowflakes
thinned and melted. In 1891, after years of cold pursuit, Hellmann
recruited the renowned Berlin photomicrographer Richard Neuhauss
to turn his skills, honed by his biomedical work, to snow, adapting
his remarkable apparatus from the laboratory to the outdoors. They
succeeded around Christmas 1892. At first, Neuhauss conceded, the
new photographs might seem hardly an advance over drawings.
"One misses in them the absolute regularity and the perfect symmetry that is so characteristic of the snow crystals of Scoresby and
Glaisher. One had become used to such a mathematical regularity in

the building of the snow crystals and is now a bit disappointed not to find it here. But it is precisely in this *departure* from ideal forms and schematic figures that we find real pictures [*reelle Bilder*] as nature presents them to us."[58] Hellmann's snowflake (figure 3.21) differed profoundly from the symmetrized crystal recorded by the Arctic explorer William Scoresby (figure 3.19). Scoresby's depictions — like the vast bulk of Nettis's — aimed to capture a perfection that eluded observers riveted by particulars.

Does the difference between Hellmann and Neuhauss, on the one side, and Nettis, Glaisher, and Scoresby, on the other, reflect no more than the fact that Hellmann and Neuhauss had a photographic camera and the others did not? Clearly not. The remarkable and much-repro- duced snowflake compendiums of Wilson Bentley, a self-educated farmer from Jericho, Vermont, make that very clear (see figure 3.20). For years, beginning around 1885, Bentley's extraordinarily beautiful white-on-black photomicrographs, taken with his bellows camera, were reproduced around the world. Neuhauss derided these images, which he took to have the appearance, but not the reality, of hands- off depiction. The black background, he lamented, was thought by naive viewers to be dark-field illumination — when, in fact, the flake images had simply been scraped out of their real background and put on black. Worse, Neuhauss regretted that "in many images Bentley did not limit himself to 'improving' the outlines; he let his knife play deep inside the heart of the crystals, so that fully arbitrary [*willkür- liche*] figures emerged."[59] Replacing the background, incising the object, snipping the edges, improving the image: these were, for Neuhauss, high crimes against objectivity. Merely using photography could not cure diseases of the will, a disorder that survives in the very construction of the German word *willkürlich*.

Idealized flakes, whether produced with or without photogra- phy, do not *refer* in the same way that Hellmann's and Neuhauss's do. While the idealized representations picked out entities not quite attached to any one particular frozen object, Hellmann and Neuhauss seized a specific — and inevitably flawed — individual. (See figures 1.2 and 3.21.) The ensuing fall from perfection startled their contemporaries. The snowflake would never be the same. "Yes," Hellmann concluded, "despite the icy hardening [*Erstarrung*] of the surroundings, these are natural pictures, warm with life."[60]

3.19. Perfected Snowflakes. William Scoresby, *An Account of the Arctic Regions with a History and Description of the Northern Whale-Fishery* (Edinburgh: Archibald Constable, 1820), classification, pp. 427–28; "mutilated," p. 431; "perfect," p. 432; "First Cause," pp. 426–27, figure in vol. 2, pl. 10. Like Nettis, Scoresby saw "mutilated and irregular specimens," but unlike Nettis, he reckoned that the greatest number were "perfect geometrical figures." Scoresby figured "the particular and endless modifications of similar classes of crystals, can only be referred to the will and pleasure of the Great First Cause, whose works, even the most minute and evanescent, and in regions the most remote from human observation, are altogether admirable." If God backed symmetry, then symmetrical snowflakes ought stand in the majority.

Fig. 3.20. Idealizing Microphotography. W.A. Bentley and W.J. Humphreys, *Snow Crystals* (New York: Dover, 1962), p. 60 (reproduced by permission of Dover Publications). The farmer-photographer Wilson Bentley spent much of his life capturing (and clipping) "perfect" snowflakes, each of which he thought was unique. Although his work was photographic, his interventions to alter the background and trim the image of the flake offended Richard Neuhauss's undying commitment to restraint in the name of mechanical objectivity (see figs. 1.2 and 3.21).

Suitably deployed, and created with iron-willed self-restraint, photographs promised objectivity. After spending years perfecting a marvelous, Rube Goldberg-style device that could produce a flash ("instantaneous") image of a falling droplet on his retina, the British physicist Arthur Worthington (the splash-man we encountered in the Prologue) could see more of this phenomenon than anyone in the world. As if frozen in time by his millisecond flash, the latent image of a drop of milk could be seen hitting water — and then Worthington could sketch the scene to abstract the ideal, underlying phenomenon from the vagaries of accident (see figure P.1). In one flash, Worthington might examine a milk drop barely touching the liquid surface. In the next burst of light, he could study a drop falling from the same height as the first but probe the impact a few thousandths of a second later in the process. By adjusting the flash to fire later and later with each subsequent drop, Worthington could track the otherwise invisible course of the splash throughout its "history."

For many years, Worthington had no particular interest in objectivity one way or the other — he was after the essence of a class of phenomena that was terrifically hard to perceive. Then, around 1894, no doubt pushed by efforts he and others saw as parallel, Worthington launched a new and intense campaign to capture the splash *objectively*. Knowing the shock of the objective, it is worth tracking Worthington's switch from retina to photographic plate with two questions in view: What were his models for this quest and its associated techniques? And how did he view the older sketched images once he had his sequenced photographs in hand?

Worthington's photography drew on shared techniques that came from near and far. By the early 1890s, all around him, Worthington could see flash photography successfully deployed to capture the physics of the very fast. In 1887, Ernst Mach, collaborating with the Austrian military photographer and physicist Peter Salcher, had captured the shadow of a supersonic bullet, using the bullet itself to trigger a bright spark. That spark cast the bullet's shadow — and even the diffraction shadow of the compressed air around it — onto a photographic plate. Mach's concerns had absolutely nothing to do with splashes but instead centered on a dispute he aimed to resolve about the damage caused (or not caused) by air compressed around the bullet's leading edge. The British, too, wanted their shadow images of

Fig 3.21. Asymmetrical Objectivity. Gustav Hellmann, with microphotographs by Richard Neuhauss, *Schneekrystalle: Beobachtungen und Studien* (Berlin: Mückenberger, 1893). For James Nettis or William Scoresby — or the author of just about any of the other compendiums of snowflake images — part of the beauty and appeal of snowflakes was that they exhibit extraordinary symmetry. It was therefore a surprise, both disturbing and bracing, that Hellmann and his microscopist-doctor collaborator, Neuhauss, found that under the cold photographic eye of the lens, a large fraction of the tiny crystals were all too asymmetrical.

bullets recorded — a problem addressed by Sir Charles Vernon Boys, who innovated by using much more sensitive photographic plates. Boys was above all a consummate instrument maker, a craftsman of such skill that he painstakingly found a value for the gravitational constant — a discovery that stood as a monument to care and precision — along with an astonishingly sensitive radiomicrometer, a much-reprinted book on soap bubbles, and, building on Mach's work, shadow photographs depicting the flight of bullets, in 1893.[61]

Meanwhile, John William Strutt, the third baron of Rayleigh, took up the spark method, making use, in 1891, of a Leyden jar to produce a faster, brighter spark — the crucial last step in the technical infrastructure that Worthington needed. (See figures 3.22 and 3.23.) It was against this background that Worthington — or, more specifically, his technically adept collaborator R.S. Cole — assembled an apparatus for photographing splash shadows. (See figures 3.24 and 3.25.) "Objective" photography (as Worthington and his colleagues understood it) moved across objects — bullets and bubbles, water spouts and droplets. The techniques and even the terminology of the objective circulated across national and disciplinary boundaries. Finally, Cole reported in 1894, it had been possible to nab "objective 'views' as opposed to shadows ... with such a very short illumination."[62] As we saw in the Prologue, in spring 1894, Worthington had succeeded in actually photographing the events he had spent so many years sketching by hand from the latent image left from the burst of light. Only with those photographs in hand did he come to see that asymmetries and faults were not merely deviations from some clear and perfect central image — that it was irregularity all the way down. No longer did it make any sense to him to continue to produce the "Auto-Splashes," those idealizations that lay behind, not in, *particular* splashes. He had passed from truth-to-nature to objectivity.

Stunned by what he retrospectively judged as a failure, despite all his previous caution, to depict nature rightly, Worthington began to introspect. How could he and others have seen for so long a perfection that had never been present? "It is very difficult to detect irregularity," he told his audience in 1895, and he went on to do a kind of *post hoc* psychological inquest into how he had gone astray. By flash projecting one of his photographs onto a screen, Worthington could test himself and others: "My experience is that most persons pro-

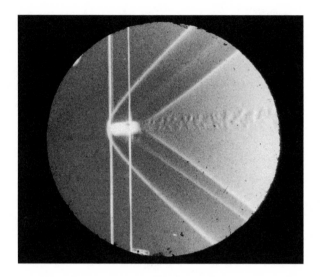

Figs. 3.22, 3.23. Instantaneous Photographs. Ernst Mach and Peter Salcher, "Brass Projectile with Hemispherical Ends," (1888), glass plate negative, Mach Nachlass, Deutsches Museum, Munich, CD52415, (*left*); Lord Rayleigh, "Some Applications of Photography," *Nature* (1891), p. 251, fig. 3 (*right*) (courtesy of Deutsches Museum, Munich). Arthur Worthington drew his technique from a wide range of contemporary attempts to photograph the evanescent. In 1887, Ernst Mach captured the flight of a bullet — and even the air disturbances around it — with a shadow photograph; later, Lord Rayleigh perfected an even faster sparking mechanism to record in a photograph the spray of a water stream (right) and the bursting of a soap bubble. Having struggled to get reflecting, not just shadow images, Worthington followed others in calling his photographic image an "objective view."

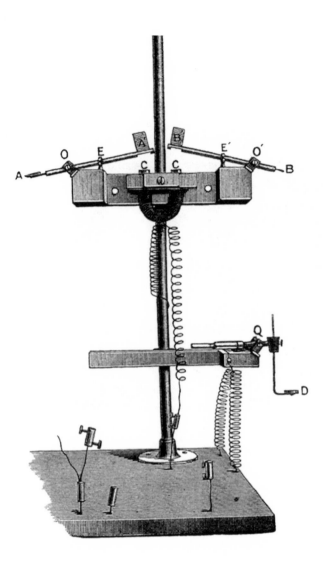

Fig. 3.24. Splash Machine. Arthur Worthington, *The Splash of a Drop* (London: Society for Promoting Christian Knowledge, 1895), p. 13. In an effort to mechanize the process of drop-impact photography (inspired by self-registering photographs of flying bullets), Worthington built this device. Pivot arms AA' and BB' are ready to tilt, but both are held in place by a strong electromagnet, C. When Worthington cut power to the electromagnet, both arms suddenly rotated, releasing a droplet of milk or mercury from watch-glass A and a sphere of ivory the size of a marble from B. Before the droplet hits the surface, the ivory sphere strikes plate D, precipitating (by means of an induction coil) a bright, very short spark sufficient to take the photograph. By varying the height of plate D, Worthington could photograph a drop any time after release — the higher the plate, the sooner the picture was taken.

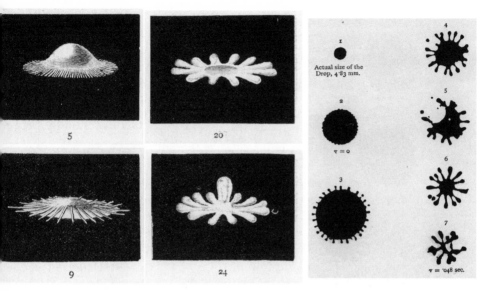

Fig. 3.25. Splash Shadows. Arthur M. Worthington, "Splash of a Drop," *A Study of Splashes* (London: Longmans, Green, 1908). Left: Worthington's drawings are from his Series I, sketched before he could make photographs. On the right are his first photographs — taken as flashes of the splash shadows. To Worthington, the identification of his older, symmetrical drawings with the new shadow photographs was clear:

drawing 5 .... shadow photograph 2
drawing 9 .... shadow photograph 3
drawing 20.......... shadow photograph 6
drawing 24.......... shadow photograph 7

The match was fine, if imperfect — until shadow photograph 7, in which the "irregularity of the last photograph almost masks the resemblance." At this point, when the phenomenon differed so dramatically from its idealization, Worthington seems to have abandoned his long-pursued hunt for the Platonic "Auto-Splash." Enter the "objective view." (Quotation from p. 152.)

nounce what they have seen to be a regular and symmetrical star-shaped figure, and they are surprised when they come to examine it by detail in continuous light to find how far this is from the truth." This was especially so, Worthington added, when "no irregularity is suspected beforehand." (His long-sought "Auto-Splash" had been perfectly symmetrical.) The psychological depiction continued: Viewers attend to a part of the image, with a preference for a part that is regular, and then tend to "fill up the rest in imagination." It was even the case, as we saw back in the opening pages of this book, that Worthington noted the discrepancy between his eyewitness perception of a splash as "quite regular" and his realization on seeing the photograph of that same event that it was far from symmetrical.[63]

In rejecting the perfected image, Worthington was not alone. Over the course of the nineteenth century other scientists — from botanists to zoocrystallographers, from astronomers probing the large to physicists poring over the small — began questioning their own disciplinary traditions of idealizing representation in preparing durable compendiums of images. Worthington's new alignment with the imperfect individual droplet was of a piece with Hellmann and Neuhauss's celebration of the individual, asymmetrical snowflake, or, for that matter, with Otto Funke's pride in depicting not-quite-rhomboid, optically distorted crystals. Here, the objectivists thought, were working objects you could count on in the long run. They cast aside the perfect, crystalline symmetry of an earlier time. Emphasizing a proud epistemic, even metaphysical idea, this widening circle of scientists relegated perfection to a chapter in the history of subjective error. Where the eye of the mind had dominated with its reasoned sight, blind sight now contested the rule.

In the rearview mirror, Worthington saw objectivity pitted against the psychological tendency to improve. Objectivity enforced the irregularity of the world on minds set to believe in the ideal regularity of nature. (See figures 3.26, 3.27, and 3.28.) Anatomists such as Jena's Karl von Bardeleben and Ernst Haeckel likewise intended their topographic anatomy atlas to be true to an *un*improved, *un*idealized nature. These makers of atlases for physiological chemistry would not, any more than those who made atlases for snowflakes, abide schematic illustrations standing in for a class or type: "The illustrations frequently have an individual character and often do not correspond to the types [*Typen*] that exist mostly in fantasy."[64] That said, the authors did not believe that photography offered the only defense against figments of imagination. Even when the Jena anatomists expanded their work ten years later, in 1904, they fiercely defended their woodcuts, adding a polemic *against* photography. More precisely, they (grudgingly) allowed that film might do for the study of exterior forms, where the goal was to capture beauty: living people, statues, bones. But when layers, complicated entities, or preparations with details were present, the woodcut, suitably colored, could not be beaten. Bardeleben and Haeckel contended that black-and-white photographs, with their limited depth of field, were simply incompetent to pick out such elements.[65] Objectivity did not imply photography; photography did not imply objectivity.

Learning to see was never, is never, will never prove effortless. For these nineteenth-century image classifiers, the shift from object-as-type to object-as-particular was long and hard, the sacrifices painful. Mathematical models, symmetry, and perfection had to be left behind; so had the hard-won knowledge of fellow scientists. The objective observer would have to renounce interpretation in the drawing. It became routine to "police" — and be seen to be policing — illustrators, lithographers, and photographers, urging them to be mindful of precise reproduction at every stage. Even instrument-produced artifacts had to be *observed* in the image. Retaining such stray effects in the pages of an atlas became a mark of authenticity, proof positive that the observer had included all that was truly at hand. The observer had to hold back, rather than yield to the temptation to excise defects, shadows, or distortion — even when the scientist or artist *knew* these intrusions to be artifacts. Mechanical objectivity aimed for this purity of observation, this new way of looking at an individual plant or particular bacterium as if liberated from the second sight of prior knowledge, desire, or aesthetics. In this blind sight lay an epochal novelty in right depiction.

Drawing Against Photography

Photography did not create this drive to mechanical objectivity; rather, photography joined this upheaval in the ethics and epistemology of the image. But once atlas makers were confronted with

$$\tau = \text{'}0021 \text{ sec.}$$

Fig. 3.26. Splash Drawing, Etched. Arthur Worthington, *The Splash of a Drop* (London: Society for Promoting Christian Knowledge, 1895), pp. 43 and 48; fig. on p. 44. Worthington's automated dropper let loose this milk drop from a height of 52 inches; the detail here, taken 0.0021 seconds after the first impact of the droplet, was one of a series of eleven images. From this height, the droplet causes the water it hits to form a hollow "shell or dome" that Worthington found "extremely beautiful." Soon (one or two hundredths of a second after this view), the return wave closes up around the original milk droplet. Sometimes the milk drop escapes, shooting upward and out; other times, the return wave encloses both the droplet and a bubble of air. "Such is the history of the building of the bubbles which big rain-drops leave on the smooth water of a lake, or pond, or puddle." Worthington quoted Robert Louis Stevenson's "Inland Voyage," in which the canoeing author sees raindrops launching water into "an infinity of little crystal fountains."

Time after contact = ˙0391 sec.

12

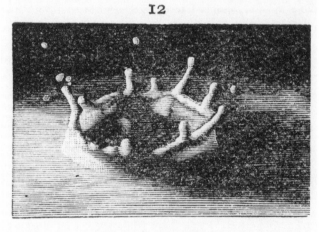

$\tau = $ ˙0391 sec.

Figs. 3.27, 3.28. Splash Photograph and Its Engraving. Photograph from Worthington, "On the Splash of a Drop and Allied Phenomena," *Proceedings of the Royal Institution* 14 (1894), opposite p. 289 (*top*); engraving from *ibid.*, image 12 of ser. 14 (*bottom*). When Worthington finally perfected a photographic system, he first took "shadow" images, modeling the procedure on the high-speed shadow photographs of flying bullets that Ernst Mach and others had managed to take a few years earlier. Those pictures — and, much more dramatically, droplet photographs — left Worthington stunned to find that the perfect symmetry of his splash drawings had been a chimera. In the 1890s, he abandoned his earlier, idealizing sight, preferring to take imperfect instances one by one. Once-beautiful crowns and domes now entered bent and broken, varying dramatically from drop to drop. At the top is an actual, spark-illuminated photograph of a splash resulting from a 16-inch droplet fall; at the bottom is an engraving of that same image.

a choice between drawings (reproduced by lithography as well as engravings and woodcuts) and photographs, debates about their relative merits ensued. Scientific artist battled scientific photographer, and in their struggle concessions were demanded on both sides: pedagogical utility, truth-to-nature, beauty, and objectivity could not always all be had at once.

The Leipzig embryologist Wilhelm His laid out the choice between the drawing (able to capture the meaning and essence of a situation) and the photograph (which could serve as a form of "raw material"):

> Drawing and photograph are complementary, without replacing one another. The advantages and disadvantages of every drawing in relation to a photograph lie in the subjective elements that are at work in its making. In every sensible drawing the essential is consciously separated from the inessential and the connection of the depicted forms is shown in the correct light, according to the view of the draftsman. The drawing is thus more or less an interpretation of the object, involving mental work for the draftsman and embodying this for the spectator, whereas the photograph reproduces the object with all its particularities, including those that are accidental, in a certain sense as raw material, but which guarantees absolute fidelity.[66]

The bacteriologist Robert Koch, whose work was key in establishing the broadly accepted criteria for naming a bacillus as the origin of a disease, held that the photograph must eventually displace the inevitably subjective drawing. After making major contributions to the study of anthrax, Koch spent some four years working on the fixing, staining, and photographing of bacteria (see figure 3.29). By 1880, he had come to view photography as essential to an objective knowledge of the microorganism: "Photographic illustrations are of the greatest significance for research on microorganisms. If anywhere a purely objective viewpoint, free of every bias, is necessary, then it is in this field. But until now exactly the opposite has occurred, and there are nowhere more numerous subjectively colored views [*Anschauungen*] and therefore differences of opinion as in the study of pathogenic microorganisms."[67]

Yet Koch conceded that much was lost in the "purely objective"

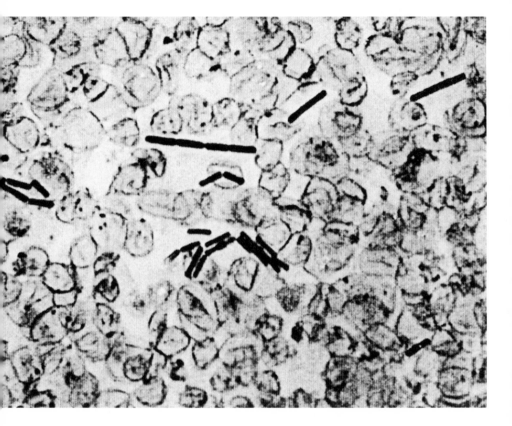

Fig. 3.29. Bacilli Photographed. Blood from two-day-old dissected corpse, magnified 700X, Robert Koch, "Untersuchungen über Bacterien VI: Verfahren zur Untersuchung, zum Conserviren und Photographiren der Bacterien," *Beiträge zur Biologie der Pflanzen* 2 (1877), pp. 399–433, table 16, no. 6. Koch used this photogram to refute Karl Wilhelm von Nägeli's "schematic drawing" of bacteria, which showed them as shorter and more "tufted" than Koch believed them to be. Against those who claimed that the appearance of bacteria could be manipulated "at will" by photography, Koch retorted that such views merely revealed complete ignorance of microphotography.

photomicrographs: the red and blue aniline dyes used to prepare samples for drawing were more pleasing to the eye than the brown ones that worked best for photography; the photograph captured even the shadow of the prepared sample and was limited to a single viewing plane; drawings of microscopic objects were always more beautiful. But all of these disadvantages paled beside the advantages of photographs, according to Koch. The photograph could discipline the microscopist "to give repeatedly an accounting of the correctness of his observation," whereas "the drawing is involuntarily already prepared in line with the subjective view of the author."[68]

Not all agreed that drawing necessarily had to be subjective. The Jena physicians' defensive apology for their woodcuts against photography signaled their own sense of being under siege. Indeed, another antiwoodcut assault came from Johannes Sobotta, a turn-of-the-century German anatomist. Sobotta's atlas of the human body remains a standard reference work in later editions. Sobotta made the importance of mechanical reproduction crystal clear when he advertised the use of photography in the preparation of his 1909 anatomical atlas — even though his own images were, in fact, drawings reproduced as multicolor lithographs. "No woodcuts have been employed, since the failure of the latter method to produce illustrations true to life has been distinctly shown by several of the newer anatomical atlases. It leaves entirely too much to the discretion of the wood-engraver, whereas the photomechanical method of reproduction depends entirely upon the impression made upon the photographic plate by the original drawing." As a further control on the discretionary power of the illustrator, Sobotta had a photograph of the designated body section taken and enlarged to the size of the intended drawing.[69] Sobotta's competitors would draw, then hand the drawing to a wood engraver. By contrast, Sobotta proposed a doubly "automated" procedure that left discretion "only" at the first stage (drawing): there would follow an automatic lithographic transfer to stone, and then a check by precise comparison of the lithograph with an enlarged photograph.

In short, the drive to automaticity was felt on both sides. There were those like Sobotta who drew their original images — but then relied on the photomechanical lithograph for reproduction, and the photograph itself as a control. And there were those who began with a photograph, like Worthington, who feared his own tendency to idealize, but who then relied on an engraver for reproduction.

Sobotta followed the same method when he turned to histology and microscopic anatomy in his 1902 treatise on that subject. Readers might worry that the samples were not representative of living tissue — that they were distorted in some way by preservation or decay. Sobotta reassured them that the vast majority of the samples came from two hanged men, several others from two additional victims of the gallows, so the "material" was still "warm" (*noch lebenswarm*). Again Sobotta had photographs made to be used as the starting point for drawings. Here, however, he noted that precision (*Genauigkeit*) should not be pushed too far — for then every disturbing accidental feature of the preparation would enter the representation. Instead, some figures were actually made on the basis of two or three different preparations. Somewhat defensively, perhaps anticipating criticism, Sobotta advised his readers that the combination was not made arbitrarily but with the careful repositioning of the camera to eliminate variation in perspective; the photographic enlargements were then cut and reassembled to reproduce a mosaic photograph against which the drawing would be judged. This, the author tells us, "would give the draftsman no possibility for subjective alterations."[70]

Sobotta's strategy thus crossed the categories of the characteristic, the *Typus*, and the ideal. By invoking specific photographs as controls on the mechanics of reproduction, he appears at first glance to follow the well-worn route to the characteristic — the individual depicted in striking detail and meant to stand in for the class. His protestations of automaticity and removal of "discretion" signal the increasing pressure of the objective. But by amalgamating fractional parts of different microscopic individuals to construct the basis from which drawings would be made, Sobotta left the domain of the purely characteristic. Is the final drawing made from the mosaic an ideal — the picture of a perfect sample one may hope one day to find? Is it a picture of an ideal that may well not exist but that represents a kind of limiting case? Or did Sobotta expect his routinized procedures to give rise to diagrams that would stand in for a *Typus*, lying altogether outside the collection of individuals past, present, and future, yet expressing an essential element of all of them? He pushed

167

such ontological questions aside; Sobotta devoted his attention instead to the procedure of controlled reproduction as a means of squelching the subjectivity of interpretation. In an earlier epoch, that of Goethe, Albinus, René-Just Haüy, and William Hunter, the atlas maker had borne an essential responsibility to resolve — one way or another — the problem of how single pictures could exemplify an entire class of natural phenomena. Sobotta's cobbled-together photographs form an apt metaphor for his uneasy authorial position, between the older desire to perfect and the newer admonition to stand aside — to keep hands off the machine-generated image.

By and large, this fear of interpretation fueled a flight from the composite image toward the individual. The very act of combining elements from different individuals appeared to many late nineteenth-century observers to leave far too much judgment to the artist. Some, however, held on to the composite — especially if it could be shown to have been assembled by means of a mechanical procedure rather than inspiration.

The British anthropometrist Sir Francis Galton shared none of Sobotta's ambivalence about amalgamation. Galton, in collaboration with sociologist Herbert Spencer, enthusiastically embraced the possibility of simultaneously eliminating judgment and capturing, in one visage, the vivid image of a group. Indeed, Galton was persuaded that all attempts to exploit physiognomy to grasp underlying group proclivities were doomed to failure if they did not use a mechanized abstracting procedure. His remedy was disarmingly simple. Each member of the group to be synthesized had his or her picture drawn on transparent paper. Exposing a photographic plate to each of these images would result in a composite image. Such a process would free the synthesis from the vagaries of individual distortion; even the exposure time of each individual could be adjusted on scientific grounds, such as the degree of relatedness, in the case of family averages. "A composite portrait," wrote Galton, "represents the picture that would rise before the mind's eye of a man who had the gift of pictorial imagination in an exalted degree. But the imaginative power even of the highest artists is far from precise, and is so apt to be biased by special cases that may have struck their fancies, that no two artists agree in any of their typical forms. The merit of the photographic composite is its mechanical precision, being subject to no errors beyond those incidental to all photographic productions."[71] What had once been a *scientific* virtue — the ability to synthesize a composite from many individuals was, for Galton, now relegated, pejoratively, to the "artistic." In the place of "pictorial imagination in an exalted degree" Galton installed a procedure with "mechanical precision."

Galton's procedure was to divide the necessary exposure for a plate by the number of faces to be included. So if the plate needed an eighty-second exposure and there were eight murderers to be synthesized, then each portrait would be photographed for ten seconds. This protocol enabled the analyst to provide a generalized picture, one that "contains a resemblance to all [its constituents] but is not more like to one of them than to another." (See figure 3.30.) Not one feature in the image is identical to any single individual, yet, Galton insisted, the composite resembles them all, one by one. He noted that the same method could be extended by weighting degrees of relatedness within a family — putting, for example, longer exposures on those most closely tied genetically to a particular person.[72]

Galton's method is a perfect instance of an image-making routine poised between our two ordinarily disjunct modes of observation: on the one side, it aimed for an ideal type that lay "behind" any single individual. On the other side, Galton's face-machine proceeded toward that ideal not with what he and others had come to see as subjective idealization (stemming from "biases," "fancies," and "judgment") but with the quasi-automated procedures of mechanical objectivity. Intriguingly, as we will see in Chapter Six, Ludwig Wittgenstein used Galton's composite as he formulated his doctrine of family resemblance.

Galton's was a scheme that would go further than merely constraining the artist's depiction of an individual; the device would remove the process of abstraction from the artist's pen. No longer would pattern recognition be left to the artists. Murderers or violent robbers could, for example, be brought into focus so that the archetypical killer could appear before our eyes (see figure 3.31). The problem of judgment, for someone like Galton, arose with the artists, and the solution lay in automated amalgamation. Here the novel mechanical aspect — the aspect that eliminated interpretation — was not in the production of the individual likeness (as in individual portraiture) or in the method of reproduction (as in lithography).

169

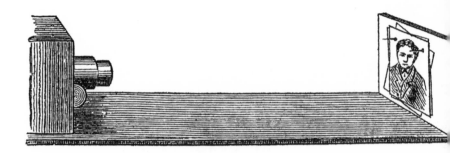

Figs. 3.30, 3.31. Galton's Physiognomic Synthesizer. Francis Galton, "Composite Portraits," *Nature* 18 (1878), p. 97 and 98. Galton had been investigating maps and meteorological charts to extract, by optical superposition, combined data. In the course of this work, he decided the same technique (fig. 3.30) could "elicit the principal criminal types" (such as murderers and violent robbers). For each photographic shot, the camera was moved so that the eyes of each particular malefactor would be aligned. If a normal exposure was eighty seconds, then, for a group of eight images, each would receive a ten-second exposure. Galton asserted that the "merit of the photographic composite is its mechanical precision." He conceded that the full composite effect (fig. 3.31) was diminished by the inevitable intervention of the woodcut engraver.

Instead, Galton had mechanized (or aimed to mechanize) the abstractive process by which one passed from individual to group. Revealingly, Galton found that his image truly was "a very exact average of its components," but that once the wood engraver (who was needed to prepare the image for publication) entered, his "judgment" altered the image. Suddenly "his rendering of the composite has made it exactly like one of its components, which it must be borne in mind he had never seen." Galton likened this seizing of the one from the many to an artist whose portrait of a child reveals the deceased father and obscures the mother (though the artist might never have met the father, and the mother's relatives might see the resemblance to her with great clarity). "This is to me," Galton concluded, "a most striking proof that the composite is a true combination." The desire, realized insofar as possible, to shift as much interpretation as possible from the artistic-interpretive to the routine-mechanical is central to objective depiction as a regulative ideal.[73]

In the late 1920s, polemics in favor of objectivity and against individual judgment were still in full bloom. The Berlin physician Erwin Christeller used his *Atlas der Histotopographie gesunder und erkrankter Organe* (*Atlas of the Histotopography of Healthy and Diseased Organs*, 1927) to caution the scientist against producing his own drawings — tempting as that might be.[74] Instead, he counseled handing the task to technicians who could produce pictures without passing through the stage of using a model; the procedure could be made "fully mechanical and as far as possible, forcibly guided by this direct reproduction procedure of the art department." Such enforced self-restraint from intervention blocked the scientist's own systematic beliefs or commitments from distorting the passage from eye to hand. This desire to extricate everyone, even himself, from the exertion of judgment extended to Christeller's advice that his fellow anatomists turn over their manuscripts to the publisher with their original anatomical preparations so the latter can be reproduced "purely mechanically" (*rein mechanisch*).[75] But pure mechanism could not proceed without a ferocious defense: Christeller insisted that the scientist's control was necessary to block others' inclinations or ignorance from interfering with the production of images: "I do not want to neglect to mention that through the whole conduct of the printing process, I maintained continuous control of the photographers and color engravers, even giving them detailed instructions and putting at their disposal my own instruments."[76] (See figures 3.32 and 3.33.)

Once so policed, and presumably only then, could the photographic process be elevated to a special epistemic status, a category of its own. In Christeller's words: "It is obvious that drawings and schemata have, in many cases, many virtues over those of photograms. But as means of proof and objective documentation for findings [*Beweismittel und objektive Belege für Befunde*] photographs are far superior."[77] This photographic superiority was inextricably attached to the removal of individual judgment. With respect to color, for example, Christeller thought that no method was perfect. Drawings carried with them an inalienable subjectivity. By contrast, photograms, made by the direct positioning of the sample on photographically sensitized paper, were tarnished only by the crudeness imposed by the limited palette of the color raster. Given the choice, the author clearly favored the crude but mechanical photographic process. Accuracy was to be sacrificed on the altar of objectivity.

So riveted was Christeller by the ideology of mechanization that he determined — as Funke had done before — to leave imperfections in his photographs as a mark of objectivity:

> With the exception of the elimination of any foreign bodies [such as] dust particles or crack lines, no corrections to the reproductions have been undertaken, so that the technically unavoidable errors are visible in some places. For example, there are small intrusions [*Überschlagstellen*] of the fibrous tissue fringes on the edge of the sections; [there is also an] absence of soft tissue components.... [I displayed these imperfections because] I believed it my obligation also, at the same time, to display with great objectivity the limits of the technique.[78]

For Christeller, the tattered tissue edge served the role of the deliberate and humbling fault in a Persian carpet. But while the carpet maker seeks to avoid the hubris of attempted perfection, Christeller's torn tissue samples, such as the one displayed in figures 3.22 and 3.23, were put forward as a testimony to objectivity: disciplined self-denial of the temptation to perfect. Their presence in the atlas was a standing renunciation of aestheticized improvement toward the ideal.

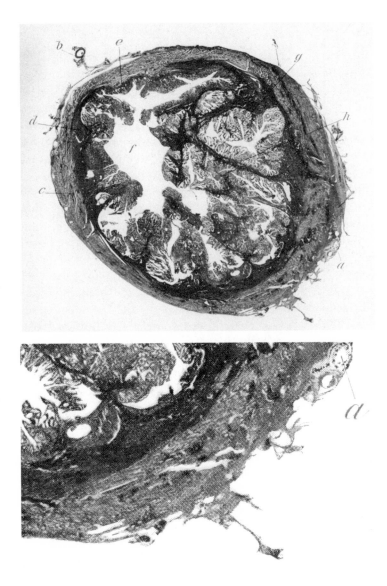

Figs. 3.32, 3.33. Tattered Objectivity, Detail. Erwin Christeller, *Atlas der Histotopographie gesunder und erkrankter Organe* (*Atlas of the Histotopography of Healthy and Diseased Organs*) (Leipzig: Georg Thieme, 1927), table 39, fig. 79. Christeller wore the imperfections of his photographic tissue sections as a badge of honor: they showed his ability to restrain from idealization. Christeller took the depicted faults — such as a misshapen snowflake, an asymmetrical milk-drop splash, and a fractured zoo-crystal — to be a central feature of the self-restrained, "purely mechanical" — and objective — image. Even the limited color palette shown here was a necessary sacrifice — hand-coloring was too subjective. This section, its edges torn in preparation, is of a polypous adenoma (benign, polyp-like tumor) of the pylorus (the passage at the lower end of the stomach) taken from a forty-seven-year-old office worker. (Please see color insert)

Self-Surveillance
Policing the artists — containing their predilection for "subjective alterations," "Zolaesque ... superfluous realism," artistic "discretion," "judgment," or "bias" by "fancy" — was only the first moment in the construction of far more encompassing set of restraints. Indeed, what characterized the creation of late nineteenth-century pictorial objectivism was self-surveillance, a form of *self*-control at once ethical and scientific. In this period, scientists came to see mechanical registration as a means of reining in their *own* temptation to impose systems, aesthetic norms, hypotheses, language, even anthropomorphic elements on pictorial representation. What began as a policing of others (artists, printers, engravers, woodcutters) now broadened into a moral injunction for the investigators, directed reflexively at themselves. Sometimes control of individual deviation could be accomplished routinely by invoking the "personal equation," a systematic error-correction term used to adjust each observer's results. In astronomy, for instance, transit observations (for example, tracking Venus across the face of the sun) required the observer to record the precise time at which a star or planet crossed a wire in a viewing device. This was accomplished by pressing a button. But the procedure was more complicated than it looked, for "a very slight knowledge of character will show that this will require different periods of time for different people. It will be but a fraction of a second in any case, but there will be a distinct difference, a constant difference, between the eager, quick, impulsive man who habitually anticipates, as it were, the instant when he sees star and wire together, and the phlegmatic, slow-and-sure man who carefully waits till he is quite sure that the contact has taken place and then deliberately and firmly records it. These differences are so truly personal to the observer that it is quite possible to correct for them, and after a given observer's habit has become known, to reduce his transit times to those of some standard observer."[79]

Adjusting for more subtle interference by the scientist's individual proclivity to impose interpretation, aesthetics, or theories was a more complex affair. But examples of the attempt abound, both in machine-dominated representational schemes that used some type of photography in one fashion or another, and in those that did not.

The ophthalmoscope, for example, provided the basis for a whole genre of atlases of the eye. One rather typical one, published by Hermann Pagenstecher and Carl Genth in 1875, clearly articulated the necessity and extraordinary difficulty of self-surveillance: "The authors have endeavoured, in these [pictures], to represent the object as naturally as possible. It cannot be hoped that they have always succeeded in this attempt: they are but too conscious, how often in its delineation the subjective view [*subjective Anschauung*] of the investigator has escaped his hand."[80] It was this betrayed hand, this escaped desire that had to be hemmed in by all means possible: "They [the authors] have kept it purely objective, describing only the conditions before them, and endeavoring to exclude from it both their own views and the influence of prevailing theories. It would have been easy to extend it considerably, and to add theoretical and practical conclusions; but the authors considered this a thing to be carefully avoided, if their work was to possess more than a passing value and to preserve to the reader the advantages of unprejudiced view and unbiased judgment."[81]

No "theoretical conclusions," no "practical conclusions" — these, the authors contended, were the necessary excisions objectivity demanded if their atlas was to become a compendium of images of record, good for the long term.

In 1890, Eduard Jaeger followed Pagenstecher and Genth, with even more urgent attention to detail. "In all these figures, there is not a single line that is arbitrarily or only approximately directed by the original." Every retinal vessel, every choroid vessel — even the smallest detail; every pathological liquid, every pigment accumulation was to have its size, form, color, and position executed under the most exact representation that "my eye can seize and my hand reproduce." For Jaeger, errors of omission were far preferable to errors of commission. That which his eye could not grasp with certainty — anything that remained unclear or poorly defined — he would rather leave out than reproduce in erroneous form. Self-restraint not only dictated the order of epistemic virtues but also governed the hierarchy of epistemic vices. Active, interventionist, speculative insertions were the worst.

As for his predecessors, Jaeger allowed that he would have to set aside modesty: his predecessors had not produced anything so faithful to nature as his plates, and it would be a good long time before anyone could deliver a similar or greater number of exact figures. In an ethical-epistemic *défi*, he demanded: Who else would sacrifice the time and effort that he had? Some figures had taken twenty to thirty, even forty to fifty sessions of two to three hours each. No, his past and future competitors would find it hard to measure up.[82] Some might claim that Jaeger's meticulous exactness was superfluous — that a less fanatical degree of resemblance would be of equal value. Or perhaps that a "genial interpretation and representation" (*geniale Auffassung und Darstellung*) of a single case or series of cases would carry an even higher value. Jaeger strenuously differed:

> As interesting and brilliant as such a [genial] representation might be, still such figures have, in relation to science, only a relative, a transitory value. Only a bit of them will endure and in later times still be valued, that which, with or without the knowledge of the depicter [*Darsteller*], is an illustration of the original [that is] faithful to nature [*naturgetreu*]. By contrast, all that which is arbitrary, that which is the expression of individual intuition in the figures, be it ever so ingenious, ever so genial, will vanish sooner or later, according to changes in opinions or the personality of the depicter, and above all in relation to progress in correct knowledge and faithful renderings of nature.[83]

Personalities change, genius or brilliance may beckon, but in the end what counts is heroic self-mastery, a surveillance of the willful self that counters genial flights of fancy with a combination of assiduousness and precision. When Jaeger's former collaborator and successor, Maximilian Salzmann, came to revise the atlas, he confessed that even he could not say he had devoted the same extraordinary effort in his figures as had his master. Morality governed his drawing table all the same. Salzmann insisted that he was proceeding with a clear conscience [*mit gutem Gewissen*], having prepared illustrations that were faithful to nature, free of schematizing or aestheticizing of even the smallest element.[84]

Pagenstecher, Genth, Jaeger, Salzmann — all were after a demanding, self-surveilling objectivity, always on the *qui vive* for traitorous interpretation. But for some scientists no drawing could ever successfully extirpate interpretation, even if it were executed with a maximum of instrumental assistance. In his microscopic studies of nerve cells of 1896, the American neurologist M. Allen Starr came down squarely on the side of Cajal — Starr bolstered the neuron doctrine, blasted the inadequacy of artistic portrayal, and supported photography: "In the most recent text-books of neurology and in the atlas of Golgi these facts have been shown by drawings and diagrams. But all such drawings are necessarily imperfect and involve a personal element of interpretation. It has seemed to me, therefore, that a series of photographs presenting the actual appearance of neurons under the microscope would be not only of interest but also of service to students."[85] By striving to eliminate "personal interpretation," "diagrams," and "drawings" altogether, Starr had to confront the difficulties associated with photographing with limited depth of field. And in abandoning the camera lucida for the photograph, Starr departed from the method of choice followed by *both* the battling future Nobelists, Golgi and Cajal.

Starr's fear of "personal interpretation" was shared by the Berlin bacteriologist Carl Fraenkel and the staff doctor Richard Pfeiffer — both at the Hygienics Institute. Intriguingly, however, the two doctors used their 1887 bacteriological atlas to present what may be the most subtle and conflicted account of them all in the great debate between drawing and photography in science. They began much the way Starr would, extolling the charms of the photographic plate and dismissing the dangers of the handmade: "A drawing can only be the expression of a subjective perception and therefore must, from the beginning, renounce the possibility of an objection-free reliability." They contended that we see not only with the eye but also with the understanding; as the difficulties mount, "simple visual perception" [*einfache Anschauung*] recedes and we come more and more to see what we believe to be the case. Inevitably, drawing reflects the understanding. "The photographic plate, by contrast, reflects things with an inflexible objectivity as they really are, and what appears on the plate can be looked upon as the surest document of the actual conditions."[86]

For Fraenkel and Pfeiffer, a "photographic eye" was not only "honest" and "unbiased" but also sharper, more precise. Photographs could capture conditions of extremely strong lighting that revealed details where the human eye would be blinded. And only the photograph allows us the possibility of showing others what we have seen without endlessly hauling out a microscope. But there was a still greater advantage to the impersonal routine of the photomicrograph. In ordinary observation (said Fraenkel and Pfeiffer), all too often the observer simply gets a general impression of the forms of bacterial colonies growing on the gelatin plate — and then, on the basis of this cursory look, declares that he is done with his investigation. In a photograph, this frequently unjustified winnowing of the "important" from the "unimportant" will not stand. Reexamining the photograph can lead the scientist to reevaluate what is actually in the image. The photomicrograph acts pedagogically by extending — in fact revising — the process of observation. In short, the photographic trace becomes an archive as a drawing could not; the photograph is a resource for further inquiry.[87]

The Hygienics Institute micrographers readily conceded, however, some serious disadvantages. First, the photographic plate could capture only a narrowly bounded fraction of the preparation. Worse, because of its limited depth of field, the photograph could show essentially a single focal plane — and at the edges of the sample, the image blurs. Old-fashioned, direct observation could see deeper into the sample; it allowed movement of the sample from side to side; it could integrate the basic facts and details; and it could make quick comparisons by moving back and forth between neighboring sites. By looking long, hard, and intelligently, the observer can sort out the structural relations and the mechanical construction of the object. The detailed accumulation of bacteria in a large-scale colony is beyond — at least beyond any *easy* — representation with photomicrography. To look at a failed plate with its blurring, its indistinct contours, its interference fringes is to see just how mangled and unrecognizable an image can become. Photography had real limits in the domain of the very small.

To counter these dangers (according to Fraenkel and Pfeiffer), one must erect the barrier of training in the use of the microscope, a study that ought to begin with the imaging of objects that have already been photographed. Where and how? In an atlas — theirs.[88] It was not that Fraenkel and Pfeiffer had no competition. In 1896, the most prolific atlas publisher of them all — Felix Lehmann, of Lehmann Verlag — persuaded his bacteriologist brother, Karl Bernhard

Lehmann, to go to press with his *Atlas und Grundriss der Bakteriologie* (*Atlas and Foundation of Bacteriology*). Like his predecessors, Karl recognized all too clearly the deep competition between the photograph and the drawing. True enough, he allowed, the photograph "is to be held in high regard for the purpose of objectively representing scientific objects, especially bacteriological objects." But that was not enough, or not always enough. First, for special kinds of biological cultures (some of which were precisely those used in diagnosing disease), drawings did better; the photograph might win in the depiction of individual entities, but for whole cultures, drawing took the honors. Secondly, drawings were superior to film images in depicting spatial depth. Here, then, is a case where the photograph was hailed as the more objective technique but nonetheless failed when stacked up against drawing as a means to prepare for the diagnosis of disease.[89]

As these image battles make clear, mechanical objectivity — self-denial coupled with the drive toward disciplined automaticity — was not for everyone, everywhere. Objectivity was costly — in different contexts, it demanded sacrifices in pedagogical efficacity, color, depth of field, and even diagnostic utility. That so many practitioners were more than willing to pay the price indicates the powerful appeal of this particular epistemic virtue. At least in their professional world, scientists at the time were quite clear about this — they had no illusion that they lived in a Panglossian world in which all the virtues pulled in the same direction. In a sense, this awareness of trade-offs in the complexity of the sciences should not surprise us. After all, in the political realm, it is no novelty that there are times and places where certain virtues dominate others — societies where the perceived virtue of egalitarianism trumps that of just reward. Or vice versa.

Objectivity figured large for the American astronomer Percival Lowell as he struggled during the first years of the twentieth century to establish the reality of the "canals" of Mars — he was willing to give up a great deal for objectivity (and yet still never persuaded the majority of his colleagues). Of one atlas-like set of sketched (and published) observations, he wrote: "Each drawing was made as if I had never seen the planet before; only twice did I allow myself even to put in afterward the snow accidentally omitted at the time. About

179

fifteen minutes only was allowed in every instance, so that each drawing does not pretend to represent all that could be seen on that night at the telescope. They were meant to get as nearly as possible impersonal intercomparable representations, — scientific data, not artistic delineations."[90]

After the fact, Lowell could see a great deal that he had omitted (see figure 3.34). But he proudly reported how he (all but twice) had resisted the temptation to reinsert the missing matter and, by so suppressing his impulse to improve, guaranteed the objectivity of his representation. These were "scientific data, not artistic delineations." Whereas artistic synthesis had previously been the guarantor of truth, Lowell in essence argued that while artistic delineations might be more complete and even more accurate, succumbing to the siren call of art would doom the objectivity of the project.

On May 11, 1905, not long after he made his sketches, Lowell and a collaborator were able to capture on film the fine lines of the planetary surface. "Thus," Lowell proclaimed, "did the canals at least speak for their own reality themselves." Speak they might, but in whispers: only one-quarter of an inch in diameter, Lowell's photographs of Mars were so blurred, gray, and puny that, at the time, they could not even be reproduced.[91] Figure 3.35 shows the pictures as they appeared in his record book, in their original blurry but unretouched form. Although the British astronomer A.C.D. Crommelin declaimed that "these photographs did a great deal to strengthen my faith in the objective reality of the canals," others looked at the same pictures and were struck by their ambiguity. Desperate, Lowell almost succumbed to artistic temptation — he considered having a neutral party (his friend and fellow Boston scientist George R. Agassiz) "retouch" the pictures so the canals would be visible in mass reproduction. Lowell's editors protested: such alteration would be a "calamity ... as it would certainly spoil the autograph value of the photographs themselves. There would always be somebody to say that the results were from the brain of the retoucher."[92] This was the by-now-familiar charge against intervention. Lowell capitulated, and in the end accuracy, completeness, color, sharpness, and even reproducibility were sacrificed to mechanical objectivity. Know as scientists might that a particular line should be there, must be there, they felt compelled, above all else, to hold back their improving hands.

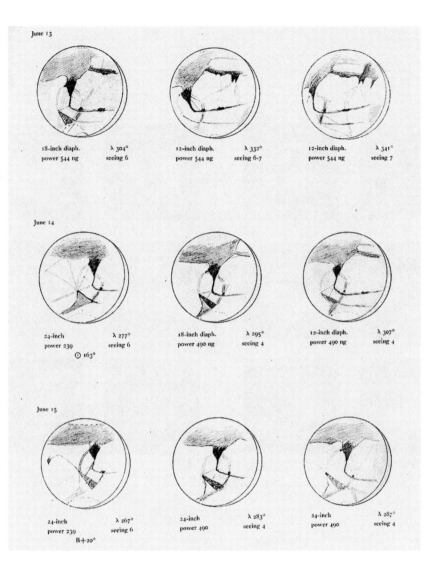

Fig. 3.34. Martian Sketches. Percival Lowell, *Drawings of Mars, 1905* (Lowell Observatory, 1906), pl. 34, June 13–15, 1905 (courtesy of Lowell Observatory Archives). The set of "impersonal intercomparable representations," of which this is one, covers about seven months. Mars itself (Lowell reported) varied in apparent size during this period, from 6.4 seconds of arc at the outset to 17.3 seconds and then back to 10.0 seconds at the end of the series. Lowell invited the reader to remove the notebook figures to the appropriate distance for these angular sizes to be replicated — he declared that the smaller apparent size drove the lack of detail in the early and late stages of variation. But of the reality of Martian canals he was sure: "Intrinsic change in many of the canals is so marked that it cannot be missed by one going through the pages." Lowell, foreword to *ibid.*, n.p.

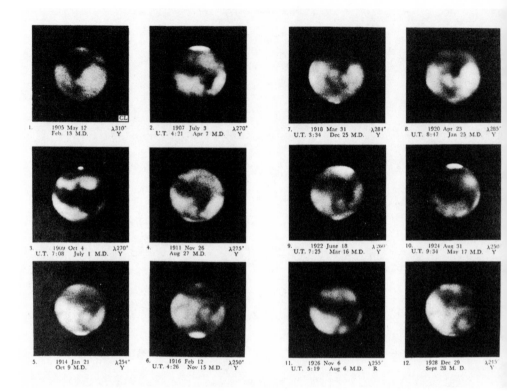

1. 1905 May 12   λ310°
   Feb. 13 M.D.   Y

2. 1907 July 3   λ270°
   U.T. 4:21 Apr 7 M.D. Y

7. 1918 Mar 31   λ284°
   U.T. 3:34 Dec 25 M.D. Y

8. 1920 Apr 23   λ285°
   U.T. 8:47 Jan 25 M.D. Y

3. 1909 Oct 4   λ270°
   U.T. 7:08 July 1 M.D. Y

4. 1911 Nov 26   λ275°
   Aug 27 M.D.   Y

9. 1922 June 18   λ260°
   U.T. 7:25 Mar 16 M.D. Y

10. 1924 Aug 31   λ250
    U.T. 9:34 May 17 M.D. Y

5. 1914 Jan 21   λ254°
   Oct 9 M.D.   Y

6. 1916 Feb 12   λ250°
   U.T. 4:26 Nov 15 M.D. Y

11. 1926 Nov 6   λ255
    U.T. 5:19 Aug 6 M.D. R

12. 1928 Dec 29   λ215
    Sept 28 M. D.   Y

Fig. 3.35. Martian Photographs. Photographs, Percival Lowell. Reproduced from William Graves Hoyt, *Lowell and Mars* (Tucson: University of Arizona Press, 1976), image pp. 180–81, text references pp. 175 and 179 (courtesy of Lowell Observatory Archives). Desperate to prove his claim that he had seen canals on Mars, Lowell pushed his junior colleague, Carl Otto Lampland, to adapt his photographic techniques to the painfully difficult task of imaging the red planet. They were explicitly seeking a "mechanism" that would allow shots to be taken through the 24-inch refractor — which had been improved through a custom-designed system of plates and filters. Of the first images, Lowell wrote: "The eagerness with which the first plate was scanned as it emerged from the last bath may be imagined, and the joy when on it some of the canals could certainly be seen." It was an uncertain certainty: astronomers and journalists pounced.

Ethics of Objectivity

Among the many late nineteenth-century scientists concerned with
the microscopic structure of the brain, Cajal (Golgi's archrival) came
to be known both for his extraordinary depictions of cell structure
and for his doctrine of the neuron's autonomy that those images supported. As a young man, Cajal had been riveted by drawing; his
father had pressed him to follow his footsteps and become a surgeon.
Together they snatched bodies from the local cemetery, and young
Cajal drew the stolen corpses with exquisite care, providing illustrations for his father's anatomical atlas. Years later, he drew his own
images on lithographic stones — and he maintained a lifelong fascination with the details of photography. Drawing in all its many forms
remained a thread for him, the outward proof of clear sight.

For Cajal, as for so many of our late nineteenth-century figures,
seeing clearly was the goal of both science and character. Clear-sightedness, both literal and figurative, lay at the heart not only of his ethical concerns, but also of his lasting contribution to neuroanatomy,
which began in the early 1890s. As we have seen, back in 1873, Golgi
had developed a staining method (using silver chromate) that made
visible the shape of individual nerve cells, and beginning in 1887,
Cajal had taken full advantage of it.[93] But the Nobel tiff of 1906 was
just the final act of a much-older rivalry: Cajal and Golgi had long
stood on opposite sides of one of the most fundamental issues of the
time. Golgi, who had worked on many aspects of the nervous system,
including insanity, neurology, and the lymphatics of the brain, argued
that neurons communicated through an inextricable net formed by
the finest branches of their axons (here he sided with many mid-
nineteenth-century neurologists in his commitment to a form of
holism). By contrast, Cajal adamantly defended the histological
autonomy of each neuron: he reckoned that Golgi and his predecessor Gerlach had committed a scientific and moral offense against
clear-sightedness — the terms of the accusation are important. As
Cajal put it, his competitors had been so "seduced by the presumed
necessity of continuous structure, they [Golgi and Gerlach] then *supposed* the existence of an anastomotic net between the axis cylinders
of different neurons." Cajal contended that such a "seduction" had
lured the weak-willed scientists away from true sight.[94]

To see without the interference of subjective haze or fog required a will bolstered by precision. With Golgi, so Cajal believed, the supposed net connecting cells achieved an "attractive structural form and even a certain appearance of being founded upon observed facts."[95] According to Cajal, where Golgi used the term "motor cells," Cajal held back ("I christened [them], so not to commit myself as to their physiology, *elements with long axons*"). Over and over, Cajal insisted that restraint was necessary, a restraint both from inference as to physiological function *and* from any temptation to succumb to the seductions of aesthetic or theoretical charm. This was a demand at once moral and epistemic: "Only by dint of evasions, irrelevances, and subterfuges could this conception [of the network by Golgi and other reticularists] be adapted to exigencies of physiology."[96]

For Cajal, Golgi's network theory was a snare and a delusion: "To affirm that everything communicates with everything else is equivalent to declaring the absolute unsearchability of the organs of the soul."[97] If one couldn't see the boundaries and thereby identify the basic objects of inquiry in the brain (so Cajal argued), then more than a neurohistological project was thwarted: the scientific project itself was doomed. Cajal desperately wanted the visual "searchability" he believed Golgi had abandoned. As a researcher, Cajal had insisted on practical procedures that led to results that could be, insofar as such was possible, *seen*. In contrast to what he viewed as the defeatist indeterminism of the network-theory advocates, Cajal identified his own efforts as objective: he took definite, well-defined entities from the world of the microscopic slide and vouchsafed their transfer to the reproduced page. "My work," Cajal argued, "consisted just in providing an *objective* basis for the brilliant but vague [neuronist] suggestions of [Wilhelm] His and [Auguste] Forel."[98] That "objective basis" meant working from the silver-impregnated tissue sample, through the camera lucida-equipped microscope, to the faithful ink trace — without willful intervention. Anything else, Cajal insisted, was a figment of overwrought imagination — an error of *subjectivity*.

Mechanical objectivity meant learning to see, twice over. First, objectivity demanded technical mastery. It was Golgi who had not only developed the original black method but also honed a faster

"Golgi method," in which he added osmium tetroxide to the bichro-mate solution, which dramatically shortened the procedure. Cajal adopted Golgi's hard-won technique but repeated the impregnation two or three times — a refinement of Golgi's staining, joined to care-ful microscopy, complemented by meticulous sketching from the projected image of the camera lucida. In principle the hand mim-icked and confirmed what the disciplined eye saw, and no more. Sec-ond, objectivity meant cultivating one's will to bind and discipline the self by inhibiting desire, blocking temptation, and defending a determined effort to see without the distortions induced by author-ity, aesthetic pleasure, or self-love. Together, for Cajal and many oth-ers, the regulation of interior states and external procedures defined objective vision.

Although mechanical objectivity was in the service of gaining a right depiction of nature, its primary allegiance was to a morality of self-restraint. When forced to choose between accuracy and moral probity, the atlas makers often chose the latter, as we have seen: bet-ter to have bad color, ragged tissue edges, limited focal planes, and blurred boundaries than even a suspicion of subjectivity. The disci-pline earlier atlas makers had imposed on their artists had been in the interests of truth, which could only be discovered by sagacious selection of the typical or characteristic. Truth did not lie on the vis-ible surface of the world. Later atlas makers, as fearful of themselves as of their artists, forfeited the typical and postponed an immediate grasp of truth because intervention was needed to produce it and because alteration of the image led all too easily to the dreaded sub-jectivity of interpretation. Could Golgi, Cajal, or, for that matter, anyone else dispense fully with *all* intervention? Of course not, and they all knew and said so. Mechanical objectivity remained an always-receding ideal, never fully obtainable. But despite being an ideal, it was not without direct and immediate effects on the lab bench, lith-ographic stone, cutting board, or microscope — in a panoply of ways, there was a continuing and insistent emphasis on moving from the interpretive to the procedural.

No atlas maker could entirely dodge the responsibility of pre-senting figures that would teach the reader how to recognize the working objects of science. To do so would have betrayed the mis-sion of the atlas itself. A mere collection of unsorted individual specimens, portrayed in all their intricate peculiarity, would have been useless. Caught between the Charybdis of interpretation and the Scylla of irrelevance, the atlas makers who pursued mechanical objectivity worked out a precarious compromise. They would no longer present typical phenomena, or even individual phenomena characteristic of a type. Rather, they would present a scattering of individual phenomena that would cover the range of the normal, leaving it to the reader to accomplish intuitively what the atlas maker no longer dared to do explicitly. As we will see in Chapter Six, researchers assiduously sought to acquire an ability to distinguish at a glance the normal from the pathological, the typical from the anomalous, the novel from the known.

Mechanical objectivity pruned the idealizing ambitions of the atlas; it also hemmed in the scientific self of the aspiring atlas maker. At the very least, the atlas maker of the eighteenth century had been a person qualified by wide experience and discernment to select and present an edition of interpreted phenomena for the guidance of other anatomists, botanists, astronomers, entomologists, or other naturalists. An exalted few had been atlas makers capable of intuiting universal truth from flawed particulars, even when scientific knowledge was meager. But even atlas makers of lesser gifts were emphatically present in their works, selecting and preparing their specimens, alternately flattering and bullying their artists, negotiating with the publisher for the best engravers, all with the aim of publishing atlases that were a testimony to their knowledge and artistic skill. Knowledge and artistry were, after all, their title to authority and authorship; otherwise, any greenhorn or untutored artist could publish a scientific atlas. Failure to discriminate between essential and accidental detail; failure to amend a flawed or atypical specimen; failure to explain the significance of an image — eighteenth-century atlas makers took these as signs of incompetence, not virtuous restraint.

Already in the early decades of the nineteenth century, however, scientists in varied fields and of very diverse methodological and theoretical persuasions began to fidget uneasily about the perils within, especially flights of interpretation and imagination. Scientists sometimes sought, not always with success, to discipline these "inner enemies," as Goethe called them, by rules of method, measurement, and work discipline.[99] But more often, and more importantly, discipline came from within: scientists confronted the "inner enemies," often conceived as excesses of the will, on their own territory. It is this internal struggle to control the will that imparted to mechanical objectivity its high moral tone. Interpretation, aestheticization, and theoretical overreaching were suspect not primarily because they were personal traits but because they were disorders of the will that interfered with faithful representation. This scientific self required restraint, a will strong enough to bridle itself. A lack of sufficient discipline indicated character flaws — self-indulgence, impatience, partiality to one's own ideas, sloth, even dishonesty — that were best corrected at their source, by assuming the viewpoint of one's own sharpest critic, even in the heat of discovery.

One type of mechanical image, the photograph, became the emblem for all aspects of noninterventionist objectivity, as two historians found self-evident by the 1980s: "The photograph has acquired a symbolic value, and its fine grain and evenness of detail have come to imply objectivity; photographic vision has become a primary metaphor for objective truth."[100] This was not because the photograph was more obviously faithful to nature than handmade images — many paintings bore a closer resemblance to their subject matter than early photographs, if only because they used color — but because the camera apparently eliminated human agency. Other advocates of mechanical, procedural, exact representation (such as Cajal) chose to draw, albeit through the camera lucida. Nonintervention — not verisimilitude — lay at the heart of mechanical objectivity, and this is why mechanically produced images of individual objects captured its message best.

The rise of the objective image polarized the visual space of art and science, just as the role of the two domains split over the role of the will. From the sixteenth century, when the illustrated scientific book originated, through the eighteenth century, the relationship between art and science had largely been one of collaboration, not opposition. Only in the early nineteenth century did Romantic artists begin to defend the willful imposition of self as the *sine qua non* of art. For their part, scientists increasingly insisted on the opposite: their images must be purged of any trace of self. Baudelaire captured the distinction when, in his "Salon of 1859," he ventriloquized the positivist painter: "'I want to represent things as they are, or as they

would be in supposing that I do not exist.' The universe without man."
Baudelaire's imagined artist replied: "I want to illuminate things with
my spirit and to project their reflection on others."[101]

Photography joined this battle between science and art, positive
recording and imaginative illumination. Richard Neuhauss, one of
the great nineteenth-century experts on photomicrography, titled a
key section of his treatise on photomicrography "Retouching the
Negative." He acknowledged that retouching was a central part of
portrait and landscape photography. In some portrait negatives, Neuhauss rather skeptically noted, the silver layer served only as a medium
upon which the colors of the retoucher would be laid. But in his corner of the world — the scientist's — this was exactly what should *not*
happen. According to Neuhauss, it is not the photographer's image
but nature's that is wanted. This was easy to say but hard to realize:
sensu stricto, every alteration of the natural ought to be forbidden.
But Neuhauss knew far too much to pretend to this ideal. Anyone
could see that two identical photographic plates, exposed in identical light conditions, could be developed to produce radically different images; one plate could show, for example, subtle, fine structures
that the other obscured.

Moreover, Neuhauss readily conceded that the gift of drawing
was not equitably distributed. Some of the best researchers had the
least skill for it. Most left the task to others, but this led to a variety
of different interpretations — a most dangerous state of affairs: "The
subjective interpretation of the artist is a point with which one must
come to terms in all circumstances. Here lies the heart of the matter:
The photogram reflects the object objectively. How does the much
celebrated objectivity appear when we take a closer look? Above all
else, the light sensitive plate copies everything that does not belong
to the object with frightening objectivity — such as the impurities of
the preparation and the diffraction edges." (Not to mention dust particles, plate defects, Newton's rings, and a host of other artifacts.)
Too much light or too little light made details vanish. Developing the
film introduced still more difficulties: membranes appeared more
than once in one image and disappeared in another. "This is the
objectivity of the microphotogram!" Neuhauss ruefully concluded.
The photomicrographer can coax details into the picture, heighten
them — or let them escape: "We can assert that a photograph can

only lay claim to objectivity if it is produced by an honest, gifted micro-photographer, working according to all the rules of the art, and richly endowed with patience and skill."[102] After forty years of scientific photography in the service of mechanical objectivity, Neuhauss knew that the photographer's art must aid science; skill was needed where automatism came up short.

By the turn of the twentieth century, faith in mechanical objectivity was unraveling. The simple promise of automaticity began to appear more ambiguous — not least to the real experts, like Neuhauss, who knew inside out all the difficulties attendant to photographing anything from bacterial cultures to asymmetrical snowflakes. Although Neuhauss and his contemporaries still upheld the ideal of objectivity, they knew it was an ideal that would not produce itself. Removing the scientists, their interfering eyes and hands, was no mean feat; it might even prove impossible.

In an 1872 address to the Versammlung Deutscher Naturforscher und Ärtzte, Rudolf Virchow reflected wryly on the challenge in the context of an attack on Ernst Haeckel's public support of Darwinian evolutionary theory:

> I am now among the oldest professors of medicine; I have been teaching my science for more than thirty years, and I may say that in these thirty years I have honestly worked on myself, to do away with ever more of my subjective being [dem subjektiven Wesen] and to steer myself ever more into objective waters [das objektive Fahrwasser]. Nonetheless, I must openly confess that it has not been possible for me to desubjectivize myself entirely. With each year, I recognize yet again that in those places where I thought myself wholly objective I have still held onto a large element of subjective views [subjektive Vorstellungen].

For Virchow, this ethico-epistemic battle against an insidious subjectivity was a never-ending struggle, one that had to be fought unremittingly against the dangerously subjective aspects of the scientific self — "my opinions, my representations, my theory, my speculation."[103] It demanded patience and more: a cultivation of the scientific self through skill and art (Geschick und Kunst). Objectivity in its purist form remained for Virchow and his contemporaries an elusive goal, a destination always just past the horizon. But even if objectivity could never be obtained in its fullness, it was not an idle bit of rhetoric. Objectivity demanded particular kinds of actions at the laboratory bench and illustrator's table.

Like Virchow, many early twentieth-century scientists increasingly concluded that subjectivity could never be extirpated. Some frankly espoused the need for subjective judgment in the production and use of scientific images; objectivity without subjectivity was, they concluded, an ultimately self-defeating ambition. Others, despairing that images would ever achieve objectivity, began to hunt for objectivity not in engravings, tracings, and photographs but in the subtle and more ethereal domain of mathematics and logic. We address these two alternatives in Chapters Five and Six. But first we must tackle a question that has already arisen in Chapters Two and Three: Who was the scientific self who sought to depict nature rightly? Taking our cue from the tight intertwining of scientific practice and character, in Chapter Four we probe the new scientific self that aspired, through a supreme act of will, to quiet the will. We want to know how it became a commonplace across such a range of sciences to say, with Cajal, that the greatest obstacle on the path to scientific objectivity was the uncontrolled, disordered will.

CHAPTER FOUR

The Scientific Self

Why Objectivity?
In the 1870s, the Leipzig embryologist Wilhelm His began a series of
attacks on his Jena colleague Ernst Haeckel's use of embryological
evidence, particularly illustrations of embryological development, to
support Haeckel's thesis that ontogeny recapitulates phylogeny (see
figures 4.1 and 4.2). His accused Haeckel of smuggling his theoretical prejudices into the illustrations (drawn by Haeckel himself in
some instances), which were intended to show the continuity of
embryological forms across species, and he came perilously close to
calling Haeckel a liar: "I myself grew up in the belief that among all
the qualifications of a scientist reliability and unconditional respect
for the factual truth are the only ones that are indispensable."[1] Haeckel
responded explosively, pointing out that his illustrations were not
intended as "'exact and completely faithful illustrations,' as HIS
would demand, but rather... illustrations that show only the essen-
tials of an object, leaving out inessentials." To call such illustrations
"inventions," much less lies, was, according to Haeckel, to drive all
ideas out of science, leaving only facts and photographs: "Wholly
blameless and virtuous is, according to HIS and other 'exact' ped-
ants, accordingly only the photograph."[2]

In his indignation, Haeckel exaggerated His's obsession with the
bare facts; His actually acknowledged the utility of drawings as well
as photographs in scientific illustration, as we have seen in Chapter
Three. But His believed that drawings always contained "subjective
elements," sometimes advantageous and sometimes not, whereas
"the photograph reproduces an object with all of its particularities,

191

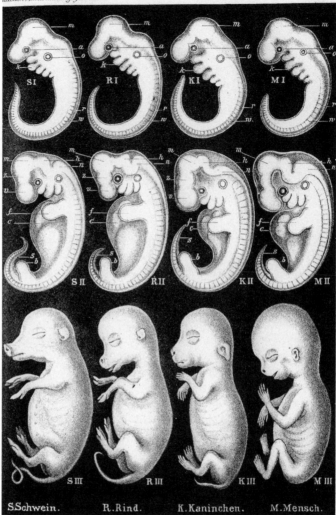

Fig. 4.1. Ontogeny Recapitulates Phylogeny. "Embryos from Three Mammals," Ernst Haeckel, *Anthropogenie, oder, Entwicklungsgeschichte des Menschen* (Leipzig: Engelmann, 1874), table 5. This plate, drawn by Haeckel himself and lithographed by the Leipzig firm J.G. Bach, shows three comparable embryological phases of a pig, a cow, a rabbit, and a human in order to make Haeckel's point about striking commonalities in early developmental stages visually. Wilhelm His was especially critical of some of Haeckel's depictions of the human embryo: he claimed that features had been exaggerated or invented to support Haeckel's claim that ontogeny recapitulated phylogeny. He fumed because Haeckel had used a camera lucida in earlier work and was therefore "not ignorant of the methods to be applied in order to obtain more exact outlines." Wilhelm His, *Unsere Körperform und das physiologische Problem ihrer Entstehung* (Leipzig: Vogel, 1874), pp. 170–71.

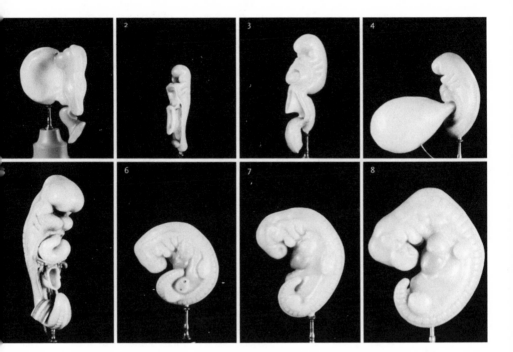

Fig. 4.2. Model Embryos. Adolf and Friedrich Ziegler (after Wilhelm His), "Human Embryos of the First Month (series 1)," in Nick Hopwood, *Embryos in Wax: Models from the Ziegler Studio* (Cambridge: Whipple Museum of the History of Science, 2002), pl. 17, p. 106 (courtesy of Anatomisches Museum, Basel). Working closely with the Freiburg scientific-model makers Adolf and Friedrich Ziegler, His commissioned this series of eight wax models (magnified forty times or twenty times), based on His's drawings in *Anatomie menschlicher Embryonen* (Leipzig: Vogel, 1880–1885), vol. 3. Each model was named after the physician who donated the original anatomical material from which the drawings were made, thus emphasizing the rarity and individuality of the specimens. (Please see Color Plates.)

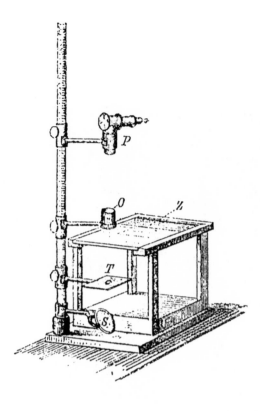

Fig. 4.3. Disciplined Drawing. Drawing apparatus, Wilhelm His, *Anatomie der menschlichen Embryonen* (Leipzig: Vogel, 1880–1885), vol. 1, fig. 1, p. 8. An object placed at T is magnified by the microscope objective O and an image is projected by the camera lucida P onto the glass drawing surface Z. An elaborate system of controls is built into the device: the drawing surface is ruled and set at a fixed distance from the object; a vertical rod graduated in millimeters allows other distances to be precisely set and replicated; the same object is sketched under different lighting conditions; sketches of embryo cross sections are then assembled next to a piece of paper marked in parallel zones that match the intervals at which the cross sections were cut. Any mismatch between drawings occasions a thorough investigation of possible causes: "In the mutual controls of the various constructions one quickly finds an exact measuring rod for the reliability of the whole process" (*ibid.*, p. 11).

including those that are accidental, in a certain sense as raw material, but which guarantees absolute fidelity."

More revealing than this bald opposition between drawing and photograph was His's own elaborate method of making images: he employed a drawing prism and stereoscope to project an image, which was then traced upon the drawing surface (see figure 4.3). These tracings of microscopic cross-sections were then subjected to a painstaking process of checking against finely lined graph paper and against one another to ascertain the exactness of the proportions. Any amendments or idealizations of the drawings or models that slipped through this system of multiple controls His equated with "conscious bungling [*bewussten Pfuscherei*]."[3] Whereas Enlightenment naturalists such as Carolus Linnaeus and Bernhard Albinus had understood it to be their scientific duty to improve drawings executed under strict constraints of empirical exactitude, His con-

demned Haeckel's intervention in drawings as tantamount to deception — even though His, like earlier atlas makers, also sought nature's
types. When Haeckel used his drawings to extract "the essential," or
what he believed to be the true idea hidden beneath potentially false
or confusing appearances, His indicted him for sinning against objectivity. Haeckel understood the charge full well. He ridiculed Rudolf
Virchow's call (discussed in Chapter Three) for objectivity in the
classroom (an explicit attack on Haeckel's passionate campaign for
evolutionary theory): if "only what has been objectively established,
what is absolutely sure" could be taught, the result would be that "no
idea, no thought, no theory, indeed no real 'science'" would ever
make its way into a lecture.[4] A sea change had occurred in science:
mechanical objectivity now confronted truth-to-nature, and hard
choices had to be made between them.

The His-Haeckel confrontation dramatizes the transformation of
scientific ideals and practices across many disciplines that we followed in Chapters Two and Three. By the middle decades of the
nineteenth century, the epistemology and ethos of truth-to-nature
had been supplemented (and, in some cases, superseded) by a new
and powerful rival: mechanical objectivity. The new creed of objectivity permeated every aspect of science, from philosophical reflections on metaphysics and method to everyday techniques for making
observations and images. In our account of the emergence of objectivity, we have focused on the latter in order to show how the airy-
sounding abstractions of truth and objectivity had their concrete
complement in the ways neurons, snowflakes, skeletons, and myriad
other natural objects were depicted on the pages of scientific atlases
in the eighteenth and nineteenth centuries. Truth and objectivity
were not merely the stuff of pious prefaces and after-dinner addresses at scientific meetings; to embrace one or the other could
translate into the choice between an exquisitely colored, sharply
outlined drawing and a blurred black-and-white photograph, or between the image of an idealized type sketched freehand and that of a
particular individual meticulously traced from a projected image. It
was a choice freighted with ethical as well as epistemological implications, as the barbed exchange between His and Haeckel shows.

Why objectivity? Why did this deep and broad change take place
when and how it did? In this chapter, we address these questions by

stepping back from atlas images to explore the ethos that made them possible. Building on the testimony from atlas makers already set forth in Chapters Two and Three, we here widen our inquiry to encompass the kind of person thought to be best suited to pursue truth-to-nature or mechanical objectivity. We have already seen how both truth-to-nature and mechanical objectivity laid heavy demands upon the atlas makers who professed these epistemic virtues: consider Albinus's Herculean labors to select, clean, pose, and then improve his skeleton, or Otto Funke's painstaking rendering of the most minute details of crystallized hemoglobin, right down to optical artifacts. These demands and the practices they imposed left their imprint on the atlas makers as well as atlas images. Truth-to-nature and mechanical objectivity molded their proponents in different, albeit equally dutiful ways: where, for example, Albinus recognized a duty to perfect, Funke bowed before a duty to abstain.

Because scientific atlases, by their very nature, had to justify the publication of a new set of definitive images in terms of the grave shortcomings of the old ones, they registered the new epistemic virtue objectivity more explicitly and forcefully than other sources. It is not a light thing to call for a wholesale change in the disciplinary eye. But the atlas makers were not alone among scientists in registering these shifting calls to duty. In the eighteenth century, geodesists and astronomers, for example, had accepted or discarded outlying data points on the basis of their best judgment about the soundness of an observation or measurement. By the 1860s, they too had come to condemn these time-honored practices as subjective and arbitrary and instead turned to objective rules to assess data, such as the method of least squares.[5] Hermann von Helmholtz's insistence on tracing the curves of muscle action by a self-registering instrument rather than using the idealized curves drawn by his predecessors similarly fostered cautious restraint.[6] In late nineteenth-century statistics, as in atlas making, objectivity also took on a moral tinge. For example, the British statistician Karl Pearson in 1892 called on enlightened citizens of modern polities to set aside their "own feelings and emotions" for the common good, on the model of the scientist who "has above all things to aim at self-elimination in his judgments, to provide an argument which is as true for each individual mind as for his own."[7] In the making of images, the taking of measurements,

196

the tracing of curves, and many other scientific practices of the latter half of the nineteenth century, *self*-elimination became an imperative.

The answer to the question "Why objectivity?" lies precisely in the history of the scientific self to be eliminated. There was nothing inevitable about the emergence of objectivity. As both an epistemology and an ethos, truth-to-nature sustained (and, in disciplines such as botany, continues to sustain, as we saw in Chapter Two) a rigorous and progressive tradition of scientific research and representation. It was and remains a viable alternative to objectivity in the sciences. Objectivity did not surpass truth, as Newtonian surpassed Galilean mechanics. Nor, as we saw in Chapter Three, did technological innovations such as photography create scientific objectivity, although the photograph became one of its principal vehicles. Eighteenth-century atlas makers such as the anatomist William Cheselden had used the camera obscura without foresaking truth-to-nature, yet the bacteriologist Robert Koch was one of the many late nineteenth-century scientists who turned to the camera obscura image fixed by the photograph to enforce mechanical objectivity. The same device could be and was turned to different epistemic ends.

Another strategy might be to seek an explanation of the advent of scientific objectivity in one of the better-known historical "revolutions" of the period — the French Revolution, the Industrial Revolution, the Second Scientific Revolution of the early nineteenth century — and these are all, no doubt, in some ultimate sense relevant. Yet the relationship is not proximate, much less intrinsic. Such an explanation would, moreover, be heterogeneous, according to a reductive base-superstructure model: one "foundational" level (the means of production, the interests of a social class, certain religious beliefs) is alleged somehow to cause an "overlaid" level of a strikingly different kind (political ideologies, taste in art, slavery). In this chapter, in contrast, we seek an intrinsic, homogeneous answer to the question "Why objectivity?" — one that puts explanans and explanandum on the same level and reveals how they interlock with each other.

Objectivity and subjectivity are as inseparable as concave and convex; one defines the other. The emergence of scientific objectivity in the mid-nineteenth century necessarily goes hand in glove with the emergence of scientific subjectivity. Subjectivity was the enemy within, which the extraordinary measures of mechanical objectivity were invented and mobilized to combat. It is no accident that these measures often appealed to self-restraint, self-discipline, self-control: it was no longer variable nature or the wayward artist but the scientific self that posed the greatest perceived epistemological danger. This untrustworthy scientific self was as new as objectivity itself; indeed, it was its obverse, its photographic negative. "Why objectivity?" becomes "Why subjectivity?" — or, more specifically, "Who is the scientific subject?"

The Scientific Subject

These questions plunge us into the history of the self, as variously studied by anthropologists, philosophers, and historians.[8] The self is entangled in a web of near synonyms and cognates in various European languages, each word embedded its own distinctive semantic field: self, individual, identity, subject, soul, *persona, le moi, das Ich*.[9] Therefore, the quarry of such a history is elusive unless pinned down to particular periods, places, and persons. We are interested here in only one specific and localized segment of this rich and capacious history, namely, the manifestations and mutations of the *scientific* self during the eighteenth through the twentieth centuries, mostly in Western Europe.

The very claim that whatever we mean by the self has a history is bewildering: how could there ever have existed a person without a self? And if selves in different times and places differ systematically from one another, how can the historian investigate these contrasting forms of selfhood, given their notorious inaccessibility to third-person observation? A great deal of the plausibility and fruitfulness of the undertaking depends on what counts as evidence and how these sources are mined. In this chapter, in addition to "ego-documents" such as diaries and autobiographies, we examine what might be called the literature of the scientific persona — collections of potted biographies and advice manuals that purport at once to describe and to prescribe the character and conduct of the scientist as a recognizable human type.

Most importantly, we pay close attention to what the philosopher-historian Michel Foucault called "technologies of the self": practices of the mind and body (most often the two in tandem) that mold and maintain a certain kind of self.[10] Following the historian of ancient philosophy Pierre Hadot, Foucault wrote evocatively of how the writing practices, the *hupomnemata*, of the Stoics and Epicureans of late Antiquity fixed and solidified a way of being in the world.[11] The kinds of practices we will be concerned with include training the senses in scientific observation, keeping lab notebooks, drawing specimens, habitually monitoring one's own beliefs and hypotheses, quieting the will, and channeling the attention. Like Foucault, we assume that these practices do not merely express a self; they forge and constitute it. Radically different practices are *prima facie* evidence of different selves. Unlike Foucault, we do not see a single self in the periods under examination here. On the contrary, we find, for example, scientific and artistic selves to be conceived and trained in diametrically opposed ways in the mid-nineteenth century.

In the case of the subjectivity that was the yin to objectivity's yang, its archenemy as well as its *raison d'être*, narrowly scientific developments intersected with broader currents in the history of the self. The career of objectivity and subjectivity extended far beyond the sciences in the nineteenth century: philosophers, artists, novelists, theologians, and intellectuals of every stripe seized on the newfangled Kantian words to pick out a novel way of being in the world that older vocabularies did not seem to capture. However divergent the philosophical and semantic reception of Kant's pair could be (and, as we saw in Chapter One, these divergences could be ludicrously wide), there was a shared sense in philosophy, psychology, and even imaginative literature that possessing a subjectivity was a different matter from being endowed with a rational soul (as Renaissance writers conceived the self) or a bundle of coordinated mental faculties (as described by Enlightenment psychology).

Because the word "subjectivity" is currently used to refer to conscious experience and its forms across cultures and epochs ("Renaissance subjectivity," "modern subjectivity"), we should make clear that we use the term here historically: it refers to a specific kind of self that can first be widely conceptualized and, perhaps, realized within the framework of the Kantian and post-Kantian opposition between the objective and the subjective. Every human being, everywhere and always, may well experience consciousness or even interiority; "subjectivity" as we shall use it is not a synonym for but a particular species of these experiences. Subjectivity is only one species of the genus self.

Consider two vivid descriptions of the self, both belonging to the genre of philosophical psychology: one was written by the French *philosophe* Denis Diderot around 1770, the other by the American psychologist William James in 1890. In Diderot's dialogue *Le rêve de d'Alembert* (*D'Alembert's Dream*), the physician Théophile de Bordeu is summoned to the bedside of the mathematician Jean Le Rond d'Alembert, who is delirious with fever, by d'Alembert's companion, Julie de Lespinasse. Bordeu interprets d'Alembert's ravings as a theory of the conscious organism conceived as a network or skein of threads, all centered on an origin, as a spiderweb is centered on the spider:

> **MLLE DE LESPINASSE:** Each thread of the feeling network can be hurt or tickled along all its length. Pleasure or pain is here or there, in one place or another, of one of those long spider's legs of mine, for I always come back to my spider. It is the spider which is the common starting-point of all the legs and which relates pain and pleasure to such and such a place though it does not feel them.

> **BORDEU:** It is this power of constantly and invariably referring all impressions back to this common starting-point which constitutes the unity of the animal.

> **MLLE DE LESPINASSE:** It is the memory and comparison which follow necessarily from all these impressions which makes for each animal the history of its life and self.

The faculties of reason, imagination, judgment, and instinct are regulated by the relation between the origin of the network and its branches. If the origin dominates, the organism is "master of himself, mentis compos"; conversely, there is "anarchy when all the ends of the network rise against their chief, and there is no supreme authority."[12] It is memory that safeguards the unity of the self over time.

In contrast to this precarious polity of the self, in *The Principles of Psychology*, James depicts the core or "spiritual" self as that which is "felt by all men as a sort of innermost centre within the circle, of sanctuary within the citadel, constituted by the subjective life as a whole." This "self of all the other selves" is that part of the stream of consciousness that endures amid the flux, and it is robust, unified, and, above all, "active":

> Whatever qualities a man's feelings may possess, or whatever content his thought may include, there is a spiritual something in him which seems to go out to meet these qualities and contents, whilst they seem to come in to be received by it. It presides over the perception of sensations, and by giving or withholding its assent it influences the movements they tend to arouse.... It is the source of effort and attention, and the place from which appear to emanate the fiats of will.[13]

James's bustling, willful self directs "this subjective life of ours" like an energetic executive: it "comes out" to meet experience with outstretched hand, "receives" thought and feeling into its office, "presides over" the clamor of perception. It is the assertive subject of subjectivity.

Between these two visions of the self—passive and active—a chasm yawns.[14] The self of Enlightenment sensationalist psychology was fragmented: atomistic sensations were combined by the mental faculties of reason, memory, and imagination to forge associations. Personal identity was as fragile as a cobweb, guaranteed only by memory and the continuity of consciousness; the sovereignty of reason at the origin of the network was always under threat from within (the vagaries of the imagination and the uprisings of the branches of the network) and without (the barrage of sensations registered by the receptive network). This was a largely passive and permeable self, shaped by its environment. The post-Kantian self, by contrast, was active, integrated, and called into philosophical existence as a necessary precondition for fusing raw sensations into coherent experience. Organized around the dynamic and autonomous will, the self acted on the world, projecting itself outward. Even perceptions were vetted, like callers at the door. This is the subjective self of Idealist philosophy, Romantic art, and, as James bears witness, early experimental psychology: a self—a "subject"—equal to and opposed to the objective world.

These two visions were admittedly advanced as speculations, albeit ones that Diderot and James each believed would resonate with the lived experience of most of his readers. They were, however, speculations that could be and were harnessed to politics, art, economics, and science. Moreover, there is considerable evidence that at least some literate elites internalized these visions and used them to describe themselves to themselves, as well as to make sense of other people.[15] During the nineteenth century, the French Revolution and the political aspirations it inspired at home and abroad, culminating in the revolutions of 1848, made new forms of political action imaginable and desirable. Flamboyantly personalized artistic styles at once documented and encouraged distinctive, private psyches. A pulsing industrial economy and educational institutions based on competitive examinations created "new men," who understood their rise to fame and fortune as a triumph of the will.

The scientific self was not simply a microcosm of these cultural macrocosms, although it shared the basic architecture of the self as lived and understood in historical context. The epistemic virtues examined in Chapters Two and Three certainly drew upon and were reinforced by attitudes, values, and social relations that operated in specific locales — among Parisian doctors or Berlin professors, American frontiersmen or London gentlemen of science. Similarly, the scientific selves explored in this chapter were doubtless inflected by local accents of class and gender: in the ethos of mechanical objectivity, for example, it is difficult to miss the Victorian admonitions to hard work or the masculine overtones of "unveiling" nature (or in the exclusionary phrase "men of science"). Yet it is equally difficult to overlook the imprint of the larger scientific context opened up and sustained by the collective empiricism described in Chapter One. The broad scope of epistemic virtues such as truth-to-nature and mechanical objectivity as reflected in atlas making stems in part from the broad mission of the atlases themselves: to establish standards for the entire disciplinary community for generations to come that would define how collective empiricism was to be practiced in a given historical context. The very existence of atlases testifies to ambitions beyond the here and now. But atlases were not the only expression of collective empiricism. Just because scientific communities were, already in the eighteenth century, dispersed in time and space, great emphasis was placed on specifically scientific values and practices that would bind its members together. The recurring and still current motif of the "other-worldliness" of scientists in anecdotes and fiction, whether it was expressed as absent-mindedness or as obsession, draws attention to loyalties that transcend (and sometimes subvert) the local and the familiar. Internalized and moralized, these loyalties stamped a distinctively scientific self, which was recognizable across a diverse range of local contexts.

Depending on which threat to knowledge was perceived as most acute at that moment, the scientific self was exhorted to take episte-mological precautions to redress the excesses of both the active and the passive cognition of nature, and to practice four-eyed or blind sight. For Enlightenment savants, the passivity of the sensationalist self was problematic; achieving truth-to-nature required that they actively select, sift, and synthesize the sensations that flooded the too-receptive mind. Only neophytes and incompetents allowed themselves to be overwhelmed by the variety and detail of natural phenomena. To register experience indiscriminately was to be at best confused and at worst indoctrinated. The true savant was a "genius of observation" whose directed and critical exercise of attention could extract truth-to-nature from numerous impressions, as the smelter extracts pure metal from ore.[16]

In contrast, the subjective self of nineteenth-century scientists was viewed as overactive and prone to impose its preconceptions and pet hypotheses on data. Therefore, these scientists strove for a self-denying passivity, which might be described as the will to wil-lessness. The only way for the active self to attain the desired recep-tivity to nature was to turn its domineering will inward — to practice self-discipline, self-restraint, self-abnegation, self-annihilation, and a multitude of other techniques of self-imposed selflessness.[17] The German philosopher Arthur Schopenhauer preached a bitter struggle with the will, on the model of Christian mysticism and the philoso-phy of the Indian Vedas, that would ultimately "rid us of ourselves" and replace the individual subject of willing and wanting with the "will-pure, eternal subject of knowing," an "unclouded mirror of the world."[18] Schopenhauer's admirer Friedrich Nietzsche detected the same mystical yearnings in the intellectual will to willessness but took a dimmer view of them. Ever suspicious of priestly pretensions of asceticism in any guise, he derided scholars who attempted to extinguish the self as a "race of eunuchs . . . neither man nor woman, nor even hermaphrodite, but always and only neuters or, to speak more cultivatedly, the eternally objective."[19] Schopenhauer and Nietzsche were on opposite sides when it came to the value of self-denying objectivity, but they were talking about the same phenomenon. By a process of algebraic cancellation, the negating of subjectivity by the subject became objectivity.

What kinds of selves meet the differing demands of truth-to-nature, objectivity, and other epistemic virtues? The term "epistemic virtues," with its ethical overtones, is warranted. Ethos was explicitly wedded to epistemology in the quest for truth or objectivity or accuracy. Far from eliminating the self in the pursuit of scientific knowledge, each of the epistemic virtues depended on the cultivation of certain character traits at the expense of others. A figurative portrait gallery of prototypical knowers of nature — the insightful sage, the diligent worker — can be reconstructed from the literature of scientific biography and autobiography, academic eulogies, memoirs, advice manuals, and actual portraits. We do not regard these accounts as faithful descriptions of the individuals they treat. Indeed, it is precisely the biographical inaccuracies, systematic distortions, and idealizations that interest us here; it is the *type* of the scientist as a regulative ideal, as opposed to any flesh-and-blood individual, that we have in our sights.[20] That these types should routinely conflate the normative with the descriptive is valuable evidence of how an ethos must be grafted onto a scientific persona, an ethical and epistemological code imagined as a self. The transformations of the scientific self are at the center of this chapter and also, in many ways, at the center of the book's overarching argument about how epistemology and ethos fuse.

But we do not believe that these scientific selves were called into being by free-floating norms and types alone. A self must be practiced, not simply imagined and admired (or castigated) as a public persona. Trading the panorama of the public portrayal of scientists for the close perspective of the *vie intime scientifique*, we then turn to the technologies of the scientific self: how doing science molded the scientist. Here we shall be especially concerned with practices of scientific observation and attention, which are essential to all branches of empirical science, intimately involved in making and evaluating images, and central to the ethical and epistemological constitution of the scientific self during the entire period under discussion, but in revealingly different capacities.

Our reframing the question "Why objectivity?" as "Who was the scientific subject?" may strike some readers as superficial, even tautologous. Where, they will ask, are the deeper underlying causes, the hidden machinery backstage, the prime mover beyond the outermost sphere? And isn't subjectivity just the necessary concomitant of objectivity, not its explanans? We must reply that superficiality is, in a certain sense, exactly the point. The kind of explanation we are after is indeed superficial, in the etymological sense of lying on the surface of things rather than hiding in conjectured depths. We reject the metaphorical (and metaphysical) reflex that, without further justification, prefers excavation to enlargement as a privileged method of understanding; instead, we suggest that in some cases an exploration of relationships that all lie on the same level, a widening of the angle of vision, can be more enlightening.[21] However, we do not regard such explanations as flimsy, in the pejorative sense of the word "superficial." They reveal patterns that show that even if a historical formation is contingent, it is not thereby a hodge-podge or chimera. Nor do we regard an explanation that reveals how the parts of these patterns fit together as tautologous. Rather, we are attempting to explain the illusion of tautology. How can two concepts, two epistemologies, two ethics, two ways of life intertwine so closely — and yet contingently, for we are within the realm of history, not necessity — that their relationship seems to be almost self-evident? This is the puzzle of objectivity and subjectivity.

Kant Among the Scientists

Immanuel Kant's philosophical reformulation of the scholastic categories of the objective and the subjective reverberated with seismic intensity in every domain of nineteenth-century intellectual life, from science to literature.[22] Whether Kant invented this idea from whole cloth or simply articulated a new way of dividing up the world is immaterial for our purposes; it suffices that he was at the very least a precocious philosophical witness to changes in conceptualizing the nature of self and knowledge that spread like wildfire in

205

the first half of the nineteenth century. Nor will we be concerned with the accuracy of the reception of Kantian philosophy in various milieus; this is already the subject of an extensive literature.[23] On the contrary, what interests us are the ways in which Kant was creatively misunderstood, or, to put it less tendentiously, adapted by scientists to their own purposes.

We begin with a brief account of how and why three influential mid-nineteenth-century scientists, each prominent not only in his discipline but also in his national context as a public intellectual, took up the Kantian terminology of objectivity and subjectivity (here understood in its broadest philosophical sense) and put it to work: the German physicist and physiologist Hermann von Helmholtz, the French physiologist Claude Bernard, and the British comparative anatomist Thomas Henry Huxley — all of whom were active during the 1860s and 1870s, the heyday of the more specifically scientific mechanical objectivity. Despite much variation in their deployment of the new philosophical language, these scientists seized on the terms as a way of articulating a turn toward epistemology and away from the metaphysics of truth-to-nature in science, in response to the ever-quickening pace of scientific advance in the first half of the nineteenth century.

By the mid-nineteenth century, dictionaries and handbooks in English, French, and German credited Kantian critical philosophy with the resuscitation and redefinition of the scholastic terminology of the objective and the subjective. Words that were once enmeshed in the realism versus nominalism debate of the fourteenth century and that had by the eighteenth century fallen into disuse except in a few treatises in logic were given a new lease on life by Kantian epistemology, ethics, and aesthetics. From the mid-seventeenth century, when Descartes still used the word *objectif* in the Scholastic sense, to refer to "a concept, a representation of the mind," to the early nineteenth century, when dictionaries began to define "objective" and its cognates as "a reality in itself, independently of knowledge," the words underwent both a 180-degree flip in meaning and a steep rise in popularity.[24] By the mid-nineteenth century, the words "objectivity" and "subjectivity" appeared, now in their substantive as well as adjectival and adverbial forms, in most dictionaries in the major European languages, often with a bow in the direction of "German philosophy."[25] When Sir Charles Lock Eastlake, the director of the National Gallery in London, translated Johann Wolfgang von Goethe's *Zur Farbenlehre (On Color Theory*, 1810) into English in 1840, he noted on the first page: "The German distinction between *subject* and *object* is so generally understood and adopted, that it is hardly necessary to explain that the subject is the *individual*, in this case the *beholder*; the object, *all that is without him*."[26]

Yet many commentators who seized eagerly upon the new/old pair "objective"–"subjective" felt that the terms did indeed require a careful and thorough explanation. Although Kant was almost universally credited with making them ubiquitous, their definitions and usage, even in philosophical and scientific circles, sometimes diverged as sharply from Kant's own as they did from medieval scholastic meanings. G.W.F. Hegel tried to sort out the confusion in his *Enzyklopädie der philosophischen Wissenschaften im Grundrisse (Encyclopedia of the Philosophical Sciences in Outline*, 1830). He pointed out that while in vernacular German the objective had now come to mean "that which is external to us and which reaches us through external perception," Kant had called "thought, more specifically the general and the necessary, the objective, and mere sensation [*das nur Empfundene*], the subjective."[27] Hegel here put his finger on the paradox of the reception of Kant's distinction between the objective and the subjective. Although mid-nineteenth-century writers — philosophers, scientists, mathematicians, novelists — found the terms irresistible, in part because of associations with Kantian profundities, they drew the boundaries between the objective and the subjective in starkly contrasting ways: between the mind and the world, the certain and the uncertain, the necessary and the contingent, the individual and the collective, the *a priori* and the *a posteriori*, the rational and the empirical. Depending on whether one read one's Kant through the philosophical lens of Samuel Taylor Coleridge's Francis Bacon or Claude Bernard's René Descartes, the German Idealist Johann Gottlieb Fichte or the French eclecticist Victor Cousin, the British polymath William Whewell or the French positivist Auguste Comte, the crucial distinction shifted its position and its import.

What was never lost in this linguistic meandering was the epistemological provocation Kant had intended in the original distinction between the "objectively valid" and the "merely subjective." In the *Kritik der reinen Vernunft* (*Critique of Pure Reason,* 1781, 1787), Kant had attacked the sensationalist philosophy of the Enlightenment as an inadequate account of knowledge, both of the world and of the self. John Locke and his successors had argued that all knowledge derived from sensation and reflection on sensation. Even the knowledge of oneself, personal identity, stemmed from the sensations represented by imagination and memory in consciousness. Kant countered that sensations alone could never cohere into an object, much less a concept. Without, for example, the *a priori* intuitions of space and time, there would be no genuine experience, only a chaos of disconnected sensations — red; loud; pungent; painful. These intuitions and, more generally, pure concepts of the understanding were therefore the "conditions of a possible experience. Upon this ground alone can their objective reality rest." Sensations such as color or odor may vary among individuals or even for the same individual under different conditions; these are artifacts of the "subjective construction" of the sense organs. Because, in contrast, the rule that every object is experienced as being in space and time countenances no exceptions, it is therefore "objectively valid."[28]

Unlike sensations, objectively valid concepts are emphatically not psychological. Nor are they metaphysical, however: the fact that all experience must be framed, for example, by causality says nothing about the ultimate reality that may or may not correspond to the representations of experience. No effort of reason, no matter how titanic, will ever reveal the essence of things in themselves, at least as they exist external to us. Kant may have discredited sensationalist philosophy as merely subjective, the stuff of psychology, but the objective validity he opposed to the "merely subjective" did not aspire to metaphysics; it was instead firmly and permanently positioned at the level of epistemology.

Kant argued that experience presupposed a certain structure of consciousness as well as of the world as represented to consciousness. Without a unified consciousness, it would not be possible to experience unified objects. What Kant called the "transcendental unity of apperception" forged helter-skelter sensations into a single, unified representation of an object, which underlies all empirical knowledge of "objective reality."[29] This was a radical break with Enlightenment sensationalist philosophy, which had envisioned the mind as a loose confederation of mental faculties more or less subordinated to reason and the integrity of the self as guaranteed by no more than the continuity of consciousness. According to the sensationalists, objects cohered and events were connected by mere juxtaposition and contingent associations, as in David Hume's analysis of causality as no more than constant conjunction. Kant's unification of the self as the necessary condition for the possibility of all "objective" knowledge was not only an alternative vision of mind but also an alternative vision of knowledge. Experience ceased to be purely sensational; it presupposed certain "transcendental" conditions that were prior to all experience.

Kant generally reserved the adjective "objective" (the substantive form appears only rarely in his critical writings) for universal and *a priori* conditions, and identified the "subjective" with the psychological or "empirical," in the sense of the empirical sensations of Enlightenment epistemology. Objective validity is determined by the necessary and universal conditions of understanding, not by the nature of things in themselves: "The object itself always remains unknown; but when by the concept of the understanding the connection of the representations of the object, which are given by the object to our sensibility, is determined as universally valid, the object is determined by this relation, and the judgment is objective."[30] Consciousness itself partook of both objective and subjective validity: the transcendental unity of apperception that fused manifold sensations into the concept of an object was "objectively valid," but the empirical unity of apperception (for example, one person's particular association of oboes with Alpine meadows) "has only subjective validity."[31] There was therefore no way to map the Kantian distinction between the objective and the subjective in any straightforward fashion onto that between the body and the soul or between the mind and the world.

The distinction between the objective and the subjective played a key role in Kant's ethics, as well as in his epistemology. The self of sensationalist psychology had been conceived as largely passive, imprinted by both external (sensations) and internal (pleasure and pain) impressions as soft wax is by a seal, to use a favorite metaphor of Locke and his followers. Overcoming this natural passivity was understood by Enlightenment thinkers as both a moral and an intellectual imperative, a gauntlet thrown down to reason to assert its control over insubordinate faculties — memory, imagination, the will and the appetites — in order to act upon the world rather than be acted upon. In contrast, the Kantian moral self was monolithic and tightly organized around the will, posited as free and autonomous (literally, "giving the law to itself"). Insofar as the will had to overcome internal obstacles, these were not rival faculties but the will itself: the "objective" side of the will, determined by the imperatives of practical reason valid for all wills, had to bridle its "subjective" side, which was responsive to the psychological motives of a particular individual.[32] Only the "good" will, which acted solely in accord with the "objective laws" ordained by reason, was genuinely autonomous; insofar as the will was also swayed by personal inclinations and interests, "as it really is with humans," it remained less than free.[33]

The mid-nineteenth-century appropriation of the Kantian terminology of objective and subjective in *science* tended to fuse the epistemological and ethical: the acquisition of knowledge was seen — and felt — to involve a battle of the will against itself. This is not to say that the epistemological was submerged into the ethical; however variously scientists interpreted the objective and the subjective, they all used the two words to identify an epistemological problem, and one very different from those that had preoccupied their predecessors in the seventeenth and eighteenth centuries. However un-Kantian scientists were in applying their Kantian language, they remained true to Kant's own militantly epistemological program. Objectivity was a different, and distinctly epistemological, goal — in contrast to the metaphysical aim of truth. And subjectivity was not merely a synonym for being prone to errors; it was an essential aspect of the human condition, including the pursuit of knowledge. But for mid-nineteenth-century scientists, this epistemological predicament was hopelessly entangled with an ethical one that was also cast in terms of the objective and subjective. To know objectively was to suppress subjectivity, described as a post-Kantian combat of the will with itself — what Schopenhauer called the will to willessness.

There was no standard scientific assimilation of Kantian terminology or philosophy. Instead, its reception was colored by indigenous philosophical traditions, disciplinary preoccupations, and individual interests. Moreover, nineteenth-century scientists stretched Kantian notions to fit new research and even new disciplines undreamed of by Kant and his contemporaries. Helmholtz, Bernard, and Huxley stand for the diversity (and creativity) of possible interpretations, but they also represent the convergent scientific dilemmas to which such interpretations were applied. They and many of their colleagues understood their specific disciplines, and indeed science as a whole, to be in a state of crisis brought on by its own advances. Scientific progress in the mid-nineteenth century struck contemporaries as faster, more violent, and less continuous than in previous generations. The headlong pace of scientific progress experienced within a single lifetime seemed to threaten the permanence of scientific truth. Scientists grasped at the new conceptual tools of objectivity and subjectivity in an attempt to reconcile progress and permanence.

In the seventeenth and eighteenth centuries, Bacon, d'Alembert, Jean-Antoine-Nicolas de Condorcet, and other reforming philosophers had contrasted the dynamic advance of modern natural knowledge with the stasis of ancient learning. But they had understood progress as expansive rather than revolutionary. New domains would be conquered — botany, chemistry, even the moral sciences would eventually find their Newtons — but old citadels — celestial and terrestrial mechanics, optics — would remain forever secure. Even Adam Smith's remarkable history of astronomy, which treated systems of natural philosophy "as mere inventions of the imagination, to connect together the otherwise discordant and disjointed phaenomena of nature," concluded with a tribute to the Newtonian system, "the most universal empire that was ever established in philosophy."[34] Between *circa* 1750 and 1840, a steady stream of histories of various sciences poured from the presses, all purporting to demonstrate the existence and extent of progress in those disciplines.[35] To continue Smith's imperialist metaphor, new territories awaited scientific conquest, but old victories remained forever safe from reversal.

Hence the British astronomer and physicist Sir John Herschel could, in 1830, still optimistically gesture toward "the treasures that remain" for the post-Newtonian natural philosopher to gather, without any hint that new treasures might devalue or replace the old. However unexpected, the new discoveries and principles would mesh smoothly with the old into "generalizations of still higher orders," revealing "that sublime simplicity on which the mind rests satisfied that it has attained the truth."[36] Discoveries accumulated; generalizations endured.

By the mid-nineteenth century, this mood of serene optimism had been ruffled by the very successes of science. It is difficult to date just when the perceived progress of science accelerated to the point of causing vertigo for its practitioners. Already in 1844, the German naturalist Alexander von Humboldt concluded the preface to his monumental *Kosmos* with a disquieting reflection on transitory science and enduring literature: "It has often been a discouraging consideration, that while purely literary products of the mind are rooted in the depth of feelings and creative imagination, all that is connected with empiricism and with fathoming of phenomena and physical law takes on a new aspect in a few decades, due to the increasing exactitude of the instruments and gradual enlargement of the horizon of observations; so that, as one commonly says, outdated scientific writings fall into oblivion as [no longer] readable."[37] Humboldt consoled himself with the familiar credo that many parts of science had, like celestial mechanics, already reached a "firm, not easily shaken foundation," and in 1867 the French astronomer Charles Delaunay declared that it was "impossible to imagine a more brilliant proof" for Newtonian astronomical theory than the discovery of the planet Neptune.[38] But by 1892, the French mathematician and theoretical physicist Henri Poincaré was calling for ever-more-precise techniques of approximation in order to test whether Newton's law alone could explain all astronomical phenomena.[39] Even celestial mechanics, that most secure of scientific bastions, was under siege.

Poincaré was caught up in what the American historian Henry Adams in 1907 called, with a shudder, the "vertiginous violence" of late nineteenth-century scientific progress. Theories succeeded one another at an ever-accelerating pace; facts pointed to contradictory conclusions. There was no firm theoretical ground safe from such upheavals: even celestial mechanics had begun to quake. The history of science would not stay written. At any moment, a theory solemnly pronounced dead might be revived, as befell the wave theory of light in the 1820s.[40] The expectations for scientific progress voiced in the early nineteenth century had not been disappointed; rather, they had been fulfilled with a vengeance. Never before had science bustled and flourished as it did in the latter half of the nineteenth century. Scientists multiplied in number, and with them new theories, observations, and experiments. But scientists themselves seemed sickened by the speed of it, and to have lost their bearings and their nerve. As Adams remarked of his scientific reading: "Chapter after chapter closed with phrases such as one never met in the older literature: 'The cause of this phenomenon is not understood'; 'science no longer ventures to explain causes'; 'the first step towards a causal explanation still remains to be taken'; 'opinions are very much divided'; 'in spite of the contradictions involved'; 'science gets on only by adopting different theories, sometimes contradictory.'"[41] It was in this atmosphere of metaphysical caution and acute awareness of the brief life spans of scientific theories (now often demoted to the status of "hypotheses") that scientists in the mid- and late nineteenth century reworked the Kantian terminology of objective and subjective.

The first and second waves of nineteenth-century positivism, launched by the writings of Auguste Comte and Ernst Mach, put scientists on their guard against rash declarations of metaphysical allegiances by pointing a warning finger toward the large and growing graveyard of discarded theories.[42] Even scientists who were critical of the positivists, as Huxley, Helmholtz, and Bernard all were, took a wary view of anything that smacked of ultimate metaphysical commitment. All three repeatedly warned that science could provide knowledge only of empirically derived natural laws, not of the ultimate nature of things. Huxley attributed the progress of modern science to an exclusive concentration on "verifiable hypotheses," regarded "not as ideal truths, the real entities of an unintelligible world, behind phenomena, but as a symbolical language, by the aid of which Nature can be interpreted in terms apprehensible to our intellects."[43] Helmholtz read the lesson of sensory physiology as applied to spatial perception as a refutation of the logical necessity of Euclidean geometry (and hence of one of Kant's allegedly *a priori* forms of intuition): no truth claim, even in mathematics, was immune from subversion by further empirical research.[44] Every theory was provisional, Claude Bernard cautioned. Scientific progress might be likened to the ascent of a high tower whose pinnacle could never be reached: "Man is made for the search for truth and not for its possession."[45]

It was against this common background of metaphysical restraint that Bernard, Huxley, and Helmholtz construed the terms "objective" and "subjective." They understood them differently from one another and from Kant, but they all used them to sort out what — if not truth — science might be about. The young Huxley, full of autodidactic zeal ("History (every morning) — Henry IV, V and VI. Read Abstract. German (afternoons) — Translate 'Die Ideale' — "), devised a classification of knowledge based on "two grand divisions" for the purpose of better organizing his studies:

> I. Objective — that for which a man is indebted to the external world
> and
> II. Subjective — that which he has acquired or may acquire by inward contemplation.[46]

Huxley assigned history, physiology, and physics to the first and metaphysics, mathematics, logic, and theology to the second, with morality straddling the divide. Bernard, assiduously working his way through Cousin's French translation of Wilhelm Gottlieb Tennemann's *Geschichte der Philosophie* (*History of Philosophy*, 1798–1819), summed up "philosophy since Kant" with the decidely un-Kantian conclusion that the "unique source of our knowledge is experience" and defined "objective knowledge" as "unconscious and as a consequence empirical," as opposed to the "rational and absolute knowledge" of relations supplied by mathematics and rational mechanics.[47] In his 1847 formulation of the principle of the conservation of energy, Helmholtz had followed Kant in the distinction between an "empirical rule" formed from subjective perceptions and an "objective law" of universal and necessary validity with respect to the unity of all forces in nature. But by the late 1860s, he had come to a considerably more agnostic view of the necessary reality of forces, as opposed to laws derived from observation.[48] Laws confronted the will as "an objective power";[49] whether the will could change a perception or not drew the boundary between the objective and subjective, a boundary discerned only by experience, not by *a priori* categories.[50]

Our point in presenting this small sampling of the ways nineteenth-century men of science turned the Kantian philosophical vocabulary of objective and subjective to their own purposes is twofold: first, to show that, although Kant undoubtedly cast a long shadow on subsequent intellectual history, his influence alone cannot explain the broad and branching ramification of the objectivity and subjectivity, both as words and as things; and second, to explain how that diffusion into the sciences followed channels cut not only by philosophy but also by the characteristic mid- and late nineteenth-century experience of ever-accelerating scientific change. This experience, much noted and commented on by contemporaries, led to an epistemological turn away from absolute truth (and indeed from all metaphysical ambitions) and toward objectivity. However variously objectivity was conceived in the sciences, it was consistently treated — in the spirit of the Kantian project, however divergent from the letter — as an *epistemological* concern, that is, as about the acquisition and securing of knowledge rather than the ultimate constitution of nature (metaphysics). It had been Kant's achievement to open up a space between epistemology and metaphysics and to set limits to the aspirations of reason with respect to the latter. This is why the nineteenth-century dictionary entries that gave thoroughly un-Kantian definitions of "objective" and "subjective" were nonetheless justified in tracing the lineage of the terms back to Kant.

Against this philosophical background, the scientists' submission to objective fact was clad in the somber language of duty. Huxley recommended universal education in science in part because it bent the will to inexorable natural laws, the "rules of the game of life."[51] Santiago Ramón y Cajal devoted an entire chapter of his *Advice to a Young Investigator* (1897) to "Diseases of the Will" among scientists; he reserved his sharpest criticism for "the theorist" who recklessly risked "everything on the success of one idea," forgetting how "many apparently conclusive theories in physics, chemistry, geology, and biology have collapsed in the last few decades!"[52] The same will that mapped the line between the subjective and the objective, which molded itself to the laws of nature, and which had to be subdued in order to become, in Huxley's words, "nature's mouthpiece," was also the essence of the self and the engine of its action in the world. Meekness and dynamism were supposed somehow to coexist in a single knowing self. Almost every aspect of the mid-nineteenth-century scientific persona was driven by this tension between humble passivity and active intervention with respect to nature. To appreciate the novelty of this persona, we must step back to survey it alongside its predecessors and successors.

Scientific Personas

Since *circa* 1700, every era has celebrated Isaac Newton as the epitome of the knower of nature, and the resulting verbal and visual portraits have been distinctive of their epochs.[53] For eighteenth-century eulogists, Newton was the scion of "one of the oldest and noblest families of the realm," his formulation of the law of universal gravitation "by far the greatest and most ingenious discovery in the history of human inventiveness," his health robust (he tended toward plumpness in old age) and his character sweet and affable, his intellect so sublime that admirers queried whether he ate, drank, and slept like other men or was "a genius deprived of bodily form."[54] (See figure 4.4.) Victorian biographers insisted that he came from good yeoman stock, led a "life that knew no ambition" and was "passed in serene meditation unruffled by conflict," solved great problems through "self-control in speculation, and his great-souled patience in the pursuit of truth," and embodied "a type on a large scale of what smaller humanity may be within its own range."[55] (See figure 4.5.) For mid-twentieth-century historians, his friendship with the young Swiss mathematician Fatio de Duillier smacked of narcissism, and his priority disputes with Gottfried Wilhelm Leibniz and other rivals were bitter and relentless, his mental health precarious, his approach to problems in mathematics and natural philosophy akin to "ecstasy" and "possession."[56]

We are not here concerned with the factual accuracy of these contrasting versions of the same life; it is, rather, their very elasticity, which allows a specific historical individual to be turned into a model of the prevailing scientific persona, that interests us. Neither treatises on ethics nor handbooks of scientific method, these portrayals are exempla of how nature should be investigated in the context of a life devoted to that end. The genre of works and lives is as old as Diogenes Laertius's collection of doctrines laced with legends and anecdotes about the philosophers of Antiquity.[57] But it is something of a surprise to encounter this genre in full vigor well after the seventeenth century, when new Cartesian doctrines of a split between knower and knowledge would appear to have made the lives of philosophers irrelevant to their works. Certainly the literary conventions of scientific publications as they gradually developed over the course of the eighteenth and nineteenth centuries do seem to have erased more and more of the author's personality and circumstances.[58] At the same time, other genres emerged and proliferated that reconnected lives and works in science: the academic eulogy, compendiums of scientific vitae, biographies and autobiographies of individual scientists, advice manuals for aspiring scientists, and psychological and medical studies of scientists as a professional group.

Each of these genres had its own conventions, modulated by nationality and period.[59] An eighteenth-century French academic *éloge* maps its subject onto a grid of neo-Stoic ideals; a nineteenth-century German autobiography narrates a scientific career as a *Bildungsroman*; a twentieth-century American advice manual includes tips on the efficient management of home and laboratory. Taken as a group, however, they testify to a growing recognition of a new type of intellectual, for whom new names began to be coined in the mid-nineteenth century: the scientist, *der Wissenschaftler, le scientifique*. (See figures 4.6, 4.7, and 4.8.) This was a persona marked out by a certain kind of character, as well as by qualifications in a particular branch of knowledge. Although the scientific persona was distinctive — and, by the mid-nineteenth century, the object of a substantial literature devoted to documenting its distinctiveness — it was framed by coeval conceptions of the self in general. Here the history of the larger genus and scientific species can be only sketched from a few representative examples from the eighteenth and nineteenth centuries. Juxtaposed, they nonetheless suffice to reveal stark contrasts in personas correlated with equally marked divergences in ethical-epistemic ways of life that frame the specific scientific practices associated with truth-to-nature and mechanical objectivity.

The Enlightenment self was imagined as at once a pastiche and a conglomerate. It was a pastiche of sensations and the traces they left in memory, combined by the principles of association and held together by the continuous thread of consciousness. It was a

Fig. 4.4. Newton Deified. "An Allegorical Monument to Sir Isaac Newton," Giovanni Battista Pittoni the Younger, 1727–1729 (reproduced by permission of the Syndics of the Fitzwilliam Museum, Cambridge). This oil painting, commissioned by the Irish impresario Owen McSwiny in 1727, the year of Newton's death, shows an apotheosis of Newton, a man deemed semidivine by eighteenth-century admirers. A beam of light shoots over a huge urn holding Newton's remains and two allegorical figures representing Mathematics and Truth and splits into prismatic colors, commemorating Newton's famous 1672 experiment on the composition of white light. Knots of sages in classical garb study astronomical instruments and weighty tomes; in the foreground, an angel and Minerva, the goddess of wisdom, lead muselike figures to Newton's shrine. (Please see Color Plates.)

Fig. 4.5. Newton Domesticated. "Newton's Discovery of the Refraction of Light," Pelagio Palagi, 1827, Galleria d'Arte Moderna, Brescia (reproduced by permission of the Civic Museums of Art and History of Brescia, Italy). The same episode in Newton's scientific career as fig. 4.4 is here commemorated, but in an entirely different setting and mood. Newton is shown as a handsome young man, richly dressed, in a domestic scene with his sister and a little boy blowing bubbles (according to the instructions of Count Paolo Tosio di Brescia, who commissioned the painting). It is the homely detail of the iridescent sheen of the bubble, not a contrived experiment with a prism, that arrests Newton's attention and prompts his discovery. (Please see Color Plates.)

Fig. 4.6. Hermann von Helmholtz in High Society. "Salon of the Countess von Schleinitz on 29 June 1874," Adolf von Menzel, in Max Jordan, *Das Werk Adolf Menzels, 1815–1905* (Munich: Bruckmann, 1905), p. 76. Menzel's now lost pencil sketch of a famous Berlin salon shows Helmholtz (*far left*), in court dress, rather stiffly rubbing shoulders with aristocrats and statesmen. As Germany's most famous scientist in the latter half of the nineteenth century, Helmholtz enjoyed great prestige, a sign of the rise of a new elite of *Wissenschaftler* alongside the wealthy and the powerful in the German Empire.

Fig. 4.7. Claude Bernard at Work. "A Lesson of Claude Bernard," Léon Lhermite, 1889, Académie de Médicine, Paris (reproduced by permission of the Bibliothèque de l'Académie de Médicine, Paris). This large painting was originally commissioned for the Laboratoire de Physiologie at the Sorbonne, in Paris. Bernard is shown at work, holding a dissecting scalpel and wearing a white butcher's apron to protect his clothes from spattered blood, flanked by eager students (who are identified by name on the frame of the painting): they are depicted as laboring scientists, with rolled-up sleeves and dirty hands. Yet amid the gore Bernard also wears the red insignia of the Légion d'Honneur awarded to him by the French government. A scribe (*right*) takes notes on the proceedings, a habit institutionalized with the laboratory notebook. (Please see Color Plates.)

Fig. 4.8. Thomas Henry Huxley Plays Hamlet. "Thomas Henry Huxley," John Collier, 1883, National Portrait Gallery, London. The comparative anatomist Huxley is depicted with a skull, an emblem of his discipline, but his offhand pose also evokes Hamlet's musings on life and death in the "Alas, poor Yorick" speech. He projects the self-confidence of an intellectual qualified by both specialized expertise and general learning (symbolized by the books upon which his left arm rests) to pronounce on the great issues of the day, from evolution to ethics to education. (Please see Color Plates.)

conglomerate of faculties, chief of which were reason, memory, and imagination. According to some widely held Enlightenment theories of mind, beginning with Locke's vastly influential *Essay Concerning Human Understanding* (1690), the self was "that conscious thinking thing," rather than some unknowable substance, whether immaterial (the soul) or material (the body): "a thinking intelligent Being, that has reason and reflection, and can consider it self as it self, the same thinking thing in different times and places; which it does only by that consciousness which is inseparable from thinking, and as it seems to me essential to it."[60] This was a self constantly menaced by fragmentation — so much so that some eighteenth-century philosophers, most notably Hume, wondered whether the sense of having a coherent self might not be illusory. Perhaps, Hume mused, personal identity was "nothing but a bundle or collection of different perceptions, which succeed each other with an inconceivable rapidity, and are in a perpetual flux and movement."[61]

On the one hand, gaps in memory or interruptions of consciousness could fission the self. Locke and his eighteenth-century readers toyed with the idea that not only amnesia but also drunkenness and even sleep might split the self.[62] On the other, the inferior faculties, most particularly the imagination, might revolt against the rule of the superior faculty of reason, causing "alienation" of the self from itself and, in extreme cases, madness.[63] The French philosopher Etienne Bonnot de Condillac went so far as to assert that all madness was due to "an imagination which, without one being able to notice it, associates ideas in an entirely disordered manner," from which everyone was potentially at risk, a power "without limits."[64] Disruptions of consciousness and warring mental faculties reinforced one another: without the metaphysical guarantee of identity and integration that had been provided in Scholastic psychology by the rational soul, there was no single, overarching framework that encompassed all the disparate aspects of mental life.[65]

During the Enlightenment, these threats to the coherent self were experienced as well as theorized. As the continuity of consciousness and memory came to replace the soul as the definition and expression of the self, introspection seemed to reveal fluid, tattered, and even contradictory identities. The French moral philosopher Charles-Louis de Secondat de Montesquieu likened the self to a spider at the center of a web of sensations and memories; should the web be torn, identity is annihilated — an image echoed by Diderot, as we have seen, to make much the same point about the "*moi*," which he saw as held together only tenuously and temporarily.[66] The Göttingen physicist Georg Christoph Lichtenberg marveled over the plurality of selves embraced by his own memory: "As long as memory holds, a group of people work together in one [person], the twenty-year-old, the thirty-year-old, etc."[67]

From the standpoint of specifically scientific virtues and vices, the Enlightenment self was susceptible to several kinds of temptation. Insufficient experience, compounded by inattention, impatience, and inexactitude, could spoil observations. An anomaly might be mistaken for the true type of nature, or a fluctuation for the constant cause. Just as moral responsibility for one's past actions depended on remembering them — connecting past and present selves — scientific responsibility for one's observations depended on recording and synthesizing them. These were the external temptations of inchoate, incomplete, and undigested sensations. A different sort of temptation waylaid the savant from within the mind. Reason might succumb to the blandishments of the imagination, that "coquette" who aimed primarily at pleasure, rather than at truth.[68] Imagination could substitute fanciful but alluring systems for genuine impressions derived from memory and sensation. Vanity seduced natural philosophers into abandoning reality for systems wrought by their own imaginations.

To read contemporary accounts of the lives and works of Enlightenment savants, whether in official academy eulogies or in novels, is to glimpse a world in which the finest minds were thought to be in constant peril of mistaking their own theoretical systems for nature. The image of castles in the air, magnificent but insubstantial, recurs in scientific censures of deluded systematists, the Don Quixotes of science. The naturalist Georges Cuvier, for example, excoriated his colleague Jean-Baptiste Lamarck for his transformationist theory of organic development, calling it one of those "vast edifices [constructed] upon imaginary foundations, resembling those enchanted palaces of our old novels that can be made to vanish by breaking the talisman upon which their existence depends."[69] Samuel Johnson's novel *Rasselas* (1759) describes how a learned and virtuous astronomer, the victim of a "disease of the imagination" exacerbated by religious sentiments of guilt, succumbed to the oppressive illusion that he and he alone could control the world's weather.[70] Condillac warned of how an abstract system in science and philosophy could "dazzle the imagination by the boldness of the consequences to which it leads."[71] (See figure 4.9.)

All these temptations stemmed from the loose organization of the Enlightenment sensationalist self and its precarious guarantees of coherence. Breaks in consciousness, lapses in memory, unruly imagination, and childhood suggestibility might all conspire to erase or distort the impressions left by experience in the fibers of the brain. The very passivity that allowed the *tabula rasa* of the mind to be imprinted by sensations also left it prey to the false ideas implanted by custom and education and to the fabrications of an inventive imagination. This self was imagined as permeable, sometimes too permeable, to its milieu, a self characterized by receptivity rather than assertive dynamism. The scientific vices to which this self was prone were a supine acquiescence to intellectual authority, surrender to the equally passive pleasures of the imagination, and insufficient care in the making, storing, and sifting of observations.

The characteristic set of eighteenth-century practices that arose to combat these temptations was as moralized as those that later surrounded scientific objectivity, but it was moralized in a different way. Habit was the shield of the virtuous, reflecting an ethics more closely linked to regimen and hygiene than to the exercise of the will.[72] In the view of Enlightenment savants, the will was, in any case, of limited efficacy in combating these epistemological vices of the impressionable self. Early education worked on the child's mind (and body) before the will could resist; in adulthood, defiance of intellectual authority depended more on penetrating critical faculties and courage than on the resolved will. As for imagination, Voltaire emphasized that neither its passive nor its active variety could be controlled by the will: "It is an interior sense that acts imperially; hence nothing is more common than to hear it said that *one is not master of his imagination*."[73] Reason and judgment alone could counter authority and rein in imagination. The will was neither the principal problem nor the solution in the quest for truth-to-nature in natural philosophy.

IMAGINATION.

Fig. 4.9. Emblem of the Imagination. Jean-Baptiste Boudard, *Iconologie tirée de divers auteurs* (Vienna: Jean Thomas de Trattnern, 1766), p. 103. Imagination is here pictured in a pose of lax passivity, her hands folded in her lap and her gaze turned inward. The wings at her temples represent the speed with which she forms images. Enthralled by the pageant going on in her mind's eye (the little figures that crown her head), she is, like the savants who lost themselves in their own systems, oblivious to the external world of experience.

Reason and judgment were also called upon to make sense of sensation, much as they were invoked to tame the imagination. Savants compared and synthesized innumerable *faits particuliers* into a stable *fait général*, which might be the type of a botanical species or anatomical organ, or the rule that unified a crowd of apparently unrelated and capricious observations about electrical or magnetic phenomena.[74] These general facts had been "carefully stripped of all extraneous circumstances" through a long series of comparative observations and experiments.[75] The French physicist Charles Dufay, for example, brought order to the bewildering field of luminescence by doing a meticulous series of experiments on substances ranging from oyster shells to diamonds, the results of which he frankly and severely pruned in order to arrive at a stable generalization. Although his manuscript notes reveal an exquisite sensitivity to the nuances and variability of luminescent phenomena, Dufay's memoirs on his experiments, published in 1730, summarized, smoothed, and omitted results, "to avoid tedious detail."[76] These judicious omissions were the textual equivalent of the visual practices deployed by Linnaeus and Réaumur, Goethe and Soemmerring, to create the reasoned images of true-to-nature atlases: schematized leaves, symmetrical insects, archetypical plants, and idealized bodily organs.

From the standpoint of both the psychological integrity of the self and the epistemological integrity of the scientific object, reason must rule with a firm hand. As Diderot had Dr. Bordeu remark in *Le rêve de d'Alembert*, the organization of a sound mind is despotic; passion and delirium correspond to anarchy, "a weak administration in which each subordinate tries to arrogate to himself as much of the master's authority as possible." Sanity is restored only if "the real self" (*cette partie qui constitue le soi*) can reassert its authority.[77] King Reason must discipline insubordinate faculties.

When, *circa* 1800, this view of the self as a fractious monarchy collided with the new Kantian views of a self unified around the will, the shock of the impact sent heads spinning. After the disastrous five-hour seminar on Kantian philosophy that he gave to a circle of prominent *philosophes* in Paris in May 1798, the Prussian philologist Wilhelm von Humboldt (the elder brother of the naturalist Alexander) wrote to Friedrich von Schiller in utter frustration:

To understand one another is impossible, and for a simple reason. Not only do they [the French] have no idea, not the slightest sense, of something beyond appearances; pure will, the true good, the self, the pure conscious of oneself, all of this is for them totally incomprehensible....They know no other [mental] operations except sensing, analyzing, and reasoning. They don't think at all about the way in which the feeling of oneself originates and don't admit that they here leave the limits of our reason.[78]

Humboldt would no doubt have been still more horrified had he read Samuel Taylor Coleridge's presentation of Kant to English readers, who were assured that the German philosopher was primarily a logician in the tradition of Aristotle and Bacon and that the *Critique of Pure Reason* would have been better titled (with a nod to Locke) "An Inquisition respecting the Constitution and Limits of the Human Understanding."[79] Yet, as in the case of the Kantian epistemology of the objective and the subjective, notions of a unified self and the supremacy of will were gradually modified and adapted to local needs and conceptions well beyond those of Kant's Königsberg. Had Humboldt returned to Paris four decades later, he would have found that not only philosophers but also most well-educated bourgeois males were wholly persuaded that they were in possession of a monolithic self, defined by an indomitable will, thanks to the institutional success of Cousin's doctrines. There was nothing mysterious about this transformation, striking though it was. A generation of French schoolboys were taught, in a philosophy curriculum coordinated by Cousin himself, to introspect and to identify their "*moi*'s" with the assertion of active will. Their cultivated individualism and voluntarism may seem diametrically opposed to self-effacing objectivity, but, in fact, subjectivity and objectivity defined poles of the same axis of the will: the will asserted (subjectivity) and the will restrained (objectivity) — the latter by a further assertion of will.[80] In Jena and Paris, London and Copenhagen, new ideals and practices of the willful, active self took shape in the middle decades of the nineteenth century.

This new conception of the self left its traces in the literature of the scientific persona. Science was no longer the rule of reason but the triumph of the will: "Much of the success in original scientific research depends on the will," wrote the author of an 1878 British guide to research methods in physics and chemistry.[81] It was the will that steeled the man of science to face the drudgery of hard intellectual labor, and it was the will that enforced the self-discipline so necessary to "the strong central authority in the mind by which all its powers are regulated and directed as the military forces of a nation are directed by the strategist who arranges the operations of a war."[82] Without a resolute will, wrote the English literary historian George Craik, the author of the well-titled *Pursuit of Knowledge under Difficulties* (1845), Newton would never have had the "self-denial, more heroic than any other recorded in the annals of intellectual pursuit," to shelve his theories when they seemed to be contradicted by the best available data on the size and shape of the earth.[83]

Will brought in its train the other dutiful virtues of patience and industry — indeed, scientific genius was nothing more than a magnification of these qualities. Take the Victorian moralist Samuel Smiles's 1869 portrait of Newton: "Newton's was unquestionably a mind of the very highest order, and yet, when asked by what means he had worked out his extraordinary discoveries, he modestly replied, 'By always thinking unto them.'... It was in Newton's case, as in every other, only by diligent application and perseverance that his great reputation was achieved."[84] Particularly in the natural sciences, the unceasing labor of individuals was understood as constituitive of the careful, empirical methods of science as a whole, in contrast to the genial inspirations of art or the dogmatic assertions of philosophy.[85]

Praise for the slow, painstaking work of scientific investigation over the lightning flashes of genius became a topos of scientific biography and autobiography in the latter half of the nineteenth century. In 1880, the French science popularizer Gaston Tissandier praised science as an exercise in patience and perseverance: "Let us listen to Newton, who will tell us that he made his discoveries in 'thinking always about them.' Buffon will cry out: 'Genius is patience.' All will speak the same language. Work and perseverance are their common motto."[86] Smiles claimed that the scientific achievements of the British chemists Sir Humphrey Davy and Michael Faraday had been realized "by dint of mere industry and patient thinking."[87] Helmholtz confessed that the ideas his admirers praised as sudden strokes of brilliance were in fact developed "slowly from small beginnings through months and years of tedious and often groping work from unprepossessing seeds."[88] Charles Darwin declared in his autobiography that although he had "no great quickness of apprehension or wit which is so remarkable in some clever men," his "industry has been nearly as great as it could have been in the observation and collection of facts."[89] Huxley preached that an educated man's body should be "the ready servant of his will" and his mind "ready, like a steam engine, to be turned to any kind of work."[90]

The doctrine of science as endless work, fueled by an unflagging will, of course echoed the platitudes of industrializing economies, and in some cases the analogy between labor in the laboratory and labor in the factory was made literal.[91] What interests us here, however, is exactly the tension between the humdrum, mechanical associations of work on the shop floor and the rather more elevated self-image of the man of science, intent on winning respect and remuneration at least comparable to that accorded to the well-established liberal professions and, in certain instances, cultural authority equal to or greater than that enjoyed by either the clergy or men of letters.[92] Why would ambitious men of the stamp of Huxley, Bernard, and Helmholtz, eager to climb the social and intellectual ladder, invite comparisons to anonymous workers and even machines? Other would-be elites anxious about déclassement in this period (for example, medical specialists and insurance actuaries) instead emphasized the ineffable tact that guided their decisions, thereby laying claim to gentlemanly status.[93] What did men of science think was ennobling about a self without subjectivity, a will without willfulness?

The key to this paradox lies in the element of sacrifice and self-denial that figured so prominently in mid-nineteenth-century scientific biographies and autobiographies. It was the distance between the brilliant and impetuous speculator and the patient drudge that measured the willpower required to hold the will in check. The prototypical men of science were not portrayed as by nature meek and mild, born for the yoke and the treadmill. They were (as the physicist John Tyndall wrote of Faraday) men of energetic, even fiery temperament.[94] Craik reserved his highest praise for Faraday's "singular combination...of the most patient vigilance in examination, and the most self-denying caution in forming his conclusions, with

the highest originality and boldness."[95] Pearson, who himself struggled to reconcile his creed of self-denial with a cultish devotion to
individualism, also singled out Faraday for special praise, as a scientist strong enough to strangle his brainchild "in silence and secrecy
by his own severe criticism and adverse examination."[96] A hero must
do battle with a worthy foe, and it was themselves whom the heroes
of objectivity met upon the field of honor. It was precisely because
the man of science was portrayed as a man of action, rather than as
a solitary contemplative, that the passive stance of the humble acolyte of nature, who (as Bernard put it) listens patiently to her answers to his questions without interrupting, required a mighty effort
of self-restraint.

In the mid-nineteenth-century literature of the scientific persona, this effort always came at the moment when the investigator
was on the brink of imposing a hypothesis upon the data. In his per-
fervid 1848 vision of science as religion, the French philologist
Ernest Renan invoked the "courage to abstain": "The heroes of
science are those who, capable of higher things, have been able to
forbid themselves every philosophical anticipation and resign themselves to be no more than humble monographers, when all the instincts of their nature would have transported them to fly to the
highest peaks."[97] To embrace mechanical objectivity was to turn the
will inward upon itself, a sacrifice vaunted as the annihilation of the
self by the self, the supreme act of will — as Percival Lowell experienced his decision not to retouch his photographs of Mars and Funke
viewed his refusal to idealize crystalline forms, as we saw in Chapter
Three.

This psychodrama of objectivity and subjectivity followed a notably different plot from the Enlightenment struggles of reason against
the seductions of the imagination. The savant who succumbed to the
counterfeit charms of a beautiful but false system thereby retreated
into the innermost recesses of the mind and deliberately shut out
reason and experience, as the infatuated lover rejects wise counsel
and common sense. His fault was too much passivity rather than not
enough. The scientist who imposed a hypothesis on the yielding data
had, in contrast, charged, not retreated. Only an act of iron will
could achieve the passivity that Schopenhauer had held up as the end
of all restless striving and the condition for knowledge — and that

231

Nietzsche had scorned as the self-mutilation of intellectual asceticism among "scientific" historians: "What, the religions are dying out? Just behold the religion of the power of history, regard the priests of the mythology of the idea and their battered knees! Is it too much to say that all the virtues now attend on this new faith? Or is it not selflessness when the historical man lets himself be emptied until he is no more than an objective sheet of plate glass?"[98] Nietzsche had caught the same Christian resonances of humility and self-abnegation that Renan and others had discerned in the new ethos of objectivity, but he condemned them as at once unmanly and traitorous to the cause of truth. Only the sick will turned inward on itself.

What are we to make of this scientific portrait album, stretching across two centuries? As the pages turn, genius migrates from well-stocked memory to steely will, as the self is reconceptualized, first as a congeries of faculties, then as a will-centered monolith. Moral imperatives shift accordingly, to combat first the temptations of the imagination and then subjectivity. Quests for truth and quests for objectivity do not produce the same kind of science or the same kind of scientist. It is the integral involvement of the scientific self in the process of knowing that accounts for the interweaving of ethos and epistemology in all these historical episodes.

But are these portraits any more than self-serving fantasies that bear as little resemblance to real science and scientists as official court portraits do to their originals? What evidence can they provide about the ways science was actually done? Are they any more than collections of stereotypes and moral lessons? These would be well-founded objections if we intended to use these personas as reliable testimony in writing scientific biographies. Our interest in them is, however, precisely *as* historically specific stereotypes and moral lessons. A stereotype is a category of social perception, and a norm is no less a norm for being honored in the breach. Because epistemology is by definition normative — how knowledge *should* best be sought — there is no avoiding its dos and don'ts. Yet in the case of the lives of the learned, including scientists, bare treatises on method have never been deemed sufficient: the pursuit of knowledge is also a way of life, to be exemplified and thereby typified. From the eighteenth through the twentieth centuries, the literature on the scientific ways of life has drawn on biographies to give flesh and blood to moral and epistemological precepts, to teach one how to be a savant, a man of science, a *Wissenschaftler*. The fact that the very same exempla are made to serve opposite purposes — Helmholtz as dutiful plodder versus Helmholtz as intuitive discoverer — is what interests us.[99] The point is that a way of life — as opposed to a methodological maxim about running control groups or doing statistical-significance tests — must be demonstrated in order to be understood. The word must be made flesh. And exempla presuppose both types and regulative ideals.

The force of these regulative ideals was felt in the daily conduct of science. When, for example, Eduard Jaeger chose, in 1890, to devote forty or fifty hours of painstaking effort to each image of his atlas (as we saw in Chapter Three), he was self-consciously plumping for a particular kind of meticulous representation. He dismissed flights of genius in scientific representation as ephemeral. Only the suppression of all subjective idiosyncracy — even individual brilliance — could produce an objective image that would endure. A century earlier, Goethe had, with equally firm conviction and care, insisted on the insight and synthetic judgment required to detect the idea in the observation. Both Goethe and Jaeger took considerable pains to uphold the highest standards of epistemic virtue, even if both — necessarily — fell short of realizing their ideals. Goethe did not fathom nature's archetypes, any more than Jaeger turned himself into a machine. But the very act of striving for truth-to-nature or mechanical objectivity can change science and self, even if these epistemic virtues, like all virtues, can never be fully realized.

Exempla and regulative ideals alone do not, however, bring selves into being. For a way of life to be realized, highly specific practices must be articulated and cultivated. In order to bridge precept and practice, our argument about the intrinsic connection between epistemology and self in general — and about the emergence of scientific objectivity along with a new kind of scientific self in particular — requires the further evidence of such technologies of the self. In the next section, we turn to one of the most central of all scientific practices, observation, and examine how its disciplines of attention simultaneously shaped scientific object and scientific self.

Observation and Attention

Observation is an enduring and essential scientific practice and is intimately bound up with the self of the observer. Observation trains and strains the senses, molds the body to unnatural postures, taxes patience, focuses the attention on a few chosen objects at the expense of all others, patterns aesthetic and emotional responses to these objects, and dictates diurnal (and nocturnal) rhythms that fly in the teeth of social convention. The practices of observation — the frozen pose of the field naturalist, the delicate manipulations of the microscopist, the observatory vigils of the astronomer, the lab-notebook jottings of the chemist — are genuine technologies of the self, often consuming more time than any other single activity. Starting in the seventeenth century, at the latest, scientific observation became a way of life. But it was not always the same way of life. The counterpoint of observation and scientific self, examined over long periods, tracks far-reaching modifications in both.

The challenge of sustaining a coherent, well-ordered self confronted Enlightenment savants in a form specific to their scientific aims and pursuits. Because they sought truth as the constants underlying fluctuating appearances — constants that could, in turn, be discerned only on the basis of prolonged investigation of a given class of phenomena — they relied heavily upon judgment exercised on the myriad impressions stored in memory. Keen senses, concentrated attention, patience, and exactitude were all required to perform reliable scientific observations, but an isolated observation, even one well made, was of no more use in synthesizing a truth about nature than an isolated impression was in forging a sense of self. In Condillac's famous philosophical thought experiment in which a statue endowed only with the sense of smell acquires human cognitive capacities one by one, the first sensation of the fragrance of a rose was insufficient to generate a sense of self; only after experiencing a number of odors that could be compared in memory did the statue become conscious of its continuity in time, of having a "*moi*."[100]

Similarly, a single observation could not reveal a truth. Nature was too variable; individual observations were always qualified by particular circumstances. Hence the importance of routinely replicating observations in eighteenth-century natural history: rarely an expression of distrust or skepticism, this practice was more often justified as necessary to stabilize the phenomenon and to extract the essential from the accidental. The Genevan naturalist Charles Bonnet, for example, urged his younger Italian colleague Lazzaro Spallanzani to repeat Bonnet's own observations and those of other naturalists before embarking on new research: "Nature is so varied that we cannot vary our trials too much."[101] The practiced observer surpassed the novice by the ability to form at a glance (coup d'oeil) "a distinct notion of the ensemble of all the parts" that captured the essence of an object or phenomenon, shorn of accidental variations.[102] Each new observation was hence a synthesis of past observations, just as the Linnaean leaf schemata discussed in Chapter One summarized observations of thousands of plant species. The integrity of the self, as well as that of scientific observations and the inferences drawn from them, depended on the continuity, exactness, and amplitude of memory.

Both forms of integrity were often safeguarded by the same practice: the keeping of a daily journal in which records of a life were kept side-by-side (sometimes on the same page) with a register of scientific observations, experiments, and reflections. Historians of eighteenth-century inner life have remarked upon the diary as an instrument of self-examination and self-consolidation, a thread connecting yesterday's self with that of today and tomorrow.[103] The day-by-day framing of one's impressions in an unbroken transcript of memory became the image of what it meant to have an intact self. When Hume sought to undermine the very idea of such a self, he did so by tearing leaves out of a metaphorical mental journal: "For how few of our past actions are there, of which we have any memory? Who can tell me, for instance, what were his thoughts and actions on the first of *January* 1715, the 11th of *March* 1719, and the 3d of *August* 1733?"[104] The self was conscious memory, and memory itself was organized like a diary. The diary was therefore more than an aide-mémoire; it shaped and spliced memories into a personal identity — or a scientific insight.

The most common such scientific records were weather diaries, kept by countless Enlightenment observers (including Locke), and the more elaborate natural-history journals, which attended to the return of swallows, the harvesting of crops, freezes and thaws, and a myriad other seasonal details of country life.[105] Scientific diaries of observations might be kept in separate notebooks from those reserved for more personal entries, as in the case of Lichtenberg's *Waste Books* and *Diaries*, or the Bern anatomist Albrecht von Haller's diary of religious soul-searching and his travel journals of the scientific capitals of Europe.[106] But sometimes the line between the two sorts of diaries blurred. On August 13, 1771, for example, Lichtenberg confided to his diary in desperation that he was beset by "terrible thoughts.... Heart head and all are infected, where shall I go?"; he also methodically noted that the barometer stood at 27″ 2‴ (according to the Paris measurement scale) at 7:00 AM after a bad storm.[107] (See figure 4.10.)

In the case of weather and natural-history diaries, the diurnal rhythms of the observer were intertwined with the observations, and the observation of self was often inseparable from the observation of nature. Even if recorded impressions could not be molded into a narrative, the bare act of transcription ensured the continuity of memory and thus the integrity of the self. When Haller faced the possibility of death, he equated the extinction of self with the emptied contents of memory: "Alas! My brain, that will soon be nothing but a bit of earth! I can hardly bear the idea that so many ideas accumulated in the course of a long life must be lost like the dreams of a child."[108]

Enlightenment savants struggled with fragmented and impressionable selves, and ministered to them with journals and regimens. But along with incoherent selves, they confronted the further epistemological problem of incoherent scientific objects. The risk of fragmenting the object paralleled that of fragmenting the self in sensationalist psychology; indeed, both stemmed from the same cause: a flood of disordered and divergent sensations registered pell-mell. The unity of the scientific self depended on memory and reason; the unity of the object of scientific observation, on the exercise of attention. Just as the private journal helped memory to guarantee the continuity and coherence of the self over time, the observational journal came to the aid of sensation in preserving the coherence of the scientific object. Attention, conceived as both a mental capacity and a scientific practice, fused myriad impressions into unified and representative objects of inquiry.[109]

Like the experiment, scientific observation has a history, with its

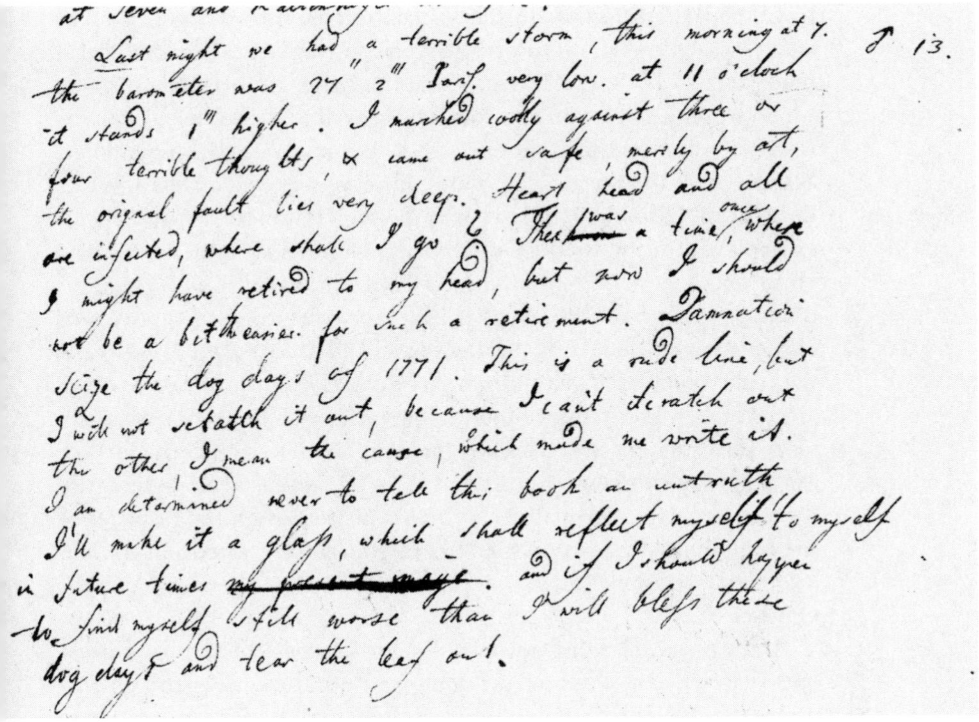

Fig. 4.10. **Storms of the Soul.** Lichtenberg's Diary, Aug. 13, 1771, Cod. Georg Christoph Lichtenberg 4, 7 (courtesy of Staats- und Universitätsbibliothek Göttingen). Lichtenberg kept separate journals: *Wastebooks* for ideas and experiments and diaries for personal experiences. But the habits of external and internal observation — and the noting down of both in entries arranged by date — were similar, as this entry recording an emotional crisis and the day's meteorological data testifies.

own record of specialized methods, instruments, and sites gradually devised and diffused. Eighteenth-century naturalists were keenly aware of the novelty of many of their techniques of observation, from the field notebook to the microscope to the tabular display of data (see figure 4.11). Observation was not only practiced but also theorized; in 1768, the Dutch Society of Sciences in Haarlem sponsored a competition on the "art of observation."[110] One of the resulting essays, by the Genevan pastor and naturalist Jean Senebier, became perhaps the best-known eighteenth-century treatment of the subject, although it did not win the society's prize.[111] Drawing on examples from the work of the most celebrated scientific observers of the age — Newton, Jan Swammerdam, Abraham Trembley, Haller, Bonnet, Spallanzani, Réaumur — Senebier celebrated the "genius of observation," which was marked by a well-stocked mind supplied with ideas garnered from objects studied from every angle: "In a given time and on a determined subject, the man of genius has many more ideas than he who lacks it, the combinations which the former can perform will come more easily to him, because he has seen the objects with all their qualities."[112] In his own essay on the faculties of the soul, Bonnet was more specific: "Genius is only attention applied to general ideas, and attention itself is nothing other than the spirit of observation."[113]

Yet a potential contradiction lay at the heart of the "genius of observation." As Senebier, Bonnet, and other eighteenth-century writers on scientific epistemology agreed, the best observations were detailed and exacting, often repeated, copiously described, and ultimately committed to the encyclopedic memory of the genial observer. However, the very detail and quantity of the observations, imprinted upon the soft-wax sensorium of the observer, threatened to dissolve the object of observation into a swarm of sensations. Prolix description exacerbated this effect. Here is Bonnet on a caterpillar he found in October 1740: "It was of a middling size, half-hairy, with 16 legs, of which the membranous [ones] have only a half-crown of hooks. The base of the color on the bottom of the body is a very pale violet, on which are cast three yellow rays, which extend from the second ring to about the eleventh [the description continues for about a page].... Yellow spots are strewn on the sides. The head is violet-colored."[114] Somewhat alarmingly, considering the

238

TABLE I.

TABLE des jours & heures auxquels font nés les Pucerons qu'enfanta depuis le premier Juin jufqu'au vingt-un inclufivement, celui qui depui' fa naiffance avoit été tenu dans une parfaite folitude.

Jours de Juin.	Nombre des Pucer. nés dans chaque j.	Nombre des Pucerons nés chaque matin, & les heures de leur naiffance.	Nombre des Pucer nés chaque après-midi, & les heures de leur naiffance.
1.	2 puc.	. . . . 0 p.	à 7 h. ½ 1 p. 9. 1 p.
2.	10 puc.	à 5 h. 2 p.* 6 1 p. 6 ½ 1 p. 7 ½ 1 p. 8 ½ 1 p. 8 ¾ 1 p.	à 12 h. ½ 1 p. 1 ½ 1 p. 6 ½ 1 p.
3.	7 puc.	à 10 h. 1 p. 11 1 p.	à 3 h. 1 p. 4 1 p.* 4 ¾ 1 p. 6 1 p. 9 1 p.
4.	10 puc.	à 5 h. 3 p.* 6 1 p. 6 ¾ 1 p.	à 12 h. ¾ 1 p. 1 ¼ 1 p. 6 1 p. 9 2 p.*

Fig. 4.11. Aphids Observed. "Table of the Days and Hours at which Aphids Were Born," Charles Bonnet, *Traité d'insectologie, ou, Observations sur les pucerons* (Paris: Durand, 1745), table 1. The naturalist Bonnet learned how to present his observations in tabular form from the mathematician Gabriel Cramer. In his study of the parthenogenetic reproduction of aphids, Bonnet watched a single aphid confined in a jar every day for over a month from *circa* 5:00 AM to 10:00 PM. This table records how many aphids were born by date and hour of the day; an asterisk indicates that Bonnet did not actually witness the birth, having momentarily left his station. The single-minded attention demanded by such strenuous observation was criticized as moral excess by contemporaries — and even by Bonnet himself, who later blamed his blindness on this relentless observational regimen.

length of his printed descriptions, Bonnet told his readers that these were only excerpts from his far lengthier journal.[115] Modern naturalists have found it difficult to make taxonomic identifications on the basis of Bonnet's descriptions, despite (or perhaps because of) their length and specificity.[116] The object as a whole shattered into a mosaic of details, and even the tiniest insect organ loomed monstrously large.

Distilling the advice of the elite of Enlightenment observers, Senebier acknowledged the necessity of detailed written reports of observations, but he also insisted that the observer be selective, so as not to confuse the idiosyncratic individual with the species under investigation. He cited with approval the example of the French zoologist Louis-Jean-Marie Daubenton, who always chose the animal with "the most ordinary proportions" for his anatomical descriptions: "As much as possible one ought to make known the mean terms [les termes moyens] which are closer to all the individuals of the species, which are the most common, and which are, so to speak, the most natural."[117] Selective attention, guided by reason, winnowed the wheat from the chaff among the raw materials gathered by the diligent observer. Only through the sustained and active exercise of attention could the observer distinguish between what was "accidental and what belonged essentially" to an object of inquiry and so avoid confusing an individual trait with a generic one.[118]

By identifying attention with active selection in observation, Enlightenment savants could even turn attention into a form of abstraction, paradoxical though the equation may seem at first glance. Although attention was, of course, directed to particulars, often minute ones, its role in assembling the generic object of inquiry from the jumble of sensations resembled the mental capacity to forge generalizations. According to Bonnet, abstraction was nothing more than attending to some traits rather than others, thereby forming "a sensible abstraction, a representative sign of all organized bodies of the [given] species which are offered to the eyes."[119] Significantly, Bonnet thought a sense of self resulted from the same process: the mind attends selectively only to those of its ideas that relate to that which perceives and appropriates sensations, thereby arriving empirically at "the notion of its own existence."[120] Attention soldered together the objects and subjects of knowledge, both assembled from the copious but fragmentary materials of sensation.

Attention was regarded by these eighteenth-century savants as primarily a matter of the appetites, a sort of visual consumption, but appetite could be retrained by habit. The remedy for squeamishness or boredom was an act not of willful self-mastery but of calculated self-deception that would become self-fulfilling: by looking long and hard enough at maggots as if they were marvels, naturalists came to believe heart and soul that they were. In his mammoth treatise on insects, the French naturalist René Antoine Ferchault de Réaumur did not chide his readers for disdaining insects; rather, he promised them surprises and enchantments to rival fairy tales and *The Thousand and One Nights*.[121] Enlightenment observation began and ended in pleasure, even under arduous mental and physical conditions. After Bonnet had observed a single aphid from 5:30 AM to 11:00 PM every day for over a month, he was disconsolate when one fine June day he lost sight of it; he was wistful for the "delights of observation" that had been his.[122] As Senebier explained, the attitude of the observer toward nature was that of "a lover who contemplates with avidity the object of his love."[123] When Enlightenment moralists commented upon the obsessive observational regimes of Réaumur, Bonnet, and other savants, they did not praise their dutiful dedication to a difficult task but reproached them for self-indulgence and lack of moderation, for appetites run amok.[124]

By the 1870s, however, psychologists writing in German, French, and English had made attention central to, even synonymous with, the exercise of will rather than the tug of appetite.[125] Volition, asserted James, only secondarily mobilizes the motor system; its first point of engagement is with a mental object: "Though the spontaneous drift of thought is all the other way, the attention must be kept strained on that one object until at last it grows, so as to maintain itself before the mind with ease. This strain of attention is the fundamental act of will. And the will's work is in most cases practically ended when the bare presence to our thought of the naturally unwelcome object has been secured."[126] As the phrases "strain of attention" and "naturally unwelcome object" suggest, the effort of attention was conceived in terms not of allurements but of onerous duty. Late nineteenth-century psychologists noted with surprise that earlier accounts of attention — for example, those of Condillac and Bonnet — had described its operations entirely in terms of the increased vivacity it lent to sensations and ideas, with little mention of the will.[127] They concluded that their predecessors had been content to study the workings of spontaneous or natural attention.

In contrast, voluntary attention was, wrote the French psychologist Théodule Ribot in 1889, quite unnatural, the product of millennia of civilization and hard work. Savages were notoriously incapable of sustained attention; so were vagabonds, thieves, and prostitutes.[128] It was only by resolutely acting against the natural human inclination to sloth, "by force of labor and pains, that man brought forth from the old foundation of spontaneous, innate attention the voluntary attention that constitutes his best instrument of scientific investigation. Out of the stubborn struggle between Nature and his nature is born the most beautiful work of man, science."[129] Voluntary attention was reclassified as work by late nineteenth-century psychologists, and with it, scientific observation. If the exertions of Enlightenment savants were labors of love, those of their successors were more often described simply as labor: they constituted the "iron work of self-conscious inference," demanding "great stubbornness and caution," as Helmholtz put it.[130]

To practice attention as an act of will and to pursue science as work with a will was consistent with the post-Kantian active self, which grasped, manipulated, and interrogated the world. But this same coiled spring of a self posed epistemological problems for scientists worried about how their own subjective projections might distort their observations. Concern about observation marred by prejudice or *esprit de système* was not new, but the Enlightenment remedy had simply been redoubled "passion for the truth"; any attempt to observe without preconceived ideas or conjectures had been dismissed as scientifically useless.[131] Yet this was precisely what scientific objectivity seemed to demand by the mid-nineteenth century. The result was an opposition between allegedly passive observation and active experimentation and a split within the scientist's own self. Insofar as Enlightenment savants had distinguished between observation and experiment, they had done so along the axis of natural and artificial conditions: observers took nature as they found it; experimenters pushed nature to its limits in the laboratory. But it was taken for granted that the experimenter was also an observer and that all observation was an active ordering of natural variety and sensations. In the 1830s and 1840s, the distinction between observation and experiment was recast in disciplinary terms, contrasting, for example, the astronomer in the observatory with the chemist in the lab.

By the 1860s, passive observation had come to be opposed to active experimentation. Bernard was among those who advanced this distinction, and he openly admitted that it was contrived: one and the same scientist had somehow both to be speculative and bold in designing an experiment to pry answers out of nature *and* to observe the results passively, as if in ignorance of the hypothesis the experiment aimed to test. The scientist was both inquisitor and confessor to nature: "Yes, no doubt, the experimenter forces nature to unveil herself, attacking her and posing questions in all directions; but he must never answer for her nor listen incompletely to her answers by taking from the experiment only the part that favors or confirms the hypothesis.... One could distinguish and separate the experimenter into he who plans and institutes the experiment from he who executes it and registers the results."[132] (See figure 4.12.) The scientist *qua* experimenter reasons and conjectures; the scientist *qua* observer must forget all reasoning and only register. This split scientific personality was the practical correlate of the tension between activity and passivity, imagined by mid-nineteenth-century scientists as an internal struggle of the will against itself.

The practices of scientific journal keeping were redesigned to hold active and passive elements of attention in balance. Whereas eighteenth-century journals had been kept not only to record but also to synthesize observations, by the mid-nineteenth century "real-time" entries were being jotted down in laboratories as events occurred.[133] Journals remained highly personal; Mach, for example, carried around pocket-sized notebooks in which he wrote down everything from experimental results to drafts of letters and reminders to buy more notebooks.[134] But just as the photograph was seen as an archive of details whose significance would be recognized only by future scientists, the lab notebook began to be imagined as a repository of raw data, unedited and uninterpreted. Exactly when entries were written down — during or after an experiment — became an issue. Faraday strongly recommended that results be noted

243

Fig. 4.12. "Nature Unveiling Herself Before Science." Louis-Ernest Barrias, 1899, Musée d'Orsay, Paris (Réunion des Musées Nationaux / Art Resource, NY). The original of this marble sculpture was commissioned by the French government for the grand staircase of the Conservatoire National des Arts et Métiers in Paris. Nature's gown, made of Algerian onyx and held up by a large green scarab, recalls the ancient mythological conflation of nature with the Egyptian goddess Isis. It blends the ancient trope of the veil of Isis, interpreted as nature's desire to hide her secrets, with the modern fantasy of (female) nature willingly revealing herself to the (male) scientist, without violence or artifice. (Please see Color Plates.)

down immediately, before subsequent results and reflections could distort memory: "The laboratory notebook, intended to receive the account of the results of experiments, should always be at hand, as should also pen and ink. All the events worthy of record should be entered at the time the experiments are made, whilst the things themselves are under the eye, and can be re-examined if doubt or difficulty arise. The practice of delaying to note until the end of the day, is a bad one, as it then becomes difficult accurately to remember the succession of events."[135]

There is internal evidence that, no matter when Faraday made his provisional lab notes (presumably on the model of these instructions), the diaries that survive were in fact written up at the end of each day, perhaps on the basis of rough notes.[136] Yet even in the redacted notes of the diaries, the cautious zig-zag between hypothesis and experimental test — and, above all, the strenuous attempts to keep the two distinct — are preserved. In a series of experiments designed to detect possible relationships between gravitational and electrical forces, for example, Faraday puzzled over whether a falling body might induce a current: "Would look like a power of affecting one end of a line and not the other. This is not likely and so is against all my suppositions, but we shall see how experiment testifies, and whether it only modifies some of my deductions and conclusions or sweeps them away altogether. Which may well be."[137] In Chapter Two, we heard Bernard on the discipline required to keep the design of experiments and the registration of results asunder, concomitantly with the active and passive parts of the experimentalist's own psyche. The practice of keeping a lab notebook had become more than an aid to memory; it was a place where hypotheses could be spun, experiments devised and described, and sharp distinctions between these activities made.

Among scientists whose careers straddled the boundary between truth-to-nature and mechanical objectivity, such as the British physicist Arthur Worthington, the tension between these two different conceptions of observation was thrown into relief. Having built his extraordinary apparatus to visualize the detailed evolution of a splash, fraction of a second by fraction of a second, he at first found it obvious that he should smooth out the irregularities, the asymmetries that seemed peculiar — and therefore negligible — in this splash or that. As we saw in the Prologue and in Chapter Three, a few years later exactly those oddities came to seem important to him: the asymmetrical images recorded by high-speed photography flaunted their objectivity. Worthington had been an active observer, intervening to extract scientific interest from what he saw; later, proud of his hard-won passivity, he aspired to let each splash draw its own lopsided portrait.

Knower and Knowledge

The divided scientific self, actively willing its own passivity, was only one possible self within the field created by the distinction between objectivity and subjectivity. Its polar opposite, equally stereotyped and normalized, was the artistic self, as militantly subjective as the scientific self was objective. For an artist to "copy nature" slavishly was to forsake not only the imagination but also the individuality that Charles Baudelaire and other antirealist critics believed was essential to great art. Subjective art invited, even demanded, the externalized exercise of the will, actively molding matter and form to fit the artist's conception.[138] As an 1885 French manual for artists put it, "If the artist neither can nor may liberate himself from the imitation of nature, his dependence has a limit . . . at the instant at which he comes to exercise his will, he arrives at the creation of a work; if not, he remains in the workaday accomplishment of a professional task."[139] For scientists, in contrast, the objective was all that resisted the external exercise of will; many of their worries about the possible interventions of subjectivity centered on the intrusions of the "arbitrary" (in the root sense of capricious acts of will) into observation and representation. Objectivity enshrined the will, but the will now exercised internally, on the self, rather than externally, on nature.

Both artistic and scientific personas spawned heroic myths, albeit complementary ones. The heroic artist was authentic, recreating the world in the image of an assertive and indelible self. The heroic scientist was disciplined, discovering the world through work. Whereas early nineteenth-century novels such as Mary Wollstonecraft Shelley's *Frankenstein, or, The Modern Prometheus* (1818) and Honoré de Balzac's *La recherche de l'absolu* (*The Quest for the Absolute*, 1834) portray once-noble protagonists who destroy themselves and their loved ones through their addictive passion for science, later fiction featuring scientists, such as George Sand's *Valvèdre* (1861), tells of wasted or warped lives redeemed by science and labor. In Sand's book, Francis, an aspiring poet nourished on novels and romantic fantasy, runs off with the bored and beautiful wife of the Swiss scientist Valvèdre, causing her death and Francis's ruin. Magnanimously forgiven by Valvèdre, Francis becomes a new man through sweaty labor as a factory metallurgist and "by steeling logic, reason, and will in severe studies."[140] In his short story "The Natural Man and the Artificial Man" (composed *circa* 1885), Cajal rang changes on the same theme: the literary Esperaindeo is lost in humanist fancies until taken in hand by his naturalist friend Jaime, who introduces Esperaindeo to "the endless work of observation." The story ends with the two friends *en route* to Jaime's paradisial electrotechnical factory, where Esperaindeo will be saved from dissolute rhetoric and fickle politics by science and hard work.[141]

Against this background, the contretemps between Haeckel and His with which this chapter opened takes on an added dimension. It was a collision between ideals of truth-to-nature and mechanical objectivity, but Haeckel cannot be dismissed as just a throwback to earlier times, an Albinus *après la lettre*. His version of truth-to-nature was altered by the very existence of—and sometimes rivalry with—mechanical objectivity. Haeckel's arguments and persona were pressed into the plane defined by the axes of objectivity and subjectivity. His spirited defense of "ideas" in images went hand in hand with an intense appreciation of the aesthetics of natural forms, most explicit in his *Kunstformen der Natur* (*Art Forms in Nature*, 1899–1904) but also clearly displayed in the exquisite plates of his earlier studies of medusae.[142] (See figures 4.13 and 4.14.) In the days of Goethe or Audubon, there would have been nothing jarring about the partnership of truth and beauty. But once framed by the opposition between objective science and subjective art, Haeckel's preoccupations made him seem eccentric—an artist in scientist's clothing. After the 1850s, something like the same puzzlement attached to the figure of Goethe: scientists like Helmholtz furrowed their brows over the apparent paradox of a great poet who was also seriously engaged in scientific research and tried to explain it away by showing how Goethe's optics, morphology, and comparative anatomy were at bottom really the expression of artistic intuition rather than scientific

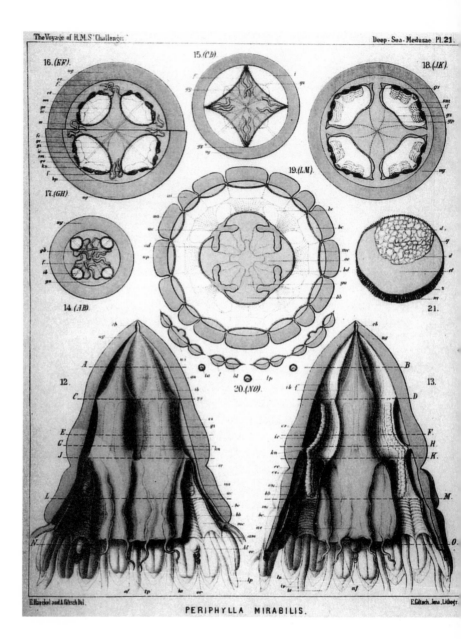

PERIPHYLLA MIRABILIS.

Fig. 4.13. The Science of Medusae. *Periphylla mirabilis*, Ernst Haeckel, *Report on the Deep-Sea Medusae Dredged by H.M.S. Challenger During the Years 1873–1876*, pl. 21, drawn by Haeckel and Adolf Giltsch, lithographed by Eduard Giltsch. The British war ship H.M.S. Challenger was converted into an oceangoing scientific laboratory and returned after three years with crates of specimens for scientists to classify, resulting in a series of fifty volumes. Haeckel's monograph on the medusae was illustrated with his own drawings, which emphasized the symmetry and elegance of these organic forms. (Please see Color Plates.)

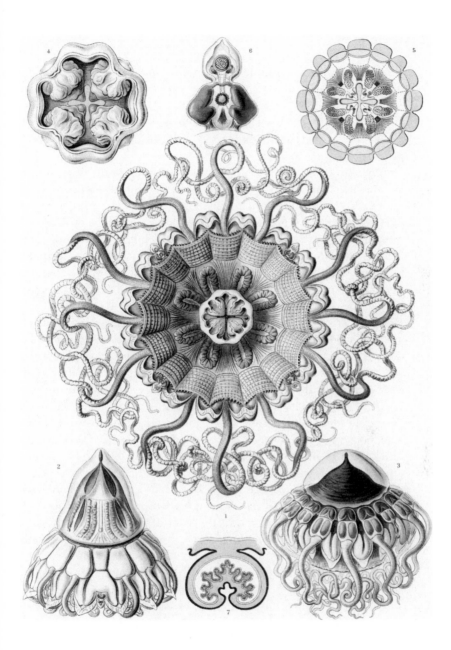

Fig. 4.14. The Art of Medusae. *Peromedusae,* Ernst Haeckel, *Kunstformen der Natur* (Leipzig: Bibliographisches Institut, 1904), table 38. These figures, also of *periphylla* medusae, are self-consciously arranged as "art forms," but the symmetries of the "basic forms" are carried over from Haeckel's earlier work in the biology of marine invertebrates, as seen in fig. 4.13. Haeckel's figures were models for many decorative works, from the monumental arch of the Paris World Exposition in 1900 (inspired by one of Haeckel's images of radiolaria) to the ornaments for Haeckel's own house in Jena, the Villa Medusa. (Please see Color Plates.)

concepts.[143] These examples make a more general point, to which we will return in subsequent chapters, about how earlier epistemic virtues are modified, though not eliminated, by later ones. The shifting relationships between scientific and artistic personas signal how the advent of mechanical objectivity changed the meaning of truth-to-nature.

Like parallel lines meeting at the vanishing point of a picture painted in perspective, objective science and subjective art converged in the dissolution of the self into its object. Nietzsche was, as we have seen, no friend of scientific objectivity; like Haeckel, he took up the cudgel for older ideals of truth in his own disciplines of philology and history. But Nietzsche made an exception for one form of objectivity, which he saw as common to the best art and science: "There is required above all great artistic facility, creative vision, loving absorption in the empirical data, the capacity to imagine the further development of a given type — in any event objectivity is required, but as a positive quality. So often objectivity is only a phrase. Instead of the outwardly tranquil but inwardly flashing eye of the artist there is the affectation of tranquility; just as the lack of feeling and moral strength is accustomed to disguise itself as incisive coldness and detachment."[144] Objectivity as a "positive quality" put back together, or so Nietzsche thought, the two halves of the self: subjective and objective, active and passive, will and world. By uniting the knower with the known in an act of "loving absorption," the will surrendered to the world without asceticism.

However illusory Nietzsche's "positive" objectivity may have been for both artists and scientists, it was proposed as a solution to a deep problem. Objectivity and the scientific self that practiced it were intrinsically unstable. Objectivity demanded that the self split into active experimenter and passive observer and that types of scientific objects be defined by atlas images of individual specimens too particularized to be typical. Nietzsche smelled the acrid odor of burnt sacrifice when the ascetic turned will against will: the objective man of science stood accused of inauthenticity, of self divided against itself. These were ethical reproaches. There were also epistemological objections to objectivity: How could an individual stand for a class without idealization or even selection? How could a universally valid working object be extracted from a particular depicted with all its flaws and accidents?

The responses to the instability of mechanical objectivity took two forms, which are the subjects of our next two chapters. On the one hand, votaries of objectivity forsook the realm of the senses altogether, fleeing from the blooming, buzzing confusion of particulars into the austere structures of mathematics and logic — there is even a tradition of mathematical atlases entirely empty of images (Chapter Five).[145] On the other hand, a new class of scientific "experts" abandoned the rigorous faith of objectivity in favor of trained judgment, taught and practiced as a skill rather than an act of will (Chapter Six). Neither answer to the internal contradictions of mechanical objectivity managed to unseat it, any more than mechanical objectivity had abolished truth-to-nature. Instead, as the code of epistemic virtues expanded, so did the potential for conflict among them.

CHAPTER FIVE

Structural Objectivity

Objectivity Without Images
In 1869, the eminent physicist and physiologist Hermann von Helmholtz lectured the annual gathering of German-speaking scientists
on the epistemological implications of the latest findings in sensory
physiology, a field to which he had made pioneering contributions.
Citing the physiologist Johannes Müller's doctrine of specific nerve
energies and his own research on color vision, Helmholtz pointed to
the gap between the external world and internal sensations. The
human eye, for example, collapsed the endlessly varied "objective
manifold of light mixtures" into only three fundamental colors;
other sensory organs were equally reductive and distorting. Helmholtz concluded that all sensations "are only signs of external objects, and in no way pictures bearing any resemblance."[1] Even the
Kantian synthetic *a priori* intuition of space was simply a "subjective
form of intuition [*Anschauungsform*], like the sensory qualities of red,
sweet, cold."[2] Yet objectivity of a sort could, Helmholtz asserted, be
salvaged from these mere signs, for they at least preserved temporal
sequences and therefore sufficed for the discovery of natural laws.
Scientific objectivity was not a matter of viewing nature as it really
was — that was impossible. Nor did it have anything to do with fidelity to sensations or ideas — these were will-o'-the-wisps generated
by the human nervous system. Instead, objectivity lay in the invariable relations among sensations, read like the abstract signs of a language rather than as images of the world.

Mechanical objectivity could be made visible. As we saw in Chapter Three, it left its signature in a multitude of scientific images. Yet

there is a form of objectivity that spurns all images, whether they are perceived by the eye of the body or that of the mind, as irretrievably subjective. Proponents of this form of objectivity, which emerged in late nineteenth- and early twentieth-century logic, mathematics, physics, and philosophy and which is still very much alive in mathematical physics and analytic philosophy, pinned their hopes instead on invariant structures; hence the title of this chapter.[3] For Helmholtz, and those who thought like him, these structures were lawlike sequences of signs; for others, they were differential equations; for still others, logical relationships. Some of the spokesmen for structural objectivity engaged in laboratory research or even engineering projects; others dwelled in the rarefied realms of mathematical logic. Their professional aspirations and enemies, their training and politics, diverged in many respects; by no stretch of the imagination can they be said to form anything like a school. But all upheld a version of objectivity (their own word) grounded in structures rather than images as the only way to break out of the private mental world of individual subjectivity. In their view, science worthy of the name must be communicable to all, and only structures — not images, not intuitions, not mental representations of any kind — could be conveyed to all minds across time and space. In a 1906 lecture, the German physicist Max Planck went so far as to suggest that this community of scientific objectivity might embrace not only other cultures and historical periods but also other worlds: "The goal is nothing less than the unity and completeness of the system of theoretical physics . . . not only with respect to all particulars of the system, but also with respect to physicists of all places, all times, all peoples, all cultures. Yes, the system of theoretical physics demands validity not merely for the inhabitants of this earth, but also for the inhabitants of other planets."[4]

All the figures treated in this chapter referred explicitly to "objectivity"; some, but not all, used the term "structures." Those who did identify "structures" as the core of objectivity understood a great variety of things under that rubric: logic, ordered sequences of sensations, some of mathematics, all of mathematics, syntax, entities that remain invariant under transformations, any and all formal relationships. Our rationale for grouping them together, despite their many striking and significant divergences from one another, is twofold: first, they diagnosed a common problem, namely, the specter of incommunicability in the sciences, and ascribed it to similar causes; second, later figures assimilated the earlier ones into a lineage when they proposed a solution, an objectivity derived from structures, however those were defined.

These intellectual genealogies were not an open-armed embrace of as many distinguished ancestors as possible, but an attempt to build upon a specific solution to an already articulated problem. Gottlob Frege may not, for example, have described his logical innovations in terms of "structures," but when Rudolf Carnap later enlisted post-Fregean logic in the service of an emphatically "structural" objectivity, he believed that he was using Fregean means to reach a Fregean end (even echoing Frege's favorite analogy between formal logic and Leibniz's *characteristica universalis*):[5] symbolic logic, as it had been developed by Bertrand Russell and Alfred North Whitehead in their *Principia Mathematica* (1910–1913) and "based on the preliminary works of Frege, Schröder, Peano, and others," would reveal the structures of an "objective world, which can be conceptually grasped and is indeed identical for all subjects."[6] Carnap recognized that Frege, Henri Poincaré, Russell, and others had understood structures in general and logic in particular somewhat differently from one another and from himself. Yet from his retrospective viewpoint, writing in the 1920s, all were bound together in a common quest for a form of objectivity that would make science communicable among all subjects, everywhere and always — Planck's interplanetary congregation of physicists.

There are further historical reasons not to insist too vehemently on an identical notion of structure, much less identical usage of the word "structure," as a criterion for inclusion among the late nineteenth- and early twentieth-century proponents of structural objectivity. It was precisely at this time, and especially in the fields of logic and mathematics, that the word "structure" acquired new meanings and intellectual glamour. Derived from the Latin verb *struere*, meaning "to build," "structure" and its cognates in the major European languages originally referred to architectural construction and were later extended to any framework of material elements (especially the human body). During the nineteenth century, the word was increasingly used (along with other architectural borrowings, such as

Bauplan) to describe how the parts of organisms were put together to make a coherent whole; it was thereafter appropriated by sociology, conceived as the study of the "social organism."[7] Around the turn of the twentieth century, "structure" became the watchword of a self-consciously innovative movement in mathematics, including set theory and the "modern algebra" of groups, rings, and ideals.[8] Philosophers, psychologists, and linguists of the 1910s and 1920s caught the "structuralist" fever. The very dynamism that made the word "structure" attractive to Carnap and others during this period also makes it an unsteady marker of intellectual affiliation.

"Objectivity," in contrast, was a word with which to conjure but also to consolidate. All the figures discussed in this chapter invoke it repeatedly, emphatically, and in the same sense: to designate the aspects of scientific knowledge that survive translation, transmission, theory change, and differences among thinking beings due to physiology, psychology, history, culture, language, and (as in Planck's fantasy) species. Their worries about mutual intellectual incomprehension were fed by mid-nineteenth-century research in history, anthropology, philology, psychology, and, above all, sensory physiology, which underscored how very differently individuals reasoned, described, believed, and even perceived. For these scientists dangerous subjectivity came to be reframed in terms of individual variability, of which the paradigmatic example was sensory experience. Unanimity on this score is our rationale for grouping them together in this chapter, under the rubric "structural objectivity."

At first glance, mechanical and structural objectivity seem to have little in common. Mechanical objectivity is about more than images: statistical techniques and experimental protocols may also be enlisted to thwart subjective projections onto nature.[9] But certain kinds of images were nonetheless central to mechanical objectivity, because they seemed to promise direct access to nature, unmediated by language or theory. Camera obscura tracings, photographs, and the inscriptions of self-registering instruments were all, at one time or another, touted as nature's own utterances. Structural objectivity, in contrast, has no truck with any kind of seeing, be it four-eyed sight or blind sight. All images must ultimately be represented to the mind of the scientist in terms of sensations and ideas, that is, via sensory, nervous, and mental processes that mid-nineteenth-century physiologists and psychologists such as Helmholtz had demonstrated to correspond only partially to external stimuli, and to be highly variable as well.

Mechanical and structural objectivity, moreover, countered different aspects of subjectivity. Mechanical objectivity restrained a scientific self all too prone to impose its own expectations, hypotheses, and categories on data — to ventriloquize nature. This was a projective self that overleaped its own boundaries, crossing the line between observer and observed. The metaphors of mechanical objectivity were therefore of manful self-restraint, the will reined in by the will. The metaphors of structural objectivity were rather of a fortress self, locked away from nature and other minds alike. Structural objectivity addressed a claustral, private self menaced by solipsism. The recommended countermeasures emphasized renunciation rather than restraint: giving up one's own sensations and ideas in favor of formal structures accessible to all thinking beings. The American logician and physicist Charles Sanders Peirce thought the submersion of self in this cosmic community guaranteed the validity even of logical inferences: "It seems to me that we are driven to this, that logicality inexorably requires that our interests not stop at our own fate, but must embrace the whole community. This community, again, must not be limited, but must extend to all races of beings with whom we can come into immediate or mediate contact. It must reach, however vaguely, beyond this geological epoch, beyond all bounds."[10]

Why, then, call both — the solitary suppression of the will and the reaching out for a communion of reason "beyond all bounds" — objectivity? Why did, for example, the mathematician Frege and the bacteriologist Robert Koch seize on the same word to describe, respectively, formalized versions of arithmetic and unretouched photographs of bacilli? Neither thought objectivity was just a synonym for external reality: Koch was painfully aware that the microscopic cross section rendered by the photograph often showed artifacts. Frege ridiculed those who thought the laws of numbers could be discovered by any kind of empirical inquiry. What mechanical and structural objectivity shared was not some claim to reveal the unvarnished facts, but a common enemy: subjectivity. Both located epistemological dangers in the self of the scientist, albeit in different facets of that self. This is why it was natural to use the same word to refer to both: objectivity is always defined by its more robust and threatening complement, subjectivity. But whereas the self restrained by mechanical objectivity was largely the creation of will-centered post-Kantian philosophy, that renounced by structural objectivity was in part the discovery of science itself, particularly the then-young sciences of sensory physiology and experimental psychology.

Using empirical methods (including some of the tools of mechanical objectivity), the post-1848 generation of physiologists and psychologists investigated the mind under laboratory conditions. What was the relationship between nerve impulses and experienced sensations? How did infants acquire Euclidean intuitions of space? Could the speed of thought be measured? Were the laws of logic simply generalizations of the laws of mental association? Armed with cameras, collimators, chronometers, and calipers, scientists studied the speed of nervous transmission, color sensations, attention spans, and even logic and mathematics as psychophysiological phenomena.[11] Some of the leading scientists of the age extended the procedures of observation-based natural science to get at the inner workings of the brain — the ganglia, tendrils, and phosphorus that they hoped would lay bare the process of thinking. Others aimed to tackle thought itself — including the ethereal realms of reason — through experimental psychology.

From the outset, the fledgling sciences of thought and sensation deployed the new Kantian vocabulary of objectivity and subjectivity as an analytical tool, to mark the division between self and world. But their own results forced a redrawing of that boundary and a remapping of the territory on both sides. On the side of subjectivity, these inquiries offered dramatic evidence of individual differences in mental processes. The methods of mechanical objectivity aimed to eliminate the distortions introduced by this or that subjective observer. Once turned upon the mind itself, these methods revealed differences in perception, judgment, and even logic. On the side of objectivity, these variations invaded science itself: in astronomy and geodesy, observers were forced to acknowledge the existence of personal equations that resisted every attempt to eliminate them by training and technology.[12] The "personality" of an astronomer's observations was discovered to be as indelibly individual as a signature.[13] Logic fared little better at the hands of the psychophysiologists. In his influential *Grundzüge der physiologischen Psychologie* (*Principles of Physiological Psychology*, 1874), the Leipzig professor Wilhelm Wundt agreed that logic was the "mental form" of science, but added: "For psychological analysis, however, the fact that psychological processes can be brought into logical form is not sufficient grounds for them to be regarded as logical judgments and inferences in their actual operations."[14] Reason itself, since ancient times upheld as uniform and eternal, threatened to shatter into the reason of this culture or that time, or even this or that individual.[15]

The response of the self-declared defenders of reason, especially philosophers and mathematicians, to these unsettling empirical claims was not to reject scientific objectivity but to deepen it. They acknowledged the variability of individual physiology and perception; they bowed to the testimony of historians and ethnologists concerning the strikingly diverse mental lives of people from other times and places; they admitted that even science was ephemeral, since new theories displaced old ones at an ever-accelerating rate, as we saw in Chapter Four. But they insisted that there nonetheless existed a realm of pure thought that was the same for all thinking beings forever and that was, therefore, genuinely objective. The objective was not what could be sensed or intuited, for sensations and intuitions could be shown to differ, and in ways that were incorrigibly private for each person. Nor was it the bare face of facts, scrubbed free of any theoretical interpretation, for today's facts might be cast in a wholly different light by tomorrow's findings. Objectivity, according to the structuralists, was not about sensation or even about things; it had nothing to do with images, made or mental. It was about enduring structural relationships that survived mathematical transformations, scientific revolutions, shifts of linguistic perspective, cultural diversity, psychological evolution, the vagaries of history, and the quirks of individual physiology.

Structural objectivity was, in some senses, an intensification of mechanical objectivity, more royalist than the king. It was no longer enough to produce an image or an instrument reading innocent of human interpretation. Mechanical objectivity had sternly jettisoned idealizations and aesthetics in scientific representations; structural objectivity abandoned representations altogether. These ascetics among ascetics aspired to a higher, purer form of knowing entirely

259

free of pictures, intuitions, or indeed any aspect of the senses; even theoretical models and geometric intuitions were suspect. Writing in 1910, the German philosopher Ernst Cassirer caught the sense of objectivity pushed ever further when he observed that science and philosophy had begun in the seventeenth century by affirming sensations as the paradigm of the objective, as opposed to the subjectivity of dreams and hallucinations. But with the advance of science, sensations expressed, at least as compared to the abstract schemata of physics, "only a subjective state of the observer." Ultimately, structural objectivity lay not in the observable facts of mechanical objectivity but only in the "final invariants of experience."[16]

Just as structural objectivity stretched the methods of mechanical objectivity beyond rules and representations, it carried the ethos of self-suppression to new extremes. Practitioners of mechanical objectivity were expected to restrain their impulse to perfect, prettify, smooth, or even generalize their unvarnished data and images. These were the facts that would speak for themselves: *res ipsa loquitur*. Nature, like Luther's Bible, should require no interpreter. Practitioners of structural objectivity went still further: one must resist the urge to believe in the contents of one's own consciousness. What had once been the prototypes of the self-evident — not merely immediate perceptions but also meticulous scientific observations, mathematical intuitions, and venerable scientific theories — were now revealed to vary from person to person and from one historical period to the next, and therefore to be subjective. The visible facts about how this particular thing looked just there, at that moment, as captured on a photographic plate, could not — *pace* mechanical objectivity — overcome the vicissitudes of individual variability and scientific change. It was, rather, structural relationships that outlived the piled-up ruins of past scientific theories and the idiosyncrasies of present scientists; these were "the only objective reality."[17]

The expression "objective reality" raises the question of the relationship between what we have called "structural objectivity" and a particular philosophical position that goes by the name "structural realism."[18] The latter has several variants, but, as the name suggests, all aim to salvage some form of scientific realism from the objections of historians, constructive empiricists, instrumentalists, social constructivists, and other critics of the claim that scientific theories are in some sense true, not just useful. To the antirealists who argue that data underdetermines theory and that an induction over the history of science indicates that all scientific theories, no matter how successful, will be eventually rejected as false, the structural realists reply that structures, understood as mathematically expressed natural laws, survive the overthrow of old theories by new. They second Poincaré here: it is structures like Maxwell's equations, not theoretical entities like the electromagnetic ether, that constitute scientific reality.

Yet the preoccupations of late twentieth-century structural realists were not those of early twentieth-century structural objectivists: the former, like all realists, were primarily interested in the justification for the claim that science was true, that it correctly described real features of the world; the latter (including Poincaré) were chiefly concerned with the justification for the claim that science was objective, that it was "common to all thinking beings."[19] Among the structural objectivists, there existed a spectrum of positions on the issue of realism and antirealism, and few, if any, of them regarded it as an urgent question — in contrast to the debate over objectivity. Among the structural realists, the only aspect of communicability that was routinely addressed was the historical continuity of scientific theories. The positions (and their proponents) sometimes overlapped, but they were not coincident. Structural objectivity, like mechanical objectivity, was first and foremost about epistemology, not ontology.

Many voices spoke out for structural objectivity in the period between roughly 1880 and 1930. Some were logicians and mathematicians, like Frege, Peirce, and Russell. Others were mathematicians and theoretical physicists, like Poincaré and Planck. Still others were scientists-turned-philosophers enthralled by the revolutionary new science of relativity theory, like Carnap and Moritz Schlick, both of whom had studied physics. They spoke in different registers and in support of different agendas. The politically conservative and devout Lutheran Frege would have had little sympathy for the engineering pragmatism of Third Republic progressive Poincaré; both would have found much to disagree with in Carnap's radical vision of philosophical and political tolerance. Frege worried about individual differences at the level of mental representations and intuitions, whereas Poincaré was concerned with salvaging permanence amid scientific change, and Carnap sought a neutral language compatible with the most diverse personal perspectives. But they converged in their articulations of an objectivity beyond mechanical objectivity — as epistemology, as ethos, and as scientific, mathematical, and philosophical practice. Indeed, it was precisely the experience of ineradicable diversity — psychological, political, historical — that made structural objectivity their holy grail.

The Objective Science of Mind

Philosophical discussions of the objectivity of mind, like almost all modern philosophical reflections on objectivity, take hold with Immanuel Kant. Near the end of the *Critique of Pure Reason* (1781, 1787), Kant offered a rough-and-ready distinction between individual subjective opinion and objectively valid conviction: "If the judgment is valid for everyone, provided only he is in possession of reason, its ground is objectively sufficient [*objektiv hinreichend*], and the holding of it to be true is entitled conviction. If it has its ground only in the special character of the subject, it is entitled persuasion." Kant described this index of the objective as "communicability [*Mittheilbarkeit*]," justifying it on the grounds that if a judgment can be communicated to other rational beings, there is a solid (though not infallible) presumption that they are talking, and talking accurately, about the same object.[20] Whether that object belonged to the world or to the mind was left open. Kant's own usage of the terms "objective" and "subjective" to describe moral and aesthetic as well as epistemological judgments suggests that he intended the widest possible construal of shared reason as well as a shared world.

But by the middle decades of the nineteenth century, a gap had opened up between the objectivity of shared reason and shared world. Scientific investigation of the world understood objectivity empirically — a word Kant had often used as almost a synonym for subjective sensation, modifying both by a disdainful "mere [*bloß*]." Moreover, empiricism in the service of scientific objectivity, in contrast to older ideals of truth, demanded that the variability of observed phenomena be carefully heeded, rather than abstracted from or idealized. The contrast between a scientific atlas of photographs versus one of drawings lay in the scrupulous rendering of each specimen in all its individual particularity, rather than as a composite of several individuals or as an idealized type. The variability that Kant had taken as the hallmark of the subjective had, in the hands of the practitioners of mechanical objectivity, become a badge of honor among the empirical sciences. Finally, by the 1860s, the objective methods of empirical science had been applied to the mind itself, as examined by physiologists, psychologists, and ethnologists alike. Laws of association, evolutionary theories of intellectual development, ethnographic reports of so-called primitive mentalities, precise measurements of reaction times and the speed of nervous transmission — all aimed to understand mental processes from perception to reasoning as natural phenomena. "Shared reason" had itself become a topic of objective empirical inquiry, rather than the standard by which objectivity was measured.

The attempts to found an objective science of mind proceeded on several fronts. Invading the Kantian heartland, Helmholtz argued that the allegedly synthetic *a priori* intuitions of Euclidean geometry in fact derived from "observable facts of experience": different experiences would generate different geometric intuitions. There was nothing transcendental about the geometric axioms and definitions that for millennia had stood as the epitome of reason; rather, they were "empirical knowledge, gained through the accumulation and reinforcement of similar, repeated impressions, not transcendental intuitions given prior to all experience."[21] Helmholtz was convinced that the same held for arithmetic. It was the task of psychology "to define the empirical characteristics that objects must have in order to be enumerable."[22] Through a combination of sensory physiology and psychology, the laws of thought would be shown to be natural laws, discoverable by the same objective methods that had led to Helmholtz's own discovery that the speed of nerve impulses was finite. As he wrote triumphantly to his father, thought itself could be made the stuff of experimental science.[23]

In his new laboratory for experimental psychology at the University of Leipzig, Wundt and his students enthusiastically extended the Helmholtzian program. The very first issue of the Wundt laboratory house journal, *Philosophische Studien* (*Philosophical Studies*), juxtaposed articles such as "On the Simple Reaction Time of a Sensation of Smell" and "Experimental Investigations on the Association of Ideas" with Wundt's own inquiry into the empirical origins of mathematics, in which Wundt unearthed traces of the "experimental beginnings" of mathematics in its earliest history.[24] Against irate philosophers, he defended his psychological approach to logic as possessing a certain "objective justification," namely an inquiry into the actual thought processes that produced knowledge. Anyone who contended that the normative force of logic derived from some abstract faculty of reason beyond the "natural law-like character" of the mental operations involved surely erred.[25] Here and elsewhere, Wundt lambasted the traditional philosophical methods of self-observation as irredeemably subjective; only experiments offered any hope of an objective science of thought. Like natural scientists, experimental psychologists would introduce controls, measurements, and mathematical analysis. Even if the contents of consciousness could not be directly measured, psychologists could avail themselves of "objective time determinations" of mental processes. To skeptics and pessimists, Wundt retorted that "there exist numerous sources of objective knowledge that promise better results than the inaccessible and deceptive [method of] self-observation, and that psychology runs no risk of running out of material, even if it restricts itself to the investigation of facts."[26]

The fundamental dimension of the new science of psychophysiology was time: the time of nervous transmission, of reaction time, of attention span.[27] Time was the dimension that submitted mental processes to measurement; time was also the dimension that connected abstract number to concrete experience, contended Wundt. Conceptions of number originally derive from intuitions of time, which in turn derive from the succession of individual sensations and representations in consciousness. Through a process of abstraction made possible by language and symbols, number concepts could achieve a generality beyond that of any specific experience. But the ultimately empirical origins and applications of these concepts required that they "be translated into concrete examples."[28]

Wundt did not doubt that advanced mathematics and the laws of thought transcended any possible experience. Abstraction succeeded in transforming "subjective" representations into "objective" concepts, which were never presented to consciousness in the form of immediate perceptions. But some form of representation was a prerequisite for even the most abstract laws of thought; hence the necessity for the symbolic representation of concepts as a substitute for intuition.[29] Although Wundt acknowledged that the trend in the history of mathematics had been toward ever greater generality and abstraction, traces of the empirical origins of its objects and concepts were still, he argued, embedded like fossils in axioms, definitions, and theorems. Indeed, it was precisely the most fundamental axioms and definitions — of number, magnitude, space — that revealed most clearly the inductive roots of mathematics.[30] The testimony of both psychology and anthropology was unambiguous: "Whenever we are in a position to trace back fundamental mathematical knowledge to its first origin, then its source is shown to be induction from experience."[31] Brandishing stopwatch and metronome, experimental psychology took up Helmholtz's challenge to anchor number concepts in experience (see figure 5.1).

The Real, the Objective, and the Communicable
It was against the new self-proclaimed objective science of mind that Frege, who taught mathematics and logic at the University of Jena, furiously defended the objectivity of thought. In an 1887 essay, Helmholtz had made the provocative claim that not only Euclidean geometry but also Frege's sacred preserve of arithmetic ultimately stemmed from experience.[32] Frege's response was characteristically acid: "Helmholtz wants to ground arithmetic empirically, come hell or high water. Accordingly, he does not ask, how far can one get, without drawing on the facts of experience? but rather asks: how can I most quickly bring in any old fact of sensory experience? ... I have hardly ever encountered anything more unphilosophical than this philosophical paper and hardly ever has the meaning of the epistemological question been more misunderstood than here."[33]

Frege's vehement distinction between the logical and the psychological is the subject of a large literature, which there is no need to rehearse.[34] Instead, we will focus on the ways his attempts to establish the objectivity of thought (especially that of logic and arithmetic) was a response to and also a critical amplification of the new objectivity of the empirical sciences of the latter half of the nineteenth century. Whereas the Kantian understanding of objectivity had extended to ethics and aesthetics as well as philosophy and science, Frege tacitly narrowed the scope of the term to apply to science

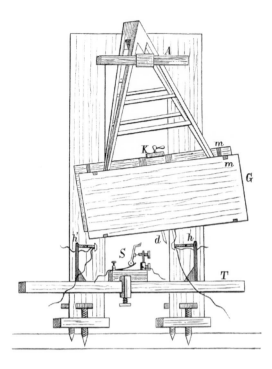

Fig. 5.1. Toward an Objective Science of Mind. Pendelmyographion, Wilhelm Wundt, *Untersuchungen zur Mechanik der Nerven und Nervencentren* (Erlangen, Germany: Enke, 1871), fig. 1, p. 7. Wundt modified Hermann von Helmholtz's self-registering instrument to measure nerve reaction times. Depending on the length of the time span to be measured, the period of the pendulum (apex at A) can be adjusted. Attached to the pendulum is a glass plate (G) upon which the electrically stimulated muscle traces out the reaction curves, without the intervention of a human hand — an instrument in the service of mechanical objectivity. Although Wundt used the apparatus mostly on frogs, the implications of the study of the speed of nervous transmission for an experimental science of human thought had already been spelled out by Helmholtz.

alone (or, rather, the more ample German *Wissenschaft*, which covers the humanities and mathematics as well as the natural sciences). Indeed, Frege made objectivity the *sine qua non* of science. And whereas previous philosophers in the Kantian tradition, including Frege's own teachers and sources, had emphasized communicability among rational beings, Frege was prompted by recent empirical investigations of the mind to focus on the *obstacles* to communicability posed by subjective mental processes. What exactly was it about subjective mind, he asked, that made it so variable, so individualized, so private?

Frege's most immediate philosophical source for his understanding of objectivity seems to have been Hermann Lotze's *Logik* (1843), in which "logical objectification" refers not to the external world but to "the common world...that is the same for and independent of all thinking beings."[35] Frege, however, accepted as genuinely objective not only physical objects such as the sun and the North Sea but also scientific abstractions about the external world, such as the

266

earth's axis. These abstractions shared objective status with purely conceptual entities, such as number: "It does no damage to the objectivity of the North Sea that it depends on our arbitrary choice which part of the general water covering of the earth we delimit and call by the name of 'North Sea.' That is no reason to want to investigate this sea psychologically. So number is also something objective. If one says, 'The North Sea is ten thousand square miles in size,' one refers neither with 'North Sea' nor with 'ten thousand' to an internal state or process, but rather one claims something wholly objective, which is independent from our representations [*Vorstellungen*] and the like."[36] According to Frege, the objective need not be physically real; rather, the real is a subset of the objective, and the objective is in turn defined as "the lawlike, the conceptual, the judgeable, what can be expressed in words."[37]

Historians of philosophy have disagreed about whether Frege was reacting against German idealism or scientific naturalism, but we have Frege's own word as to which specific empirical studies of logic and mathematics he found objectionable.[38] Some of his targets were philosophers: he was contemptuous of John Stuart Mill's attempts to derive number concepts from the experience of counting pebbles.[39] Others were scientists: he indignantly rebutted the Vienna physiologist and histologist Salomon Stricker's claims that number concepts were acquired via the muscular sensations of eye movements while counting.[40] And he dismissed the ethnologist Thomas Achelis's view that the "generally valid norms of thought and action cannot be won by a one-sided, merely deductive abstraction, but rather through an empirical-critical definition of the objective, fundamental laws of our psychophysical organization, which are still valid for the broader popular consciousness [*Völkerbewußtsein*]." This "empirical-critical" definition of the norms of thought would come, Achelis insisted, not from philosophy but from ethnology and psychology as pursued by Wundt and his students.[41]

Mill, Stricker, and Achelis were spokesmen for an empirical approach to logic and mathematics; they had inspired or been inspired by the Wundtian program for an objective science of mind, but they were themselves neither logicians nor mathematicians. Frege, however, also detected dangerous defections to the empirical camp among his own colleagues.[42] He upbraided Hermann Hankel, the author of a book on complex numbers, for suggesting that key concepts might be defined by an appeal to empirical intuition [*Anschauung*], and even reprimanded Georg Cantor (whose mathematical theory of the transfinite Frege otherwise applauded, because it was so obviously remote from any possible experience) for having incautiously invoked "internal intuition [*innere Anschauung*]" when he ought to have provided a rigorous proof.[43] He accused the logician Benno Erdmann of conflating "the laws of thought [*Denkgesetze*]" with "psychological laws."[44] Frege was not even prepared to make concessions on pedagogical grounds. Chiding Ernst Schröder, the author of a textbook on arithmetic and algebra, for conflating concept formation with abstraction from a concrete object, Frege rejected induction as a means for deriving and defining mathematical entities like unity: "A concept does not stop being a concept even if only one thing falls under it, which is thus fully determined by [the concept]."[45]

By the time Frege took on these opponents in *Die Grundlagen der Arithmetik* (*The Foundations of Arithmetic*, 1884), a debate had been raging for at least a decade about whether mathematics and logic could withstand the onslaught of scientific physiology and psychology. Paul Du Bois-Reymond, who was the brother of the physiologist Emil Du Bois-Reymond and who had been cited with approval by Helmholtz in the article on arithmetic so noxious to Frege, tried to sum up the state of the controversy in his *Die allgemeine Functionentheorie* (*General Theory of Functions*, 1882). The "idealists" "posited a world that is not somehow subordinated to our representations [*Vorstellungen*], or even our most remote intuitions and concepts, but that nonetheless, beyond these representations, possesses a real content of which we are deeply conscious, even if [it is] humanly unimaginable." The "empiricists" countered: "We are not justified in assuming entities and in weaving them into mathematical thought processes from which we have and could not have any representation."[46] Mathematicians, psychologists, physiologists, ethnologists, and philosophers were involved in the debate, and Frege attacked them, one and all. He might attack, in one sentence, Mill for his philosophical naïveté; in the next, Helmholtz for his physiological presumption; in the one after that, Schröder for his psychological leanings. All fell afoul of Frege for conflating subjective representations with objective concepts.

What were the psychological entities that Frege found so threat-ening, and to which he opposed objective entities, both real and con-ceptual? Two categories, both derived from experience and both somehow visible to the mind's eye, defined subjective mind for Frege: representations (*Vorstellungen*) and intuitions (*Anschauungen*). Both of these terms carried venerable Kantian pedigrees in nine-teenth-century German philosophy, and their meanings had, by the latter half of the century, been further ramified by the empirical studies of the psychologists and physiologists (many of whom also took Kant as their departure point, or at least as their foil).[47] Frege's usage, indebted to both traditions, was roughly the following. Rep-resentations were mental pictures of objects formed either by sensa-tion or by imagination; intuitions were also somehow "picturable" but were more deeply rooted presuppositions about the spatial, tem-poral, and causal order of experience. Both were irretrievably subjec-tive, according to Frege. What made them subjective was not their failure to correspond to something in the external world; Frege's no-tion of the objective-but-not-real also failed the correspondence test. Rather, they were subjective because they were privately "owned," as opposed to objective thoughts, which were the common property of all rational beings: "Representations need a bearer [*Träger*].... To be the content of my consciousness belongs so essentially to each of my representations that every representation of another is indeed as such different from mine."[48]

Frege was aware that his use of the term "representation" to refer solely to the subjective deviated from standard usage, especially in contemporary psychology and physiology. In *Grundzüge der physiolo-gischen Psychologie*, Wundt had routinely distinguished between "ob-jective" representations, such as sensations, which are produced by stimulation of the nerve endings of sensory organs, and "subjective" representations, which are generated by the activities of conscious-ness. Even objective representations may not actually resemble the stimuli, but they are nonetheless causally linked to external stimuli.[49] Helmholtz had made a similar distinction in the context of sensory physiology: objective sensations referred to the external world, sub-jective ones to the sensory apparatus itself.[50] Yet Frege explicitly avoided the phrase "objective representations" as confusing and con-signed all mental pictures entirely to the realm of the subjective.

Anything that was picturable, subject to the laws of association, and above all private was *ipso facto* "psychological" and could not be modified by the adjective "objective."[51] Nor could it be scientific: "Thus, I can also acknowledge thoughts as independent of me; other men can grasp as much as I; I can acknowledge a science in which many can be engaged in research. We are not owners of thoughts [*Gedanken*] as we are owners of our ideas [*Vorstellungen*]."[52]

Over and over, in different ways and with different emphases, Frege argued that arithmetic is not particular to one person or another. Representations of the individual mind were inadequate to capture the concept of number. "If number were an idea, then arithmetic would be psychology. But arithmetic is no more psychology than, say, astronomy is.... If the number two were an idea, then it would have straightaway to be private to me only.... We should have to speak of my two and your two, of one two and all twos." Frege's opposition to psychology, both as scientific discipline and as subject matter, was at root hostility to empiricism as the ground of concepts. If representations and intuitions ultimately stemmed from experience, as the empirical philosophers and psychophysiologists claimed, then they could have nothing to do with logic and arithmetic. "In arithmetic," Frege concluded toward the end of *Die Grundlagen der Arithmetik*, "we are not concerned with objects which we come to know as something alien from without through the medium of the senses, but with objects given directly to our reason and, as its nearest kin, utterly transparent to it.... And yet, or rather for that very reason, these objects are not subjective fantasies. There is nothing more objective than the laws of arithmetic."[53]

Frege hoped to eliminate what he regarded as sins against the objectivity of arithmetic and logic by introducing new practices for proving theorems. Although he conceded to the psychologists that all rational beings known to us seem to require some "sensory perception" for "intellectual development," he maintained that mental pictures and intuitions smuggled into logical and mathematical demonstrations wrought havoc with rigor. Such elements derived from experience led to just the sort of sloppy inductions that Wundt had described as the origins of all mathematics and to gaps in demonstrations where appeals to intuition and ambiguous language replaced watertight arguments. The antidote would be a purely symbolic language of logical proof, the *Begriffsschrift* ("concept-writing"), which would purge the mind of both words and images: "In order that nothing intuitive can infiltrate [the proof] unnoticed, the seamlessness of the chain of inferences must be assured at all costs."[54] Frege likened the relationship between the *Begriffsschrift* and ordinary language to that between the microscope and the naked eye. The eye was more convenient for ordinary use, but only the microscope was suited for "scientific purposes."[55] Just as precision instruments had advanced the natural sciences and thereby revealed the errors of the unaided senses, the *Begriffsschrift* would, Frege hoped, free logic and arithmetic from the deceptions of intuitions and words, which were also tainted by the senses.

He admitted that the *Begriffsschrift* yielded no new results. Moreover, even his most sympathetic readers, such as his Jena physicist colleague and patron Ernst Abbe, found the symbolism rebarbative and the project eccentric; Russell confessed that he had possessed the book for years before he understood it, and then it became comprehensible only after "I had myself independently discovered most of what it contained."[56] Meant to guarantee the communicability and therefore the objectivity of arithmetic and logic, the *Begriffsschrift* itself proved opaque. Frege nonetheless insisted on the scientific utility of his symbols, which he saw as the partial realization of Leibniz's dream of a *characteristica universalis* and as potentially extendable to other sciences, such as mechanics and physics.[57] The *Begriffsschrift* would be a tool of structural objectivity, a shield to protect logic and arithmetic from both the psychological and the psychologists — at one point, he feared psychology would swallow up all sciences.[58]

Built into the symbolism of the *Begriffsschrift* was Frege's fundamental distinction between a "mental representation" (*Vorstellung*) of a certain specific content or state of affairs and a "judgeable" (*beurtheilbar*) conceptual content. Only the latter could be affirmed or negated and thus qualify for logical treatment. The "mere representation" was written in the *Begriffsschrift* as

$$- A;$$

and the judgeable proposition was written as

$$| - A.$$

271

If, for example, | − A signified the judgment that "opposite mag-netic poles attract each other," then − A signified "merely the men-tal representation of the attraction of opposite magnetic poles called to mind in the reader." Frege himself regarded this possibility of dis-tinguishing between content and judgment as key. When critics complained that Frege's *Begriffsschrift* was simply a more unwieldy version of George Boole's logical algebra, Frege retorted that the novelty of his symbolism lay in the possibility of representing "con-tent through written symbols in a more exact and comprehensive manner," not just in recasting logic into algebraic formulas.[59] In order to make the *Begriffsschrift* still more independent of the vaga-ries of intuition and language, Frege abandoned the ancient logical distinction between subjects and predicates. Although judgments might be differently formulated, all that mattered in the *Begriffss-chrift* was their "conceptual content," that is, the inferences that could be deduced from them. Frege noted further that while Aris-totelian logic identified a number of kinds of inference, all of them could be translated into his one principal form. But he emphasized that the preference for his one form over Aristotle's many had noth-ing "psychological" about it, being "only a question of form in the sense of the greatest functionality."[60]

Frege conceded that words and other symbols were an improve-ment over the particulars of sensation and memory, but he con-tended that they were still insufficiently general or precise for the formation of concepts, which must express what specific things have in common. Analogous to the human hand and the naked eye, natu-ral language was a flexible instrument but ill suited for the rigor demanded by science. What was needed was a specialized, deliber-ately unhandy tool: "And how is this exactitude made possible? By the very rigidity, the permanence of the parts, the absence of which makes the hand so all-around skillful." The *Begriffsschrift* would com-plete the mind's liberation from "the restless flow of our actual thought movements" by substituting a world of pure concepts and the logical relationships among them.[61]

The price of objectivity in logic and arithmetic, as set forth in the relentless formalism of the *Begriffsschrift*, was rigidity and strict con-trol, which "would permit no transition that did not follow the rules set forth once and for all."[62] The implication was that the temptation

272

to break the rules by an illicit appeal to sensation, intuition, or language would be otherwise irresistible. Like the photograph that checked the impulse to project sharp outlines and pleasing symmetries onto an imperfect specimen, the *Begriffsschrift* held all seductive pictures and equivocations at bay. Both served as sentries against subjectivity, but the one embraced images while the other repudiated them. (See figure 5.2.)

For Frege, the battle against subjectivity was not based in Platonic contempt for appearances or Cartesian distrust of bodily sensations but was rooted in the struggle to transcend the privacy and individuality of representations and intuitions. To understand why he and other advocates of structural objectivity could take for granted that sensations, representations, and intuitions *were* individualized, contrary to earlier epistemological assumptions, we must turn once again to the emergent sciences of physiology and psychology. Frege and his contemporaries were well aware that color sensation had, through the investigations of the sensory physiologists, become the foremost example of privatized subjectivity. Color sensations were emblematic of what structural objectivity was not: individualized, incommunicable, impermanent. How can I communicate what I see when I see red?

The Color of Subjectivity

By the late nineteenth century, color had become a paradigmatic example of private, incommunicable subjectivity. Despite the tendency of modern histories of epistemology to trace a continuous arc from seventeenth- through twentieth-century philosophical discussions of color, nineteenth-century reflections on the subjectivity of color were not just a variation on the early modern distinction between primary and secondary qualities.[63] Although that distinction received rather different formulations by, say, Descartes and Locke, it can be roughly summarized as the distinction between what the world is really like and our perceptions of the world. We humans infer that objects in the world are yellow or red or green because we see them as such, but in reality the colors are phantasms created by the interaction of our perceptual apparatuses with certain kinds of particles of different shapes and speeds. As Descartes puts it in his treatise on *Optics* (1637): "And first of all, regarding light and

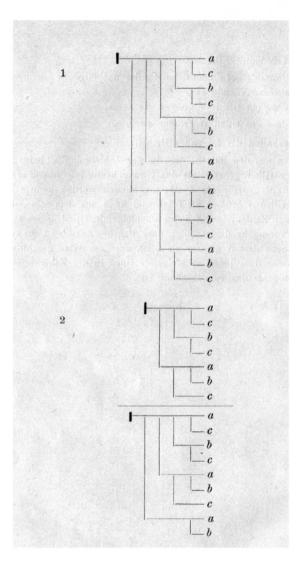

Fig. 5.2. Pure Thought. "Representation and Derivation of Some Judgments of Pure Thought," Gottlob Frege, *Begriffsschrift, eine der arithmetischen nachgebildete Formelsprache des reinen Denkens* (Halle: Nebert, 1879), p. 30. It took a full page of notations to express the principle of transitivity in the case of a series of numbers A, B, C ... in which each successive term is larger than its predecessors: if M is greater than L, then N is also greater than L. Frege himself realized that readers would find the details of his notation tedious. But precisely because his *Begriffsschrift* was so opaque and cumbersome, in contrast to diagrams that aimed at clarity and efficiency, Frege hoped that it would counter subjective intuitions.

color ... it is necessary to think that the nature of our mind is such that the force of the movements in the areas of the brain where the small fibers of the optic nerves originate cause it to perceive light; and the character of these movements cause it to have the perception of color: ... there need be no resemblance between the ideas that the mind conceives and the movements that cause these ideas."[64] This is a problem of representational accuracy: the contents of perception do not look like the things in the world, although perceptions and light stimuli may be (and usually are) reliably correlated with one another.

Now consider a characteristic expression of the problem of color as understood in the late nineteenth century, again by a philosopher-scientist, Poincaré. For Poincaré, the problem was one of the irredeemable privacy of sensation: "The sensations of another will be for us a world eternally closed. [Whether] the sensation that I call red is the same as that which my neighbor calls red, we have no way of verifying." This was enough to disqualify color as objective: "Nothing is objective but that which is identical for all; hence one cannot speak of such an identity unless a comparison is possible, and can be translated into a 'coin of exchange' that can be transmitted from one mind to another."[65] What was at stake here was not whether red was a property of the world or only the human way of perceiving the world but whether all minds perceived red the same way. It is the correspondence among minds rather than that between a mental picture (in any mind whatsoever) and the world that is at issue.

Poincaré deployed the post-Kantian, modern vocabulary of objectivity; insofar as Descartes used the words, it was with their old Latinate, scholastic meanings (and never to describe the problem of color).[66] But the contrast between these two framings of the problem of color runs deeper than terminology. Descartes was not particularly worried about the privacy of color sensations. Although he recognized that certain bodily disorders (for example, jaundice) may produce deviant color perceptions, he assumed that all normal minds perceived red in the same way. Nor was he concerned with verifying this assumption, finding a suitable way to communicate and compare his sensation of red with that of his neighbor. He was, in short, not moved by the modern dilemma of the gap between the objective and the subjective, as exemplified by the problem of color. He had other epistemological fish to fry, namely the unreliability of perceptions as opposed to clear and distinct ideas. Poincaré, for his part, no longer deemed Descartes's problem of color a philosophical problem at all; it was, rather, a fact of sensory physiology, exhaustively investigated by scientists, who, for example, had matched wavelengths of light measured in millimicrons to the perception of spectral yellow.[67] For Poincaré, the problem of color was one of individual variability and (as for Frege) communicability. Only pure relations (such as quantity), the invariants underlying the fluctuations of experience, were shared by all minds and therefore constituted "the sole objective reality . . . common to all thinking beings."[68]

It would be misleading to suggest that Poincaré, Frege, and other leading spokesmen for structural objectivity were particularly interested in the sensory physiology of color — they were not. Yet the late nineteenth-century science of color — a powerful combination of physics, physiology, and psychology — raised in sharpest form the difficulty that did exercise them: Could there be an objectivity of mind, and if so, how would it be related to the objectivity of the external world, on the one hand, and to the subjectivity of mental processes, on the other? More pointedly, what would be its relation to the most promising contenders for an objective science of mind, those new sciences known variously as sensory physiology, psychophysics, and physiological psychology? Was the objectivity of the empirical sciences of mind compatible with the objectivity of mind? It was in this context that mechanical objectivity provoked the reaction of structural objectivity.

These questions were new to the mid-nineteenth century and were prompted by the latest scientific developments. When, in the 1780s, Kant had discussed what was too subjective to be communicable to other rational beings, his examples were opinions and beliefs about such matters as the existence of God and an afterlife.[69] Among philosophical and scientific empiricists, reports of sensory experience, including scientific observations, had since the late seventeenth century been regarded as the most reliably communicable material — as thousands of pages in scientific journals and treatises bear witness. The association between experience and incommunicability was forged by the emerging experimental sciences of the senses in the first half of the nineteenth century.

Sensory physiology and philosophy were tightly intertwined, especially in Germany. Physiologists such as Müller and Helmholtz attempted to turn philosophical claims for the spontaneity of consciousness or the existence of the synthetic *a priori* into empirical research programs. Philosophers responded to the discoveries of the physiologists with challenges of their own.[70] The science of color in particular pioneered the use of the newfangled Kantian terminology of "objective" and "subjective" to describe both methods and subject matter. Already in 1810, when the words had scarcely entered German dictionaries in their new, Kantian sense, Johann Wolfgang von Goethe used them to organize the series of optical experiments reported in his treatise *Zur Farbenlehre (On Color Theory)*. In Goethe's usage, subjective effects are those that originate in the eye itself; objective effects originate in an external light source, usually the sun. Ideally, objective and subjective versions of the same experiment should be paired.[71] For Goethe, objective and subjective phenomena were complementary and equally essential to the science of colors. They differed in their locus (internal or external to the observer) and their duration (fleeting or more durable), but not their reality. Even among later scientists who disapproved of Goethe's anti-Newtonian tirades and found his methods too phenomenological, *Zur Farbenlehre* was praised as a treasure trove of "subjective" visual phenomena that attracted a new generation of researchers.[72]

Sensory physiologists soon anchored the new terminology of "objective" and "subjective" phenomena in practices of inquiry developed to explore the distinction. One of Goethe's most remarkable disciples, the Czech physiologist Jan Purkinje, refined self-observation and experimentation on what he, following Goethe, called subjective visual phenomena to the point where he could observe his own retina, as well as the blood vessels in the eye, and control the movements of the eyeball (see figure 5.3).

Most difficult of all, according to Purkinje, was the trained ability to separate objective from subjective visual impressions, which required the scientist to progress through a series of ever-more-demanding exercises in self-observation, until complete visual passivity was attained, so as to see "as the primitive [*Naturmensch*] sees a painting, as a mere surface of various colors. Through this abstraction, which is simultaneously the most specialized empiricism, one

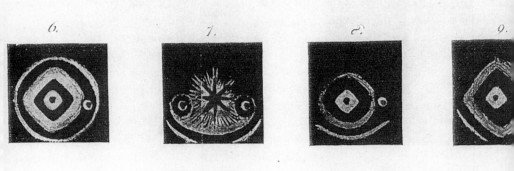

Fig. 5.3. **"Galvanic Light Figures."** Johann Purkinje, *Beobachtungen und Versuche zur Physiologie der Sinne* (Berlin: Reimer, 1823–1825), vol. 2, table 1, figs. 6–9. Dedicated to Goethe, Purkinje's account of his self-experimentation from "a subjective perspective" made distinctions between objective and subjective phenomena fundamental to sensory physiology. These figures were what Purkinje saw when he electrically stimulated his eyeball (6), his forehead (7), and the middle (8) and tip (9) of his eyebrow. Such perceptions were the fruit of discipline and practice: "It surpasses all imagining, how gradually the attention increases ever more in subjective experiments on sight and perceives phenomena that vision — usually lost in the external world — could otherwise never succeed in making sensible" (*ibid.*, p. 74).

enters into the sphere of the organic living subject-object, in which every material process is at once an ideal, subjective one."[73] As Purkinje and other sensory physiologists realized, such virtuoso feats of self-observation accentuated individual differences in sensory acuity and discipline. In his magisterial *Handbuch der physiologischen Optik* (*Handbook of Physiological Optics*, 1856–1867), Helmholtz paid tribute to these feats of observation but noted that some of the effects observed by Purkinje had yet to be achieved by other physiologists and suggested that perhaps they had derived from "the individual peculiarities of his organ [his eyes]."[74]

Even among subjective visual effects that numerous researchers, after some practice, could train themselves to see, individual variation persisted. This was often the case for phenomena of color vision. Helmholtz reported that he saw polarization figures "not just in homogeneous green, yellow, red, nor even in mixed, but rather in the saturated gradations of these color tones that colored glasses give."[75] Even for more mundane, objective visual phenomena, physiologists reported significant individual differences. The Prague professor of physiology Ewald Hering was surprised to discover through a series of exacting experiments in 1885 that he and his two assistants, Wilhelm Biedermann and Edgar Singer, diverged in their identification of spectral colors and mixtures thereof. All three were experienced and acute observers, a necessary precondition for such experiments, as Hering stressed, and all three tested normal by the usual standards for full color vision. Yet, reported Hering, "[a] green that appeared pure to me, was seen as decisively yellowish by B., and that which appeared to him as pure green seemed bluish to me: Between S. and B. there was an analogous and still more striking difference."[76] On the basis of these and numerous other divergences, Hering concluded that normal color vision was anything but uniform. Some cautious sensory physiologists and psychophysicists published individualized data, explicitly so labeled, for their own eyes (see figure 5.4).

Data poured in from other sources attesting to the individuality of color experience. Helmholtz's and Hering's experiments documented variability in the color vision of normally sighted and highly trained observers. Better known to the public at large were findings concerning color blindness and other deficiencies in color vision. In April 1876, a catastrophic train accident in Sweden was blamed on the color blindness of a railway employee who had fatally misread a signal. Of the 266 Swedish railway employees subsequently tested, 19 were pronounced color-blind. These findings created a sensation in the European press and, along with several important publications on the sensory physiology of color vision by Helmholtz and Hering, stimulated a burst of scientific research on the subject after *circa* 1875.[77] Not all this research was physiological; historical and ethnological studies examined the allegedly deficient color sense of archaic and primitive peoples. The Wroclaw opthamologist Hugo Magnus argued on the basis of philological evidence that the ancient peoples who had produced the Sanskrit Rigveda, the Hebrew Bible, and the Homeric epics could distinguish only the bright colors of red and

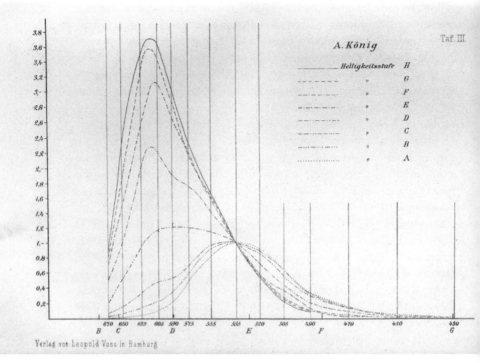

Fig. 5.4. Subjective Color, Objective Light Intensity. Arthur König, "Über den Helligkeit-swert der Spektralfarben bei verschiedener absoluter Intensität," in Arthur König (ed.), *Beiträge zur Psychologie und Physiologie der Sinnesorgane: Hermann von Helmholtz als Festgruss zu seinem siebzigsten Geburtstag* (Hamburg: Voss, 1891), pp. 309–88, table 3. The sensory physiologist König measured the perceived brightness of colors as a function of wavelength (given in micrometers on the abcissa) and absolute light intensity (the levels designated on the right by the filled and broken lines). These values are for König's own eye only; values for other experimental subjects (each individually designated with a name or cipher) were given in additional graphs.

yellow, while darker colors at the other end of the spectrum, such as blue and violet and perhaps even green, were designated and perceived as an undifferentiated dark hue.[78] Ethnologists jumped into the fray, testing so-called *Naturvölker* from equatorial Africa to the far reaches of North America with multicolored swatches to try to distinguish between genuine differences in color perception versus simply a scanty color vocabulary.[79] These and other well-publicized controversies in the 1870s and 1880s over the causes and frequency of color blindness and the historical and cultural development of color vision made the perception of color a paradigm of individual differences in mental representations. (See figure 5.5.)

For the most part, Frege, too, used color sensations as an obvious example of subjective mind — of mental representations that notoriously varied from person to person, like pain: "Whereas each [person] can only feel his pain, his desire, his hunger, can only have his sound and color sensations, numbers can be the common object for many, and indeed are exactly the same for all, not just more or less similar inner states from various [people]."[80] But there were other passages in which, repeatedly albeit fleetingly, Frege suggested that certain aspects of color might take on an objective — that is, structural — aspect. Notably, Frege enlisted the example of color blindness, an extreme example of variant color sensation, to make his point. Although color-blind people cannot distinguish between sensations of red and green, they can, Frege asserted, make the same linguistic distinctions that those with normal color vision do: "The color-blind person can also speak of red and green, although he cannot distinguish these colors in sensation. He recognizes the distinction from the fact that others make it, or perhaps by a physical experiment. Hence the color-word indicates often not a subjective sensation, about which we know nothing as to whether it agrees with that of others — for obviously the same name in no way guarantees this — but rather an objective property."[81]

The use of color words, rather than the experience of color sensations, could be made a matter of public agreement and therefore, Frege suggested, objective. This tentative strategy on how to make color objective was later pursued by Frege's student and admirer Ludwig Wittgenstein.[82] As in the case of number, Frege attempted to reconquer as much scientific territory as possible from the private

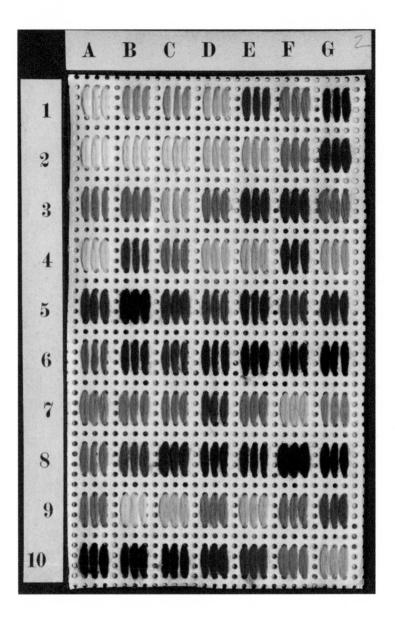

Fig. 5.5. Testing Color Sensations. A. Daae, *Die Farbenblindheit und deren Erkennung*, trans. from Norwegian by M. Sänger (Berlin: Dörffel, 1878). These colored yarn samples were used by ophthalmologists to test for color blindness and, more generally, the refinement of color perception. The test was originally introduced for signalmen on ships or trains, to make sure they could distinguish between red and green. It was subsequently also used for ethnographic inquiries into the color sense of non-European peoples, providing further evidence for the diversity of color experience. (Please see Color Plates.)

realm of the subjective — here venturing into what the late nineteenth-century sciences of mind had staked out as the inner keep of
the incommunicable. Other proponents of structural objectivity,
however, accepted the psychophysiological account of color as the
ineffable personal experience par excellence and sought a science
that could open windows for the closed-in self. It was no accident
that Poincaré chose to epitomize the privacy of the subjective by the
sensation of the color red.

What Even a God Could Not Say

Poincaré's account of what made science objective could be con-
densed into the lapidary motto "Pas de discours, pas d'objectivité."
This eliminated all sensations, including one's own. Psychophysiology taught that the "sensations of others will be for us a world eternally closed. We have no means of verifying that the sensation I call
red is the same as that which my neighbor calls red."[83] If I have the
color experience A when I spy a cherry, and someone else has the
color sensation B, we may both use the label "red," but the inner reg-
istrations of A and B are not comparable. The moment we want to
rely on color sensations or any other immediate experience, a veil of
solipsism descends, isolating us one by one. Here Poincaré confronted the same problem as Helmholtz and Frege. But if raw experience was not communicable, Poincaré continued, relations were.
"From this point of view, all that is objective is devoid of all quality
and is only pure relation. Certainly, I shall not go so far as to say that
objectivity is only pure quantity (this would be to particularize too
far the nature of the relations in question), but we understand how
someone could have been carried away into saying that the world is
only a differential equation."[84] All his life, Poincaré looked to these
equations to capture the elements of mechanics that, in either their
older Newtonian form or their newer incarnation, grasped the world
rationally. These compact forms were everything Poincaré liked:
they organized relations among phenomena; they held their distance
from any single interpretation; and they could be compared to locate
the simplest one that did the work to hand.[85]

   Poincaré's defense of "the objective value of science" was a battle
fought on two fronts. On the one hand, he opposed the traditional
metaphysics of truth with his philosophy of conventionalism. Simple

structures were, for Poincaré, the goal of scientific work, for it was precisely in this collective simplicity that convenience lay: convenience not just for you or me but for all people, for our descendants. This could not be just by chance. A quadratic equation was simpler than a cubic one, come what may and to whom it may. "In sum, the sole objective reality consists in the relations of things whence results the universal harmony. Doubtless these relations, this harmony, could not be conceived outside of a mind that conceives them. But they are nevertheless objective because they are, will become, or will remain, common to all thinking beings."[86] Yet objective reality did not equal truth from the viewpoint of a god. Science would never penetrate the true essence of things, not even with the aid of divine revelation. For how could these deepest truths be transmitted to human minds? "If any god knew it, he could not find words to express it. Not only can we not divine the response, but if it were given to us, we could understand nothing of it; I ask myself even whether we really understand the question."[87] Truth failed the test of communicability.

On the other hand, Poincaré resisted the radical empiricism of the Austrian physicist Ernst Mach, the American psychologist William James, the French philosopher Henri Bergson, and their followers. At this moment, *circa* 1900, some scientists, mathematicians, and philosophers abandoned lived experience as hermetically subjective, and others embraced it wholeheartedly: the really real, claimed the radical empiricists, is the phenomenological surface of things.[88] All speculation about what lay behind or between these sensations was the airiest of metaphysics. Physics, psychology, and physiology would, Mach asserted confidently, soon converge into a single science of the analysis of sensations. "For us, therefore, the world does not consist of mysterious entities, which by their interaction with another, equally mysterious entity, the ego, produce sensations, which alone are accessible. For us, colors, sounds, spaces, times... are provisionally the ultimate elements, whose given connexion it is our business to investigate. It is precisely in this that the exploration of reality consists."[89] Even the abstract concepts of physics and mathematics could ultimately be traced back to "the sensational elements on which they are built up."[90] These sorts of proclamations were sufficiently alarming to Planck for him to wage a sustained campaign against what he called Mach's anthropomorphism, but he never doubted Mach's loyalty to the scientific enterprise.[91]

More effusive devotees of radical empiricism, such as the French mathematician and philosopher Edouard Le Roy, plunged into the stream of experience headfirst, leaving science behind on the shore. True understanding, wrote Le Roy in his paeans to Bergsonian philosophy, meant immersion in the world of sensation, not in the dictates of modern science, "conceived of late under much too stiff and narrow a form, under the obsession of too abstract a mathematical ideal which corresponds to one aspect of reality only, and that the shallowest."[92] Borrowing the language of convention from Poincaré, his former teacher, Le Roy contended that scientific laws and facts were artificial, the fabrication of the scientist, and that science supplied nothing more than rules for practical action. As Poincaré himself paraphrased Le Roy's Bergsonian philosophy, "there is no reality except in our fugitive and changing impressions, and even that reality vanishes as soon as one touches it."[93]

Faced with Le Roy's corrosive "nominalism," decked out in the colors of his own conventionalism, Poincaré sought to articulate a form of objectivity that would be proof against such threats to the validity of science. No recourse to Truth with a capital T was possible; Poincaré had early and often rejected anything so metaphysical. Instead, he had espoused laws of science that resembled conventions for the international establishment of the meter more than they did the eternal forms in Plato's heaven. His highest praise for a scientific theory was that it revealed relations that stood the test of time, whether the entities it posited — electrons, the ether — were real or not.[94] Theories about the true nature of electricity or life were nothing but "crude images," images that were always temporary, in a perpetual state of flux in which one picture gives way to another. Nor would the analysis of sensations, *pace* the radical empiricists, suffice to guarantee the objectivity of science: how could anything so evanescent and ineffable be made common to all thinking beings? Instead, Poincaré found his answer to Le Roy and other doubters in the intellectual "coinage of exchange" that could be transmitted from one mind to another. No picture, whether theoretical or sensory, could fill this bill. All that many minds could hold in common were the relationships that "cemented" together groups of sensations. "Hence when we ask what is the objective value of science, this means not, Does science lead us to know of the true nature of things? but rather, Does it lead us to know the true relations of things?"[95]

This "indestructible cement" of relations persisted when particular theoretical schemes and experience faded. Science was for Poincaré a classification, and classifications were not true or false, only convenient or inconvenient.[96] Classifications laid bare hidden structures. At the heart of Poincaré's mathematics, for example, lay a fascination for the qualitative rather than quantitative study of differential equations.[97] That is, instead of trying to approximate the solutions to such equations by numerical series, he wanted to study the kinds of behavior that the solution curves exhibited. Did many solutions cross at a specific point ("node")? Did only two solution curves intersect at that point, with all others approaching it asymptotically ("saddle point")? Or did the solutions terminate in a single point ("focus") or orbit around one point ("center")? Using this division, he could classify the solution curves, prove that certain characteristic relations were true of the number of nodes, foci, and saddle points on surfaces such as the sphere. And in the application of such concerns to physical systems, he could distinguish between orbits of planets that stably remained within certain regions of space and those that would, in the fullness of time, wander off to infinity. When Poincaré turned to images, he typically depicted the topological (qualitative) — not the metrical (quantitative). He was after the relational, the structural (see figure 5.6).

Poincaré's injunction to heed enduring relations rather than ephemeral theories was not merely a historical lesson or philosophical adage; it shaped every aspect of his teaching and treatises. In his Sorbonne lectures on electricity and optics, delivered between 1888 and 1899, for example, he systematically reviewed the electrodynamics of André-Marie Ampère, Wilhelm Eduard Weber, Helmholtz, and Hendrik Antoon Lorentz. For each theory, he set out its principles and assumed entities; he developed the mathematics and then, crucially, extracted those commonalities among the theories that were in accord with experiment. Some theories opted for two electrical fluids, others for a single kind — as far as Poincaré was concerned, the key fact was that both could be rendered compatible with the observed laws of electrostatics. From the standpoint of a

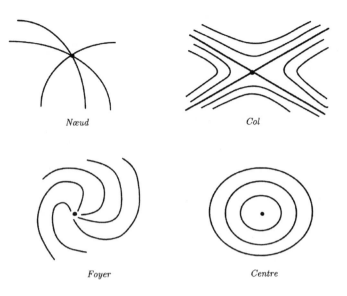

Nœud   *Col*

Foyer   *Centre*

Fig. 5.6. Poincaré's Relational Images. Simplified from Henri Poincaré, "Mémoire sur les courbes définies par une équation différentielle," *Journal de mathématiques* 8 (1882), pp. 251–96; this figure is from June Barrow-Green, *Poincaré and the Three Body Problem* (Providence, RI: American Mathematical Society, 1997), p. 32, fig. 3.2.i (reproduced by permission of June Barrow-Green and the American Mathematical Society). Poincaré developed a qualitative, topological approach to the study of differential equations. In physical terms, he imagined a plane drawn through the solar system so that an orbiting planet would puncture the plane each time around the sun. He could then study this map of successive punctures (consequents) — classifying the resulting curves by whether they formed nodes (*noeuds*), saddle points (*cols*), foci (*foyers*), or centers (*centres*). Poincaré often used images, sometimes very complex ones — but almost always images of this relational rather than representational type.

mechanical model, such details were a matter of indifference, for no mechanical explanation couched in terms of differential equations could be unique. Citing the controversy in optics between Agustin-Jean Fresnel, who had claimed that light vibrations were perpendicular to the plane of polarization, and Franz Neumann, who had contended that they were parallel to it, Poincaré concluded: "If a phenomenon permits one complete mechanical explanation, it permits an infinity of others that accord equally well with all the particulars revealed by experiment."[98]

It was not just a positivist distrust of metaphysics that drove Poincaré to treat the realist pretensions of scientific theory as so much ontological rococo. His researches on the quickly changing landscape of electromagnetic theory had impressed upon him how short-lived even the most promising theories often were. After judiciously weighing the claims of open- and closed-current theories in electrodynamics in light of the latest experimental findings, he was nearly ready to consign the open-current theories of Ampère and Helmholtz to the history books in favor of Maxwell's closed currents — yet the latest experiment by the French physicist Victor Crémieu had once again thrown everything into confusion. "I will not risk a prognostic that could be contradicted between the day on which it goes to press and that on which the volume appears in bookstores."[99] This was also the lesson taught by the history of science: Descartes had sneered at the pre-Socratic natural philosophers; Newtonians had mocked Descartes; no theory lasts forever.[100] On the first day, theories are born; on the second, these beautiful images of the world are all the rage; on the third, they are the classic, venerable theories of the world; on the fourth, they become superannuated; on the fifth, they are all but forgotten. Only relations endure. "If one of them has taught us a true relation, this relation is definitively acquired, and it will be found again under a new disguise in the other theories which will successively come to reign in place of the old."[101] Scientific objectivity, for Poincaré, was more than a matter of overcoming the privacy of subjective sensation; it was also the cord of continuity that connected scientists across generations.

Late in life, Poincaré mused over the moral import of science. Although he rejected any attempt to ground morality in science, he did entertain the possibility that doing science might nourish certain sentiments that could be harnessed to moral ends. Among these was the submersion of the self in a greater whole: "And science renders us another service; it is a collective work and cannot be otherwise; it is like a monument the construction of which requires centuries and to which each brings his stone; and that stone sometimes costs him his life. It thus gives us the sentiment of a necessary cooperation, of the solidarity of our efforts and those of our contemporaries, and even that of our predecessors and successors."[102] The stones in this grand edifice were neither facts nor theories, neither images nor truths; they were the relations that for Poincaré constituted objectivity. Relations intelligible to all thinking beings wherever or whenever they lived created a community that, like Planck's vision of interplanetary physics, knew no bounds. The objective cut across the particular or local; it went, ultimately, beyond even that which was human to embrace "all thinking beings." Russell echoed these capacious sentiments in a 1913 essay: science made the solitary individual "a citizen of the universe, embracing distant countries, remote regions of space, and vast stretches of past and future within the circle of his interests."[103] Objective thought might not capture much in the world, but it was the basis for science, community, and whatever hope of immortality anything human might aspire to: "Thought," observed Poincaré, "is only a gleam in the midst of a long night.... But it is this gleam which is everything."[104]

Dreams of a Neutral Language
The gleam of shareable thought caught the eye of Rudolf Carnap when he was a university student in Jena. Following courses on Frege's *Begriffsschrift* and philosophy of mathematics, Carnap found inspiration in the new view, espoused not only by Frege but also by Russell, Whitehead, and other mathematical logicians *circa* 1900, that concepts could be correctly understood only through symbols. Like Frege, he saw in the new symbolic logic a realization of Leibniz's *characteristica universalis*, now interpreted as a "theory of relations" [*Relationstheorie*] that would be applicable to all sciences. In his studies of philosophy and physics, Carnap came to realize that such a *scientia generalis* could not hope to unite the content of the various sciences.[105] His doctoral dissertation, *Der Raum* (*Space*, 1922), showed how physicists, philosophers, mathematicians were all after different things when they spoke about space: formal space, intuitive space, and physical space. Carnap experienced this perspectival diversity over and over again — as he moved from his religious home to a wider, more ecumenical university environment, fought in the bloodied trenches of the First World War, struggled for postwar socialism, and pressed for the adoption of new Esperanto-like languages. In philosophy, perspectival diversity reigned as well: "With one friend I might talk in a language that could be characterized as realistic or even as

materialistic; here we looked at the world as consisting of bodies, bodies as consisting of atoms.... In a talk with another friend, I might adapt myself to his idealistic kind of language.... With some I talked a language which might be labeled nominalistic, with others again Frege's language of abstract entities of various types, like properties, relations, propositions."[106] Carnap adamantly held to what he called his "neutral attitude," which he soon elevated to an ontological (and political) "principle of tolerance." The theory of relations he advanced in his magnum opus, *Der logische Aufbau der Welt* (*The Logical Construction of the World*, 1928), aimed to overcome "the subjective departure point of all knowledge in the content of experience" by constructing "an intersubjective, objective world...identical for all subjects."[107]

Objectivity, for Carnap, was deeply associated with this very particular way of abstaining from particularity while maintaining a commitment to the structural integrity of shared knowledge. To explain what he meant by a structure, Carnap asked his readers to imagine a map of the Eurasian railway network. Distances might not be represented to scale; the names of towns might have been omitted; all other geographical features might have been erased. Yet just by studying topological features such as the nodal points of the network — how many lines came in and out of a station — one could begin to identify stations. Should this structural feature be insufficient to differentiate all the stations — two or more might, for example, be the nodal points at which eight rail lines met — then other features (for example, telephone lines, the number of inhabitants of a town) could be used. If two places could not be differentiated by any such structural features, then they were "scientifically" identical: "That they are subjectively different from one another, in that for example I find myself in one place rather than the other, does not objectively signify a distinction."[108] (See figure 5.7.) Such structures were "neutral" as regarded the wearisome debates of idealists versus realists, being neither "produced" nor "simply recognized" by thought, but "constructed."[109] Following this ontologically and experientially neutral stance was key to Carnap's assembly of the *Aufbau*. To build from elementary bits to higher and higher forms, as in geometry, offered a *structure*, one that could be assembled in different ways using different starting points and (as in Hilbertian geometry) different contents.[110] Objectivity depended on structure alone; everything that pertained "not to structure, but to material, everything that is referred to concretely, is in the final analysis subjective" — and hence unfit for science.[111]

Carnap's neutralist stance toward structure involved more than logical quantifiers. For him and his Vienna Circle colleagues, it was also a moral stance, a way of life, in conscious defiance of traditional philosophy: "The new type of philosophy has arisen in close contact with the work of the special sciences, especially mathematics and physics. Consequently, it is the strict and responsible orientation of the scientific investigator that will be aimed at as the basic attitude in philosophical work, while the attitude of the traditional philosopher is more like that of a poet. This new attitude not only changes the style of thinking but also the task. The individual no longer undertakes to erect in one bold stroke an entire building of philosophy." Instead, the work would more closely resemble that of the physicist or historian who collaborates in the collective building-up of knowledge. "In slow, careful construction, one bit of knowledge after another will be secured; each contributes only what he can endorse and justify before the whole body of his coworkers. Thus, painstakingly, stone will be added to stone, and a safe building will be erected upon which each following generation can continue to work."[112] The practice of science and philosophy would, Carnap believed, find "inner kinship" with other movements in entirely other domains of life: architecture, education, and, more broadly still, in "meaningful forms of personal and collective life." These reforms overflowed the narrow confines of philosophy; nothing less was demanded than a new kind of person, a new "style of thinking and doing . . . the mentality that seeks clarity everywhere."[113]

This engineering-scientific ethos of a collective *Aufbau* in philosophy was what Carnap's fellow Vienna Circle enthusiasts — the physicists Philipp Frank and Moritz Schlick, along with the sociologist Otto Neurath and other like-minded colleagues — wanted as well.[114] The collaborative nature of their venture was built into the very typography of some of their texts. Carnap's *Aufbau* and *Logische Syntax der Sprache* (*The Logical Syntax of Language*, 1934) teem with references to others' work, not buried in endnotes but written into the flow of text, set off as discursive "references," parenthetical

Fig. 5.7. Structural Map. "Pneumatic Post Network," in Paul Kortz (ed.), *Wien — Am Anfang des XX. Jahrhunderts: Ein Führer in technischer und künstlerischer Richtung* (Vienna: Gerlach & Wiedling, 1905–1906), vol. 1, fig. 103, p. 153. This diagram of Vienna's steam-driven pneumatic postal system (whose average speed was 1 kilometer per minute) might have been the kind of structural network Rudolf Carnap had in mind. Not drawn to scale, the individual stations are distinguished only by their function (for instance, machine or air station) and their relative position as nodes within the network. The engineering of modern cities — building networks of trams, telegraphs, electricity, pneumatic tubes, and water pipes — made such diagrams emblematic of new ways of conceiving even old cities like Vienna as abstract spaces structured by relationships rather than by distinctive places and landmarks.

remarks, and asides. Together, in the late 1920s, the members of the Vienna Circle dissected texts, some sentence by sentence. Politics and values were to be checked at the door; "In logic, there are no morals," Carnap proclaimed. Anyone could construct his own logic — his own language — as he pleased. What one could not do was fudge syntactical rules or methods; unverifiable "philosophical arguments" were banished. Even in the apodictic realm of mathematics, Carnap applied this abstemious edict, opposing anything that smacked of dogmatism. In the debate over the foundations of mathematical functions, some mathematicians in the intuitionist camp demanded that all functions be actually exhibited explicitly — no arguments by contradiction would be allowed. Carnap wanted tolerance there — so long as both intuitionists and anti-intuitionists obeyed the rigorous rules of the new game that demanded all statements stay within the bounds of proper logical syntax and experience. Carnap went still further, extending his brief for tolerance to language, politics, and ontology.[115] Suppress unshareable experience, withhold absolute ontological commitments, reject universal procedural demands, leave dogmatic values and politics at home. Such abstinence would be realized in shared procedures, shared rules, shared constructions — the essence of communicable thought, all to be reformulated in terms of structures. Here lay objectivity.

In one such moment of wall building with fellow masons, Carnap insisted emphatically that "for science, it is possible and at the same time necessary to restrict itself to structure statements." Then he paused for one of his frequent in-the-text interventions, launched with an upper-case "REFERENCES." Here he tied his view back precisely to those passages of Poincaré's works that defined scientific objectivity in terms of relations, and bringing in the work of Russell:

REFERENCES. Considerations similar to the preceding ones have sometimes led to the standpoint that not the given itself (viz., sensations), but "only the relations between the sensations have an objective value." [Carnap cites Poincaré's *Valeur de la science* (*Value of Science*, 1905).] This obviously is a move in the right direction but does not go far enough. From the relations, we must go on to the structures of relations if we want to reach totally formalized entities. Relations themselves in their qualitative peculiarity, are not intersubjectively communicable. It was not until Russell [Carnap cites Russell's 1919 *Introduction to Mathematical Philosophy*] that the importance of structure for the achievement of objectivity was pointed out.[116]

Russell was quite explicit that the nature of a particular relation was of no importance, only the class of objects ordered by it mattered. "Father" picks out the ordered class of objects (x,y) such that x is the father of y. Having abstracted from the particular relation in question, Russell went further. Suppose *ab*, *ac*, *ad*, *bc*, *ce*, *dc*, and *de* are ordered relations of arbitrary terms *a* through *e*. Then this network of relations could be captured by a map (see figure 5.8) that would stand for a common structure corresponding to any number of particular realizations in the phenomenal world of experience (particular values of the elements). Personally as well as philosophically, Russell brushed aside the significance of particulars. When James wrote to him in 1908 urging that he give up mathematical logic in order to hold fast to "concrete realities," Russell coolly replied: "But on the whole, I think relations with concrete realities a barrier to understanding the general characteristics which different things have in common, & the general interests me more than the particular."[117]

Russell argued that structure relations would recover the "objective counterparts" to subjective phenomena, including those of space and time. "In actual fact, however ... the objective counterparts would form a world having the same structure as the phenomenal world, and allowing us to infer from phenomena the truth of all propositions that can be stated in abstract terms and are known to be true of phenomena."[118] If the human phenomenal world has three dimensions, then so should the objective structure to which it corresponds; if the phenomenal world is Euclidean, then so must be the objective world of structure. Philosophers, Russell lamented, have all too often sought the ontological bedrock by driving a wedge between experience and reality. The few thinkers who had tentatively proposed a correspondence between phenomena and the real had been too timid, fearful of conflating phenomena and noumena. By Russell's lights, however, these difficulties vanished if the analogies between the worlds of experience and reality were articulated in terms of structure rather than content: "Every proposition having a

294

Fig. 5.8. Russell's World Structure.
Bertrand Russell, *Introduction to Mathematical Philosophy* (1919; London: Allen & Unwin Ltd., 1924), figure on p. 60 (the Bertrand Russell Peace Foundation Ltd.). For Russell, a map of relations reveals its *structure*. For example, the map of fig. 5.8 picks out ordered couples connected by arrows. The "field" may be changed without changing the structure (swap a new entity *q* for old *d* but keep the arrows the same). Conversely, the field can remain the same but the structure can be altered (add an arrow from *a* to *e*, for example). Russell took there to be two corresponding worlds with the same structure: a phenomenal (subjective) one and abstract (objective) one — because of this correspondence, Russell contended we can, in fact, *know* the objective world through experience. Any communicable proposition must be true of both worlds or neither.

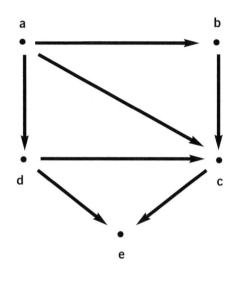

communicable significance must be true of both worlds or of neither: the only difference must lie in just that essence of individuality which always eludes words and baffles description, but which, for that very reason, is irrelevant to science."[119] Like Poincaré, like Carnap, Russell located the scientific in the structural and the communicable — and opposed both to the *je ne sais quoi* of individuality

For these philosophers and scientists, structure safeguarded communicability — among generations of scientists, among cultures, even among species and planets. This was a lesson that Schlick — the Vienna Circle's unofficial leader — took to heart. He remarked that Albert Einstein had relied on coincidence to define events — when the train approached a clock in Einstein's 1905 paper on special relativity, for example, and again to define an event in space-time in his 1915 general theory of relativity. Schlick asked: "Why exactly do we make use of this procedure? The only correct answer is, because of its objectivity, that is, because of its inter-sensual and inter-subjective validity." When one moved the tips of one's fingers together, contact was perceived as both a tactile and a visual coincidence of events. Just as these two senses were registered independently of each other in a single person, Schlick reasoned, other observers would confirm by their own visual sense that the two fingers touched.[120] "In general objectivity obtains only for those physical propositions which are tested by means of coincidences, and not for propositions which are concerned with qualities of colour or sound, feelings such as sadness or joy, with memories and the like, in short, 'psychological' propositions."[121]

Like Frege and Poincaré, Schlick defined objectivity in terms of independence from the physiological and the psychological, conceived in terms of individual variation. This epistemological flight from a certain kind of body endowed with certain kinds of sense organs sometimes left the realm of the human altogether. Kant had sought knowledge valid for all rational beings, even for angels. In the still-new twentieth century, Schlick began imagining bizarre, surgically created monsters for whom objective knowledge ought still to be valid:

> We now imagine that by means of an operation the optic nerve is connected to the ear, while the auditory nerve is joined to the eye. We should then hear all light-impressions as sounds, whereas all tonal impressions would be seen as colours or shapes. A painting would produce on us the impression, say, of a musical composition, while a piece of music, conversely, would appear to us as a coloured picture. The world of our experience would thus be utterly and entirely different ... but there is no doubt that a man ... if only he had enough intelligence, would come to establish exactly the same natural laws as we do, and his description of the universe would coincide perfectly with our own.... He would paint his world — utterly different in content from ours, but yet it would somehow display exactly the same abstract order or structure.[122]

What had begun as a quest to transcend the idiosyncrasies of individual human experience as documented by the psychophysiologists had grown into an ambition to cast off even the constraints of species.

The Cosmic Community

Schlick's cross-wired monsters hint at just how cosmic the community of structural objectivity had become by the early twentieth century. The fantasies of the scientists and philosophers chimed in with those of *fin-de-siècle* novelists imagining extraterrestrial life. *Circa* 1900, the aliens of science fiction ceased to be human-animal hybrids (bird-men, frog-men, and so on) and became truly other in physical form and sensory apparatus: geometric figures pulsing with light; faceless, antlike moon dwellers; gelatinous, cyborg Martians.[123] Schlick's monsters look neighborly by comparison. In an 1896 article on the possibility of intelligent life on Mars, the novelist H.G. Wells asserted that "there is every reason to think that the creatures on Mars would be very different from the creatures on earth, in form and function, in structure and in habit, different beyond the most bizarre imaginings of nightmare ... even granted that the unimaginable creatures of Mars had sense-organs directly comparable with ours, there might be no common measure of what they and we hear and see, taste, smell, and touch."[124]

Yet in turn-of-the-century tales of earthling-extraterrestrial encounters, intelligences connect where all analogies of anatomy, sensation, and passions fail. In Wells's own novel *The First Men in the Moon* (1901), the scientist Cavor, in the hands of Selenite captors, cannot repress a shudder of horror at their utterly inhuman appearance. In conversation with their leader, a huge brain attached to a shriveled, insectile body, Cavor, however, gradually overcomes his revulsion, as mind interacts with mind: "I found something reassuring by insensible degrees in the rationality of this business of question and answer. I could shut my eyes, think of my answer, and almost forget that the Grand Lunar has no face."[125] However strange the form taken by the aliens was and however malevolent or inscrutable their emotions and intentions were, their capacity to communicate with humans as well as with one another was largely taken for granted. The luminous cones and cylinders of J.H. Rosny's *Xipéhuz* (1888) aim to massacre the human race; their outward forms and senses resemble those of no other life form known to the nomad Bakhoûn, who warily observes them. Nonetheless, he soon discovers that they communicate by means of symbols and are capable "of exchanging ideas of an abstract order, probably equivalent to human ideas."[126] In Camille Flammarion's *Fin du monde* (*The End of the World*, 1894), when Martians send Earthlings a telephotogram to warn of an impending collision of the earth with a comet about to land somewhere in Italy ("Get out of Italy," the message helpfully concludes), little skepticism arises regarding the existence of Martians or their ability to communicate ideas intelligible to humans; rather, debate ensues as to whether they really know Italy by name — one of those subjective particulars left out of Carnap's structural railway map (see figure 5.9).[127] In the cosmic community imagined by these writers, wholly different senses, wholly different emotions, even wholly different bodies offer no impediment to communication among intelligent beings. All one has to do is close one's eyes, to block out the distracting, distorting, disturbing images — and think of objectivity.

These were, of course, the extravagant scenarios of science fiction, not the workaday experience of scientists. Yet at a more earthbound level, the wave of international collaborations — and international rivalries — that swept over late nineteenth-century science created practical problems of communicability that plagued even polyglot scientists. Large international congresses resembled convocations of diplomats wrangling over treaties, complete with national delegations, competing interests, long memories for past slights and honors, and Byzantine protocol. For example, the correspondence surrounding the mammoth star-mapping project known as the *Carte du ciel*, launched by an international congress held in Paris in 1887, is full of intrigues, spats among national delegations, and efforts to secure a linguistically gifted chairman (Otto Wilhelm Struve, the director of the Pulkovo Observatory in Russia, was chosen). An entire folder bulges with the careful preparations for evening dinners at the congress, down to the minutely planned seating charts.[128] Even among scientists assembled in pursuit of a common goal — the mapping of the heavens, the determination of the gravitational constant, the standardization of the meter — smooth communication could not, as a practical matter, be taken for granted.[129]

Moments of mutual incomprehension must have become a routine part of scientific life in the latter half of the nineteenth century, as travel from lab to lab, congress to congress, university to univer-

Fig. 5.9. Get Out of Italy. "The Martian dispatch is projected on the screen," Camille Flammarion, *La fin du monde* (Paris: Flammarion, 1894), p. 133. A packed public session of the Paris Académie des Sciences scrutinizes a photophonic message in a kind of hieroglyphic code sent by Martian astronomers in Flammarion's apocalyptic novel. Although Mars has not previously communicated with earth, no one has any trouble deciphering the message; a map identifies Italy (more specifically, Rome; more specifically still, the Vatican) as the point of the comet's impact. In *fin-de-siècle* futuristic fantasies like this one, communication with aliens seldom poses a problem; thinking beings, however strangely formed, are assumed to understand one another.

sity intensified. Even stay-at-homes like Charles Darwin had to wrestle with publications in foreign tongues.[130] Perhaps such encounters were the backdrop for the comparative psychologist C. Lloyd Morgan's claim that understanding the mind of one's dog was only quantitatively, not qualitatively, different from trying to fathom that of a foreigner.[131] When Frege claimed that "the more strictly scientific an exposition is, the less noticeable the nationality of the author will be [and] the easier it will be to translate," he was expressing not only an ideal of objective thought but also a rule of thumb by which to gauge it, one all too familiar to scientists of the day.[132]

Seen in the light of both extraterrestrial fantasies and globe-trotting practicalities, the community of all thinking beings postulated by the mathematicians and logicians looks positively cozy. Certainly some of its would-be members derived comfort from the comradeship they imagined they would find there, conversing about structures across time and space. Russell, exhausted and isolated by his demanding work on mathematical logic, wrote to a friend about the consolations of his "imaginary conversations with Leibniz, in which I tell him how fruitful his ideas have proved, and how much more beautiful the result is than he could have foreseen; and in moments of self-confidence, I imagine students hereafter having similar thoughts about me. There is a 'communion of philosophers' as well as a 'communion of saints,' and it is largely that that keeps me from feeling lonely."[133] Einstein also sought solace in what he called a "paradise" beyond the personal, populated by "friends who cannot be lost," "people of my type [who are] largely detached from the momentary and the merely personal and who devote themselves to the comprehension of things in thought."[134] The practice of mechanical objectivity had been a solitary and paradoxically egotistical pursuit: the restraint of the self by the self, of will affirmed by the very act of will denied. In contrast, structural objectivity demanded self-effacement — or at least self-narrowing, stripping away all but thought in order to enter a community.

Some, such as Poincaré and Carnap, experienced this obliteration of individuality as a sacrifice. But others, including Russell and Einstein, welcomed it as a liberation, "an escape from private circumstances, and even from the whole recurring human cycle of birth and death."[135] Still others, like Peirce and the British statistician Karl

Pearson, couldn't make up their minds. Both thought that what Pearson called "self-elimination" would come at the cost of a heroic struggle with egotism in the name of duty — for Peirce, the duty to strive for logical validity; for Pearson, the duty of the ideal citizen (already exemplified in his opinion by the man of science) to "form a judgment free from personal bias."[136] Yet both sometimes also wrote of the attainment of impersonality in and through science as the zenith of self-cultivation, a flight from "that tiresome imp, man, and from the most importunate and unsatisfactory of the race, one's self."[137]

But the practices of structural objectivity — Frege's *Begriffsschrift*, Poincaré's panoramic surveys of theories, Carnap's ardent neutrality, Russell's "communion of philosophers" — were not solely concerned with the suppression of subjectivity. They expressed a yearning, as well as a fear: longing for a common world, and one that can be communicated, not just experienced. For the proponents of a certain kind of objectivity, if even godlike knowledge of the nature of things failed the test of communicability, it could not be science. The struggles of nineteenth-century scientists to dampen the pathologies of the will were not enough. Self-restrained image making could not satisfy many early twentieth-century physicist philosophers and mathematician philosophers whose worries went far deeper. They were suspicious of their own psychology, dubious about the deceptiveness of naive visualization, dismissive of worldviews and school philosophies. Structural objectivity may not, in the final analysis, serve truth so much as the cosmic community, Poincaré's "universal harmony."

A curious parallel between self and world governed conceptions of structural objectivity. On the side of the world, all that mattered were structures — not phenomena, not things, not even scientific theories about things. Observed phenomena and conjectured mathematical models, primary and secondary qualities, were all on a level as far as structural objectivity was concerned: not so much unreal as irrelevant. On the side of scientific self, only that small sliver of the thinking being counted, purified of all memories, sensory experience, excellences and shortcomings, individuality *tout court* — everything except the ability "to provide an argument which is as true for each individual mind as for his own."[138] Structural objectivity did not so much eliminate the self in order to better know the world as remake self and world over in each other's image. Both had been stripped down to skeletal relations, nodes in a network, knower and known admirably adapted to each other. The German mathematician Hermann Weyl captured this parallelism in a metaphor that has proved singularly tenacious: invariants under transformation. Attempting to explain Johann Gottlieb Fichte's and Edmund Husserl's notion of "the absolute ego [*das absolute Ich*]," Weyl reached for an analogy from projective geometry. The points stand for objects in the world; the ordered triples locate points in a coordinate system for subjects. If the ordered triples are regarded only as numbers, "the experience of a pure consciousness," these numerical relationships will be unaltered by a change of coordinate systems — that is, by any arbitrary linear transformation. Under such transformations, all the subjective egos "have equal rights" so long as only the objective relationships are considered, as opposed to the geometric points that preserve indelible individuality.[139]

For Weyl — as for Carnap and Cassirer — Einstein's special theory of relativity provided inspiration for a new scientific philosophy, with structural objectivity as its centerpiece. Einstein was preoccupied throughout his career with the meaning of objectivity in physics. His view could be and was taken as a form of structural objectivity. But a close reading of his reflections on objectivity with regard to relativity reveals a more subtle position.

Einstein charged that in Newtonian theory "the present" identified points in time uniquely for all reference frames: "Silently assum[ing] that the four-dimensional continuum of events could be split up into time and space in an objective manner — i.e., that an absolute significance (a significance independent of observer) attached to the 'now' in the world of events."[140] Einstein took special relativity to shatter the objectivity that seemed to characterize time by itself. Time could be defined objectively only alongside space. Using the notion of a rigid body (a body that could move but not change state), Einstein contended, we build the idea of space. In particular, a rigid ruler could lay out the spatial coordinates of Euclidean geometry. How could one similarly define a public (shared) idea of the "now"? Einstein's May 1905 solution to that problem, the final step in his construction of special relativity, set a procedure for non-arbitrarily defining "the same time" at distant points A and B. Put identical clocks at A and B: Einstein coordinated them by sending a light signal from A to B, bouncing it off B, and measuring the round trip time back to A. If the round trip took, say, two seconds, then the one-way trip could reasonably be taken to be one second. So if clock A sends a light signal at noon, clock B gets set to noon plus one second when the flash arrives. In this way, Einstein had what he considered a criterion for "objective" time — two events were simultaneous in a frame of reference if they occurred at the same time as measured on synchronized clocks.[141]

Here's the first rub: in the special theory of relativity two events simultaneous in one constantly moving reference frame are not simultaneous in another: as Einstein says, "'Now' loses for the spatially extended world its objective meaning. It is because of this that space and time must be regarded as a four-dimensional continuum that is objectively irresoluble, if it is desired to express ... objective relations without unnecessary conventional arbitrariness."[142] In other words, two observers will disagree as to the separation of two events in both space and time — there is no unique division between differences in space and differences in time that will be shared by all observers. An analogy helps. In ordinary Euclidean space, the "difference in x" and "difference in y" between two spatial points are arbitrary; those differences depend on the orientation of the coordinate system. But Pythagoras tells us that the *distance squared* between the two points $[(\Delta x)^2 + (\Delta y)^2]$ is fixed no matter how the coordinate system is rotated. If it is two miles from my house to yours, that's that. Einstein insisted (using the language of the mathematician Hermann Minkowski, in units where the speed of light is one) that a similar situation held in relativity: the "space-time distance squared" $[(\Delta t)^2 - (\Delta x)^2]$ does *not* depend on the inertial reference frame even though the different, constantly moving observers will disagree about the difference in time (Δt) or the difference in space (Δx) separately. Or, as Minkowski put it, "Space and time are doomed to fade away into mere shadows and only a fusion of the two will remain." Einstein called that fusion "objective," though in his general relativity theory Minkowskian space-time was but a special case.[143]

The second rub: Einstein did not take objectivity itself to be purely objective. In a 1949 essay written in honor of Einstein, the philosopher Henry Margenau offered a view that was — and remains — quite common among philosophers: objectivity was that which remained invariant under changes of perspective, often characterized as group transformations. Like all structural objectivists, Margenau protested that the world of the senses could never, on its own, vouchsafe objectivity — it could not truly be, as he put it, "independent of the observer." Objectivity "must have as few anthropomorphic traits as possible. One might mean thereby that reality must appear the same to all, appear, that is, in sensory perception. But this can certainly never be assured in view of the intrinsic subjectivity of all our sensory knowledge."[144] Nor (according to Margenau) is the desired interpersonal aspect of theories captured simply by making correct predictions. Rather, "the criterion of objectivity lies somehow within the very structure of theory itself . . . that is within some formal property of the ideal scheme which pretends to correspond to reality." The question for Margenau was, What property of this "structure" or "ideal scheme" (his terms) could be objective? Ordinary distance by itself wasn't objective — that differed from one moving observer to another — only relativistic invariants were. Generalizing, Margenau asserted: "Objectivity becomes [for Einstein] equivalent to *invariance* of physical laws, not physical phenomena or observations."[145] For Margenau, theories were structures — a necessary criterion for objectivity — but only invariance secured that status.

Einstein bridled at Margenau's interpretation, which he found far too constraining: "This discussion has not convinced me at all. For it is clear *per se* that *every* magnitude and every assertion of a theory lays claim to 'objective meaning'," but that objectivity exists only "within the framework of the theory." Only in theories that claim that "the same physical situation" holds — under different descriptions — does the problem of group invariance arise: "It is . . . not true that 'objectivity' presupposes a group characteristic, but that the group-characteristic forces a refinement of the concept of objectivity." True, invariance under a group is heuristically useful because it radically limits possible theories. In that case, as in relativity, the idea of invariance *is* a valid constraint on what is truly shared (objective) in the mathematical-physical structure. But invariance under transformation was not, for Einstein, a *sine qua non* of objectivity in general.[146]

Einstein's caution against identifying group invariance and objectivity was but one caveat to Margenau and the philosophers. Objectivity went to the heart of Einstein's understanding of Kant, whose adage moved him: "The real is not given to us, but put to us [*aufgegeben*] (by way of a riddle)." Einstein wrote: "We represent sense-impressions as conditioned by an 'objective' and by a 'subjective' factor. For this conceptual distinction there is no logical-philosophical justification. But if we reject it, we cannot escape solipsism. It is also the presupposition of every kind of physical thinking.... The only justification lies in its usefulness.... The 'objective factor' is the totality of such concepts and conceptual relations as are thought of as independent of experience, viz., of perceptions."[147] In the case of relativity, Einstein took subjective time to be the beginning of our construction of objective, coordinated time. That subjective starting point, alongside what he always insisted was a *conventional* method for coordinating clocks, showed very clearly how inextricable the subjective and objective were within a theory. Einstein's synchronization was not simply "given to us" as an unavoidable bit of "raw data," nor was it a logical necessity. Yet the justification of the convention was nonetheless achieved through the success of special relativity as a whole.

So was Einstein a structural objectivist? Yes and no. Yes, he was relentless in his hunt for theoretical structures that "conditioned" our sense impressions. Yes, within the relativity theories he sought invariance — in many ways, this was his life's work. But, at the same time, Einstein insisted over and over that as indispensable as objectivity was, physics did not come to it element by element or even symmetry by symmetry. Instead, objectivity issued from the integrity of a theory like relativity taken as a whole, complete with principles, observations, and conventions. For Einstein to take invariant structures as objectivity was far too narrow. But to identify mathematical-physical structure *per se* with objectivity was far too broad: Einstein took each theory, with its peculiar combination of conventional and nonconventional elements, to pick out the objective.

Einstein's protests to Margenau notwithstanding, the subtlety of his theory-specific holistic approach to scientific objectivity left little trace in later philosophical views on objectivity. Transmitted to analytic philosophy via the writings of Frege, Carnap, Poincaré, Schlick, and Russell, structural objectivity retains its hold within contemporary epistemology. The suspicion of the individual, the private, the sectarian, and the ineffable has, if anything, deepened; the positive ideal of objective knowledge as that which remains invariant under the transformations of any and all perspectives is still current. In a particularly striking formulation, the philosopher Thomas Nagel called this kind of objectivity "the view from nowhere":

> A view or form of thought is more objective than another if it relies less on the specifics of the individual's makeup and position in the world, or on the character of the particular type of creature he is. The wider the range of subjective types to which a form of understanding is accessible — the less it depends on specific subjective capacities — the more objective it is. A standpoint that is objective by comparison with the personal view of one individual may be subjective by comparison with a theoretical standpoint still farther out.... We may think of reality as a set of concentric spheres, progressively revealed as we detach gradually from the contingencies of the self.[148]

The knower who moves outward through Nagel's concentric spheres undergoes a winnowing in which "the contingencies of self" — but not the thinking essence — are stripped away. The philosopher Robert Nozick adapted Weyl's metaphor of mathematical transformation to make much the same point. He defined an objective fact as "one that is invariant under (all) admissible transformations"; the title of the book in which this definition appears — *Invariances: The Structure of the Objective World* (2001) — rings all the changes on the theme.[149] Only structures, according to these philosophers and their predecessors survive the vicissitudes of many minds (human, angelic, Martian), of many worlds (physical, chemical, biological), and, above all, of the many theories that litter the history of science. By x-raying the object of knowledge into structures and distilling the subject of knowledge into a thinking being indistinguishable from all other thinking beings, objectivity is preserved — or, at least, that is the hope.

It was, however, a hope purchased at a high price, as far as empirical scientists were concerned. Although they acknowledged the limitations of mechanical objectivity, they were not prepared to abandon the world of sensory experience or the scientific images that aimed to represent it. Nor were they ready to surrender representations and intuitions in order to achieve the kind of scientific self that could be inducted into the cosmic community dreamed of by the mathematicians and logicians. Instead, they plunged back into the visual, into sensations and images.

In the twentieth century, scientists still committed to knowledge of the eye produced atlases on everything from stellar spectra to ganglia that proudly proclaimed their subjectivity. In explicit defiance of the canons of mechanical objectivity, they championed judgment and intuition. Neither genius nor labor would reveal the right image; what was needed was self-confident expertise. This was a scientific persona openly guided by unconscious intuition and perceptual habit, anathema to advocates of both structural and mechanical objectivity. In Chapter Six, we trace this second, opposed reaction to mechanical objectivity and explore the epistemic virtue it called into existence: trained judgment.

CHAPTER SIX

Trained Judgment

The Uneasiness of Mechanical Reproduction

In 1905, the radiologist Rudolf Grashey and a number of his contemporaries could no longer contain the many in the one. For them, the link to the multitude of variants could not be held in any single representation, be it ideal, typical, or characteristic. Instead, the most a picture could do was serve as a signpost, announcing that this or that individual anatomical configuration stood in the domain of the normal. By the 1930s, Grashey was relentless in his analysis of errors that could be produced through the naive use of the x-ray. But beyond any particular problem of distortion or spurious juxtaposition, Grashey attacked a more fundamental difficulty associated with the use of individual photographs to demarcate the normal from the pathological. The problem is this: If one is committed, as was Grashey, to the mechanical registration of images of individuals, then how can one distinguish between variations within the bounds of the "normal" and variations that transgress normalcy and enter the territory of the pathological? Grashey's own solution was to elevate the most striking of such rare deviations to a place of honor (*Ehrenplatz*) in the x-ray laboratory.[1] They would then serve as boundary posts of the normal, guiding the diagnostician away from false attributions of pathology. In the early 1900s, moreover, the metaphysical position underlying Grashey's view was widespread: no single scientist could capture a category, whether it was composed of normal skulls or just about anything else. The implicit nominalist metaphysics that had prevailed under mechanical objectivity for much of the late nineteenth century was destabilizing (see figure 6.1).

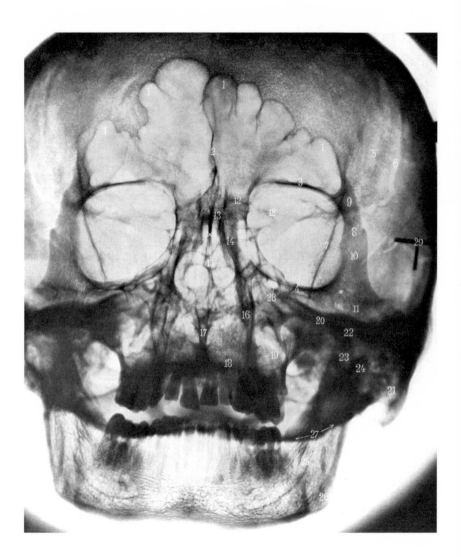

Fig. 6.1. A Normal Variant. Rudolf Grashey, *Atlas typischer Röntgenbilder vom normalen Menschen*, 6th ed. (Munich: Lehmann, 1939). Grashey transferred classification from author to reader by publishing a series of "wanted posters" (*Steckbriefe*) that illustrated the far reaches of the normal and thereby distinguished the normal — with all its variations — from the pathological.

This chapter describes how the ambition to produce an objective image mechanically came to be supplemented by a strategy that explicitly acknowledged the need to employ *trained judgment* in making and using images. Slowly at first and then more frequently, twentieth-century scientists stressed the necessity of seeing scientifically through an interpretive eye; they were after an *interpreted image* that became, at the very least, a necessary addition to the perceived inadequacy of the mechanical one — but often they were more than that. The use of trained judgment in handling images became a guiding principle of atlas making in its own right. Where the eighteenth-century atlas maker took it as obvious that idealization was precisely what was called for, by the mid-nineteenth century many scientists considered idealization anathema. But the history of epistemic virtues did not stand still. In the early twentieth century, a widening circle of scientists in diverse disciplines began to chafe under the constraints of the mechanical image, even while the old forms of scientific sight persisted. In short, a new possibility arose: judgment-inflected vision as a goal for scientific sight.

Along with this new form of seeing and new status of depiction came a different way of cultivating the scientific self. Self-denial and actively willed passivity were intrinsically conscious; therein lay their moral worth, as deliberate sacrifices made to scientific objectivity. Yet by the 1920s, after an efflorescence of psychologies of the unconscious of which Freudianism was only the most famous, scientists writing about how to live the scientific life no longer envisioned it as a conscious inward struggle of the will against the will.[2] Indeed, it was not a struggle at all, or at least it was not a struggle that promoted scientific achievement, *pace* nineteenth-century claims to the contrary.

Now it could be said, as Sir Peter Medawar, a winner of the Nobel Prize for Physiology or Medicine, did in the 1970s, "A scientist's life is in no way deepened or made more cogent by privation, anxiety, distress, or emotional harassment."[3] Nor was the most important intellectual work necessarily even conscious, for discovery and insight depended on hunches that erupted suddenly from the inaccessible mental depths. Such "leaps of the imagination" were thought to result from long incubation and rest and to occur "at a time when the investigator is not working on his problem." After telling several stories of such thunderbolt inspirations in science, the

Harvard neurologist and physiologist Walter B. Cannon in 1954 likened the process to a not-too-well-supervised factory: "The operation going on in an industry under the immediate supervision of the director is like the cerebral processes to which we pay attention; but meanwhile in other parts of the industrial plant work is proceeding which the director at the moment does not see. Thus also with extraconscious processes."[4] There was nothing to be gained by dogged perseverance; better to put the problem aside or, better still, as the endocrinologist Hans Selye assured readers in 1964, get a good night's sleep.[5]

Great scientific accomplishment was no longer essentially a matter of patience and industry, but neither was it a Promethean gift of divine fire. Although brilliance could not be taught, intuitive thinking could, even if no one understood exactly how it functioned. "The mere empirical application of observations concerning the stimuli that we found to promote or impede creative thought can help, even if we do not understand how these factors work. Even a process that must go on automatically in the unconscious can be set in motion by a conscious, calculated effort."[6] But the involvement of the will began and ended with that first effort; volition was *ipso facto* excluded from the unconscious. Nor was the will required to bend body and mind to duty, for science had ceased to be dutiful. It was now superfluous to exhort would-be scientists on the necessity of never-ending work; sloth among scientists was rare,[7] and in any case, no one should consider a scientific career (advised Medawar), "until he discovers whether the rewards and compensations of a scientific life are for him commensurate with the disappointments and the toil ... Once he has felt that deeper and more expansive feeling Freud has called the 'oceanic feeling' that is the reward for any real advancement of the understanding — then he is hooked and no other kind of life will do."[8] The will had no place in a psychology of destined vocation — or addiction.

At the juncture of hypothesis and data, that crossroads at which the nineteenth-century researchers had confronted the choice between objective virtue and subjective vice, a wide range of mid-twentieth-century successors counseled trained judgment and trained instincts. Hypotheses, like hunches, were universally acknowledged as essential guides to research and explanation. Yet mistakes of inter-

pretation were accepted as inevitable. How to know when a hypothesis was not a beacon but a fata morgana? The French physiologist
Charles Richet suggested in 1923 that "to know when it is necessary
to persevere, to know when it is necessary to stop oneself, this is the
gift of talent, and even of genius."[9] In some cases, perseverance could
be a positive hindrance, tempting the scientist down endless blind
alleys.[10]

Here there were no rules, much less mechanical procedures, to
guide the scientist — only the expert, trained intuitions that had
become a new form of right depiction. But although the concern
with judgment, unconscious assessment, and protocol-defying expertise was made in explicit criticism of mechanical objectivity, it
was *not* the same critique leveled by Gottlob Frege and his logico-
philosophical allies. The structural objectivists were suspicious of an
objectivity grounded in reference and experience; they preferred
relations bound into structures that could be unproblematically
shared. According to Frege, concepts of numbers do not derive
directly from their reference, but are defined by identity: the "same
number" maps the two sets of objects, element by element. Nor was
the claim "I see red" a direct allusion to an individual's inner response; rather, it was associated with a color located between others
on a spectrum. Henri Poincaré certainly valued the kinds of intuitions afforded by geometry, topology, and the curves of functions,
but at the end of the day, structural objectivists as a group were dubious about the direct value of empirical, referential picturing.

This chapter's narrative concerns another kind of doubt raised
about mechanical objectivity, one that clung to images (suitably
reinterpreted) and came from deep inside the community of empirical scientists. This twentieth-century struggle aimed to maintain the
scientific image while recognizing the corrosion of faith in an objectivity vouchsafed by an aspiration to an automatic transfer from
object to paper. It is about a newfound confidence among scientists
in the twentieth century, a confidence born in professional training
that let them take on board the new developments in instrumentation and image production, but that left them far from self-abnegating. It is about a faith, also new, that assessments of images could be
made in ways that relied on a scientific self, one reducible to neither
failures nor victories of the will.

If the makers of the objective image had had a slogan, it might have been: Where genius and art once were, there self-restraint and procedure will be. The shift from reasoned images to "objective" images opened up the space of depiction beyond general objects (type, idealization), to include specific objects (individuals, mechanical images). In the twentieth century, as the limits of procedure-governed mechanical objectivity became more apparent, one atlas maker after another insisted that objectivity was not sufficient — complex families of visible phenomena needed trained judgment to smooth, refine, or classify images to the point where they could actually serve any purpose at all. Instead of the *four-eyed sight* of truth-to-nature or the *blind sight* of mechanical objectivity, what was needed was the cultivation of a kind of *physiognomic sight* — a capacity of both maker and user of atlas images to synthesize, highlight, and grasp relationships in ways that were not reducible to mechanical procedure, as in the recognition of family resemblance.

Under the new possibilities of trained judgment applied to image making and reading, a new, less centrally directed scientific self finds articulation in the opening years of the twentieth century. At one level, this should not be surprising. By 1900, a wide range of models of the unconscious proliferated in the sciences of the mind. Perhaps more surprisingly, unconscious criteria — "tacit," "sophisticated," "experience"-based pictorial judgments — came to be seen as a crucial component of day-to-day scientific routine. From the classification of skulls to the development of mathematical understanding, scientists and even mathematicians began to invoke and celebrate "intuitive" criteria for sorting and solving. This positive formulation of trained expert assessment was a far cry from the understanding of a scientific self predicated on the will to willessness.

Machines were hardly abandoned among those scientists who argued that mechanical objectivity was not enough: in fact, some of the *most* sophisticated instruments (electroencephalograms, for example) were, as we will see in a moment, the site for the greatest discontent with rigid protocols. *Trained judgment* came increasingly to be seen as a necessary supplement to any image the machines might produce. Nor was this a return to truth-to-nature. For Johann Wolfgang Goethe in 1795, the depiction of the *Typus* did represent something in nature (though not something apparent from this or

that individual). For Bernhard Albinus in 1747, the "true" representation of a subject referred to nature not only because it borrowed
from several individuals but also because it improved on any single
one of them. For William Hunter in 1774, the link to the general
occurred through a particular individual, chosen precisely so that it
might represent (in both senses) a whole class. Different as they
were, all three views took it for granted that a single representation
could stand in for (and behind) the myriad variations of nature.

When atlas makers no longer claimed self-evident generality for
their images, a gap opened between the atlas images and the objects
that atlas users actually encountered. This was Grashey's problem:
with only one sketch, the guide book no longer resembled the scen-
ery. Closing that chasm would take effort — and could not be accomplished by the image maker or the image alone. The *user* of the atlas
became, therefore, quite explicitly key to making the collection of
images work. Many instances were needed to convey the extent of
the normal, since the normal spanned a space that even in principle
could not be exhausted by individual representations, each differing
from the rest. The German nuclear physicists Wolfgang Gentner,
Heinz Maier-Leibnitz, and Walther Bothe worked for years, beginning in 1938, to produce remarkable cloud-chamber pictures of
many different kinds of nuclear interactions.[11] When they published
their *Atlas of Typical Expansion Chamber Photographs*, they included
multiple examples of alpha particles ionizing a gas, beta particles
scattering from different substances, and positrons annihilating electrons (see figure 6.2). Collectively, the physicists hoped, these "typical" images would evoke patterns in the minds of their readers. At
the height of mechanical objectivity, the burden of representation
was supposed to lie in the picture itself; as the twentieth century un-
folded, however, this responsibility fell increasingly to the scientific
readers. Judgment by the author-artists joined the psychology of pattern recognition in the audience.

Caught between the infinite complexity of variation and their
commitment to the specific simplicity of individuals, mid-nineteenth-century atlas authors invoked a philosophical psychology.
Enlightenment atlas makers had taken selection and distillation as
their principal authorial tasks; now they shed these, relying instead
on the eyes of the audience. Such a solution preserved the purity of

315

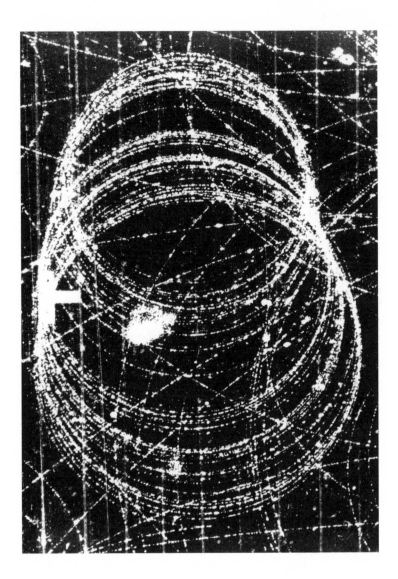

Fig. 6.2. Spiraling Electron. Wolfgang Gentner, Heinz Maier-Leibnitz, and Walther Bothe, *An Atlas of Typical Expansion Chamber Photographs*, 2nd ed. (London: Pergamon Press, 1954), p. 51. Here an electron with an initial energy of 16.9 million electron volts is created in a pair with a positron. The point of pair creation is visible as a left-opening fork at the beginning of the spiral just above the middle of the bottom of the image. The positron arcs down to the left and out of the image; the electron spirals approximately thirty-six times in the magnetic field, drifting upward due to a slight increase in the magnetic field near the center of the chamber. As the electron progresses, it loses energy due to two processes: collisions (ionization) shed 2.8 million electron volts, and the emission of a photon (visible as a sudden jump in the spiral's diameter in the seventeenth circle) causes the loss of an additional 4.5 million electron volts. This kind of detailed interpretation accompanied every image in the atlas — an example of using theory, but theory that was, at the time of publication, considered shared and well established.

blind sight at the cost of acknowledging the essential role of the readers' response: the human capacity to render judgment, the electroencephalographers would cheerfully allow, is "exceedingly serviceable." For Grashey, the problem occurred in shadows of bone, not ink tracings, but the weight of nature's diversity was similarly felt: "One must know these variations," Grashey insisted. "We need an all-points bulletin issued for them. A series of pictures in this atlas is devoted in part to spreading widely 'wanted posters' [*Steckbriefe*] for them."[12] Images of the human face served as the model for grasping Grashey's x-ray images.

But, as we have seen, the rise of mechanical objectivity produced new kinds of instabilities. The triumph of the individual over the generic avoided the problems associated with spurious idealizations, but depictions of individuals made it much harder to handle (that is, identify or teach) the "normal variations" that could arise in a species. Is this star or starfish the same as the one depicted? Two new responses emerged. The first, as we saw in the last chapter, was to develop a notion of *structural* objectivity, an objectivity that, in its emphasis on structural relations rather than objects *per se*, was both a rejection of mechanical objectivity (by turning away from empirical images) and an intensification of objectivity on another scale (by pushing even harder for a knowledge independent of you and me). Structural objectivists bypassed mechanical objectivity because they reckoned that mimetic representations of even the most carefully taken photograph would never yield results truly invariant from one observer to another. Of course, invariant structural accounts worked particularly well for the topologist or philosopher, but it was not a solution to the morphological problems facing biologists, microscopists, or astronomers. Instead of rejecting the empiricism of the image, trained experts sought to make use of a "sophisticated" or "trained" eye to put back together what a radical nominalism risked tearing apart.[13]

Whether they were classifying stellar spectra or electroencephalograms, the atlas makers who believed in expert judgment were self-consciously aiming to use their atlases to identify groupings of objects. Wanting *neither* merely to collect an assortment of isolated individual occurrences (the risk of mechanical objectivity) *nor* to provide idealized entities that entirely lay behind the curtain of appearances (the risk of truth-to-nature), the proponents of trained judgment employed a variety of metaphors. Most prominently, they turned to facial similarities — families, as it were. While there were no explicit, strictly procedural rules for sorting, say, a particular kind of star on the basis of its spectrogram or identifying a petit mal seizure from its electroencephalogram, one could *learn* how to identify and group them, just as one learned to recognize this or that set of people by the subtleties of their appearance. In a sense that Ludwig Wittgenstein made famous but did not originate, family resemblances (partially overlapping features without a necessary and sufficient "core" set of properties) picked out concepts and classes like "game" or "number."

Skull A may have certain features in common with skull B; skull B may have different features in common with skull C — but skulls A and C may share no common defining properties. Object families are recognized by trained judgment; no simple rule-based procedure leads us easily from normal skull with variation number one to normal skull with variation number twenty-three. We have arrived at the third of the historical alternatives that have risen against the regime of depiction pursuing truth-to-nature (with general, idealized objects, revealed by genial intervention). Truth-to-nature (types) is positioned against mechanical objectivity (individuals), but then mechanical objectivity is addressed by structural objectivity (relational invariants), and trained judgment (families of objects). This division does not mean that each replaced the former in sequence: on the contrary, each new regimen of sight supplements rather than supplants the others. Structural objectivity intensified the search for a world without us — but it did so by stepping away from the empirical, mimetic capture of objects and toward relations and structures. And although both truth-to-nature and trained judgment opposed mechanical objectivity, the enemy of my enemy is not necessarily my friend: trained judgment differs from truth-to-nature precisely because the scientists invoking judgment to form their atlas images in the twentieth century had already taken on board or worked through mechanical objectivity. Sequence matters — history matters.

The novelties are as striking as the continuities. In the early twentieth century, scientific atlas makers began to issue explicit and repeated warnings about the limits of objectivity and to make accompanying calls for judgment and interpretation. Within the first third of the twentieth century, new possibilities emerged as the self-abnegating scientists and their various modes of automatic registration began to yield to scientists who worked with highly sophisticated instruments but were, nonetheless, proud of their well-honed judgments in the formation and use of images. Instead of our imagined nineteenth-century slogan "Depict as if the observer were not here," the twentieth-century atlas writers might have said, "At the end of procedural depiction begins trained judgment." As we have emphasized throughout, elements of older strategies for the depiction of nature persist after new forms emerge. There is no "programmatic," "paradigmatic," or "epistemic rupture" here. Even after the great efflorescence of atlases espousing mechanical objectivity occurred in the mid-nineteenth century, for example, one sees instances of the eighteenth-century "truth-to-nature" that, it was supposed, could only be discerned by the sage or genius. Similarly, mechanical objectivity never died. Some atlas writers embraced a vision of an unemended mechanical objectivity deep into the twentieth century. Our argument is not that mechanical objectivity, in an instantaneous break, suddenly vanished during the first third of the twentieth century. Rather, it is that during this period the ethical virtue of self-eliminating pictorial practices was confronted by a new form of epistemic ethic associated with active and highly trained judgment.

For a concrete instance of the persistence of mechanical objectivity, consider the following excerpt from Henry Alsop Riley's 1960 *Atlas of the Basal Ganglia, Brain Stem, and Spinal Cord*, an excerpt that perfectly illustrates the goal of mechanical, automatic reproduction safe from interpretation: "This process [of hand-based illustration], however, makes the illustration a purely selective presentation and therefore the user of the atlas is often uncertain of the exact outline, relations and environs of the structures illustrated. The advantage of a photograph . . . seems to be self-evident. The photograph is the actual section. There is no artist's interpretation in the reproduction of the structures."[14]

For Riley, authorial overselection was a vice to be resisted. Allowing the scientist or artist interpretive autonomy would throw into doubt the reliability of the object depicted. Riley contended that hardly anything need be said to defend the superiority of photographs. So tightly did the photographic image bind itself to the object that he could conclude, "The photograph is the actual section." Resemblance became identity.

By this late date in the mid-twentieth century, however, given all the attention that had been devoted to the limits of photographic reliability, a pure, unblinking faith in the photograph could not be completely sustained. Riley readily conceded that staining was not completely targetable to specific parts of the specimen—his photographs revealed the irregularity of even the best and most technically skilled staining. Alas, even occasional scoring (from dissection) of the samples could be detected. Nonetheless, for Riley, the game was worth the candle—his procedure ensured that "the accuracy and reliability of the photographs makes up for at times an inartistic appearance," where being inartistic was a right-handed criticism (rather than a left-handed compliment).[15]

Like Riley, the authors of the 1975 *Hand Atlas* dismissed the artistic in favor of mechanically objective reproduction: "The authors have provided more realistic illustrations by substituting the surgeon's camera for the artist's brush."[16] According to some of those who espoused the mechanical-objective view, realism, accuracy, and reliability all were identified with the photographic. Nature reproduces itself in the procedurally produced image; objectivity is the automatic, the sequenced production of form-preserving (homomorphic) images from the object of inquiry to the atlas plate to the printed book. Photography counted among these technologies of homomorphy, underwriting the identity of depiction and depicted.

But if mechanical objectivity survived into the twentieth century, it also came to be supplemented across a myriad of scientific fields. Our interest is not in extrascientific attacks on objectivity (romantic literary, artistic, or mystical blasts against the scientific worldview) but in the practices used within laboratory and field inquiry to establish matters of pictorial fact about the basic objects of many scientific fields. The atlases, handbooks, surveys, and guides we have seen thus far chart a central territory of science. In these compendiums of pictures, the simple (even simplistic) nineteenth-century model of images grounded in the protocols of mechanical objectivity came under the fire of scientifically trained judgment.

Do we mean to imply that the practitioners of mechanical objectivity did not exert judgment? Their protestations to the contrary, *of course* the rubbings, projections, and even photographs never extirpated judgment in some absolute and transhistorical sense. As we saw in Chapter Three, sophisticated image makers such as Richard Neuhauss knew perfectly well that photography never could function without skill — as he acerbically noted, the photograph, wrongly handled, could reveal objects that weren't there and hide those that were. But for these scientists, mechanical objectivity was a regulative ideal, a shaping ambition that conditioned whether and when practitioners sought to improve what they did on the page, in the field, and at the laboratory bench. Our argument is that, increasingly during the first half of the twentieth century, the espousal, celebration, and cultivation of trained judgment — as a necessary supplement to objectivity — became a new kind of regulative ideal, one that, in its own ways, reshaped what scientists wanted from their working objects — and from themselves.

Accuracy Should Not Be Sacrificed to Objectivity

During the early decades of the twentieth century, first slowly, then faster, scientists began to stop preening over their self-abnegation, over those tools and practices that had let them present nature "in her own language." Gone, too, was the prevalence of the ferocious denial of any peculiarly human assessment of evidence. In field after field, atlas makers articulated a new stance toward depiction, one that frankly set aside the hard-won mechanical objectivist ideals of absolute self-restraint and automaticity. For example, Frederic A. Gibbs and Erna L. Gibbs launched their compendious *Atlas of Electroencephalography* (1941) with the proclamation that "this book has been written in the hope that it will help the reader to see at a glance what it has taken others many hours to find, that it will help to train his eye so that he can arrive at diagnoses from *subjective* criteria."[17] Surely there are exceptions to every rule (as we saw in Chapter Three, for example, His worked to find a place for subjective drawing), but in the history of late nineteenth-century scientific atlases one finds *very* few scientists in 1850 or 1870 or 1890 who explicitly espoused the subjective as a necessary, central component of making and using scientific images of record. (See figure 6.3.)

Could it be that Gibbs and Gibbs simply did not understand the way "objective" and "subjective" had been deployed by the mechanical objectivists of the previous hundred years? Could they be "talking past" those who deplored the subjective? No, the Gibbses understood full well the pictorial practice of mechanical objectivity. And they emphatically rejected it, as is clear from the continuation of their racial-facial explanation:

> Where complex patterns must be analyzed, such [subjective] criteria are exceedingly serviceable. For example, although it is possible to tell an Eskimo from an Indian by the mathematical relationship between certain body measurements, the trained eye can make a great variety of such measurements at a glance and one can often arrive at a better differentiation than can be obtained from any single quantitative index or even from a group of indices. It would be wrong, however, to disparage the use of indices and objective measurements; they are useful and should be employed wherever possible. But a "seeing eye" which comes from complete familiarity with the material is the most valuable instrument which an electroencephalographer can possess; no one can be truly competent until he has acquired it.[18]

In this context, "indices" and "objective measurements" are closely connected. Fourier transforms, autocorrelations, and other attempts to parameterize the complex spikes and wave patterns of the electroencephalogram were positioned precisely as alternatives to the "subjective" criteria. The Gibbses' vaunted subjectivity is not, however, a return to the older epistemic virtue of truth-to-nature. Where in the mid- to late 1800s mechanical objectivity was counterposed to genial intervention in nature in order to idealize, perfect, or average, the procedure accompanying interpreted images was to be far different. Instead of Goethean genius (which discerns the *Urpflanze* behind the earthly plant) and in place of automatism and self-denial, beginning in the 1930s and 1940s an increasing number of scientific atlases invoked trained judgment based on familiarity and experience. Two opponents of mechanical objectivity should not be conflated: the sage revealed the true image of nature, and the trained expert possessed and conveyed to apprentices the means (through the "trained" or "seeing" eye) to classify and manipulate.

322

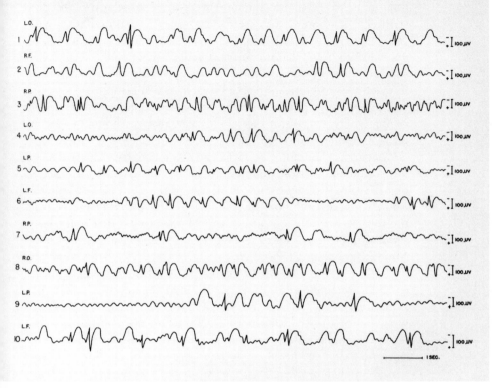

Fig. 6.3. Electroencephalographic Judgment. Frederick A. Gibbs and Erna L. Gibbs, *Atlas of Electroencephalography* (Cambridge, MA: Addison-Wesley, 1941), p. 75. In advocating the use of "subjective criteria," a "seeing eye," and distinctions made "at a glance," Gibbs and Gibbs explicitly argued for a form of scientific sight that would distinguish different neurological conditions. They argued that the blind sight of rule-governed mechanical objectivity was useful but needed supplementation. Required beyond measurements was a form of trained physiognomic sight which, when applied to the electroencephalographic traces, could analyze them the way "the trained eye" can so effectively distinguish "an Eskimo from an Indian."

Some years later, in 1950, the Gibbses produced a new edition of their atlas, expressing in the new preface the same anti-objectivist sentiment in somewhat different language: "Experimentation with wave counts . . . and with frequency analysis of the electroencephalogram . . . indicate[s] that no objective index can equal the accuracy of subjective evaluation . . . if the electroencephalographer has learned to make those significant discriminations which distinguish between epileptic and nonepileptic persons. Accuracy should not be sacrificed to objectivity; except for special purposes analysis should be carried on as an intellectual rather than an electromechanical function."[19]

"Accuracy should not be sacrificed to objectivity." This astonishing statement — astonishing from the perspective of mechanical objectivity — is the epistemic footprint of the new, mid-twentieth-century regime of the interpreted image. How different this is from the reverse formulation of mechanical objectivity: that objectivity should not be sacrificed to accuracy. Recall an example of the opposite decision from Chapter Three: Erwin Christeller's insistence in his *Atlas der Histotopographie gesunder und erkrankter Organe* (*Atlas of Histotopography of Healthy and Diseased Organs*, 1927) that "[it] is obvious that drawings and schemata have, in many cases, many virtues over those of photograms. But as means of proof and objective documentation to ground argumentation [*Beweismittel und objektive Belege für Befunde*] photographs are far superior."[20] In the search for such *objektive Belege*, advocates of mechanical objectivity, roughly starting in the 1830s to the 1920s, were willing to sacrifice the color, sharpness, and texture of scientific representations for a method that took the brush from the artist's hand and replaced it with instruments. In their time, Lowell's tiny, blurry, black-and-white photographs of Mars had counted for more than artistic renderings, even if the latter would have been in color, sharper, more complete, and reproducible. For such a mechanical objectivist, photographs or procedure-driven images said it all. For advocates of rigorously trained judgment such as Gibbs and Gibbs, however, it was equally obvious that the "autographic" automaticity of machines, however sophisticated, could not replace the professional, practiced eye.

We are hit here by the full force of the contrast between the scientific sight of mechanical objectivity and that of trained judgment.

Hellmann's snowflake (figure 3.19), presented as an individual in all its delicate asymmetry, functions very differently from the Gibbses' electroencephalogram (figure 6.3). If making truth-to-nature images required four-eyed sight (that of the naturalist directing that of the artist), Hellmann's technology was a joint enterprise between himself and Neuhauss, an accomplished expert on microphotography. The Gibbses' atlas demanded a new kind of collaboration with the active, subjective electroencepalographer-in-training — they had used their own trained eyes to classify the traces, and their goal was to provide others with that same ability. Mechanical objectivity alone would not suffice (a perfectly administered electroencephalogram was not enough); a rigid adherence to rules, procedures, and protocols was insufficient. The electroencephalographer had to cultivate a new kind of scientific self, one that was more "intellectual" than algorithmic.

In their radical devotion to mechanical means and their protestation of innocence against the charge of intervention, the nineteenth-century atlas writings betray a certain defensiveness, a nervousness before the charge that the phenomena were not actually out there, but instead were mere projections of desires or theories. For Gibbs and Gibbs, that acute anxiety is absent; they did not worry about the possibility that the phenomenon might be a "mere projection." This confidence in scientific judgment was rooted in the changing contours of the scientific self, and this new kind of scientist in turn resided in a much-changed environment. Increasingly, there was a sense that scientists could rely on the cognitive capabilities of less-than-conscious thought. There were technical difficulties — such as those in interpreting electroencephalograms — that defied easy subordination to simple, shareable rules. Finally, a huge growth in the size of the scientific community was facilitated by a remarkable expansion and transformation of scientific pedagogy in Europe and North America during the period roughly between 1880 and 1914, especially in Germany, France, Great Britain, and the United States. A few examples and statistics must suffice to sketch the scope and magnitude of these changes.

Whereas in the 1840s the German physicist Franz Neumann had had to convert his house and garden into a makeshift laboratory in order to teach experimental physics to his students at the University of Königsberg, between 1870 and 1920 twenty-one well-equipped physics institutes were built in Germany (to say nothing of institutes for chemistry, experimental psychology, geology, and physiology).[21] In 1876, there were 293 students enrolled in science faculties at French universities; by 1914, their numbers had swelled to 7,330.[22] An 1899 report by Alexandre Ribot urging the modernization of French education led in 1902 to the introduction of a separate curriculum for the sciences in secondary education.[23] The Royal Commission on Scientific Instruction and the Advancement of Science in Great Britain (also known as the Devonshire Commission) concluded in 1875 that "the Present State of Scientific Instruction in our Schools is extremely unsatisfactory ... little less than a national misfortune," and strongly recommended the establishment of doctoral programs in the sciences at Cambridge, Oxford, and the University of London.[24] The Cavendish Laboratory was founded at Cambridge University in 1874; between 1870 and 1910, the number of science graduates at English universities increased sixty-fold.[25]

Beyond these bare numbers and official reports lay the reality of new spaces, new instruments, and, above all, new ways of training advanced science students to see, manipulate, and measure — a calibration of head, hand, and eye perhaps unprecedented in its rigor and range. Seminar teaching, first introduced by philologists in German universities in the early nineteenth century, was adapted by scientists to the needs of their own disciplines; the pedagogical innovation soon spread to other countries.[26] Instead of listening passively to lectures, students were actively inducted into the craft and standards of their specialties — in the laboratory, the botanical garden, the observatory, and the field, as well as in the seminar room. Aspiring scientists first honed their skills by repeating exercises that were already part of the repertoire of the discipline. Fledgling chemists were set to synthesizing known compounds; young physicists replicated well-established results and re-solved old problems; stripling zoologists practiced classification on models and specimens of known species. Discipline and duty figured prominently in these exercises, whether the members of Neumann's physics seminar were learning to make a precision measurement or a class of Edinburgh medical students was being drilled "in the use of the microscope until every man knew his instrument as a trained soldier knows his rifle, and until in the handling of it he was as perfect as the veteran in the manual of arms."[27] Models of everything from medusae to embryos stocked the shelves of leading university institutes from Leipzig to Boston. In the case of Friedrich and Adolf Ziegler's extraordinary wax embryos (as in figure 4.2), the models radically decontextualized their objects, greatly enlarged them, and turned the transparent wisp with blurry boundaries under the microscope into "huge and memorable shapes."[28] Other models aimed at *trompe l'oeil* verisimilitude, so that they could stand for, and even replace, natural specimens, as in the case of the glass botanical models commissioned by Harvard University from the Dresden craftsmen Leopold and Rudolph Blaschka.[29]

This late nineteenth-century explosion in pedagogical innovation blazed a path to scientific formation that contrasted starkly with what had preceded it. The vast majority of eighteenth-century savants came to their science as autodidacts and practiced it as lone individuals. Uniformity in a field was enforced by the authority of a towering practitioner (such as Linnaeus) or an institution (such as the Paris Académie Royale des Sciences). In the last quarter of the nineteenth century, training became collective and standardized — and the number of people involved in one or another aspect of science also increased dramatically. The annals of science in the middle decades of the nineteenth century are full of complaints about the difficulty of enforcing some sort of uniformity, some common direction in the vast volume of research being conducted by many different people in many different places and published in many different forms. This was not just the familiar complaint of information overload but an expression of concern about the divergence of results and, still more alarming, the objects of scientific inquiry.

One response to this impending chaos was top-down, in the form of magisterial review articles by figures of the stature of Sir John Herschel and James Clerk Maxwell that surveyed recent developments from an Olympian height, separated dross from gold, and offered signposts for the direction of future research. But far more effective was the new mode of seminar instruction, in which students internalized and calibrated standards for seeing, judging, evaluating, and arguing. These were the habits of mind and body that by the early twentieth century had been instilled and ingrained in a generation of scientists. The training-based self-confidence of Frederic A. Gibbs and Erna L. Gibbs and similar atlas makers was new, deriving not only from the enhanced standing of science in society and the professionalization of science as a viable career, but also from a scientific pedagogy that had succeeded in forming self-assured experts.

In the Gibbses' 1941 *Atlas of Electroencephalography*, we can see traces of the emergence of this new, more confident scientific self, breaching the boundaries set by mechanical objectivity. The Gibbses explicitly opposed their "intellectual" approach to an electromechanical one. Such a clash again signals a changed vision of who the scientist is. Neither eighteenth-century sage nor nineteenth-century lay ascetic, the scientist of the twentieth century entered as an expert, with a trained eye that could perceive patterns where the novice saw confusion. The "practiced eye" was as significant to geology as to electroencephalography –for example, in atlases such as Oskar Oelsner's 1961 mineralogical study, which trained the budding geologist to sort microscopic ore samples. Reflectivity, Oelsner noted, depended crucially on the polishing of the surface, so "beginners using it can often make gross errors." Color, too, was susceptible to "remarkable misinterpretations" until the neophyte had acquired a "very experienced eye."[30]

Emphasizing the activity demanded of the picture user, the Gibbses went on to liken the development of skills needed to "read" an encephalogram to those required to read a new language using an unfamiliar alphabet and a different script. True, they acknowledged, encephalography is not simple to master, but with three months of practice, they promised, an average (scientific) person would be able to achieve 98 percent accuracy.[31] The expert (unlike the sage) can be trained and (unlike the machine) is expected to learn — to read, to interpret, to draw salient, significant structures from the morass of uninteresting artifact and background. As an encephalographic atlas from 1962 strikingly put it, "The encephalogram remains more of an empirical art than an exact science."[32] This "empirical art" does several things: first, it identifies that portion of the wave train that is "regular" — unlike automatic methods that must ploddingly examine each fragment, the eye quickly assesses some portion of the signal as "regular" or "typical." Second, even the unaided eye finds "patterns" (the author's quotation marks).

This frank admission of the craft nature of encephalogram reading dovetails — and may have absorbed — a debate over judgment and objectivity in clinical medicine. For example, quite a number of inter-war "patrician" British clinicians aimed to subordinate instruments and scientific standard measures to guard the primacy of their own individual judgment. On this depended not only their status but also their livelihood. For such elites, the celebration of bedside assessment was defensive, a rear-guard and increasingly ineffective interwar attempt to preserve their earlier preeminence at a time when they were being squeezed out by laboratories, tests, and medical scientists. Instruments and laboratory procedures — mechanical objectivity — were for these elites a threat, a direct challenge to their hard-won authority and their place within the upper reaches of society.[33]

Though the medical patricians and the Gibbses' atlas both challenged the triumph of objectivity alone, their reasons for doing so were quite different. Gibbs and Gibbs did not pretend to any (real or virtual) patrician status, and their stance toward instruments was altogether different. Far from opposing high-tech medicine as a threat to their status, they embraced it: they were, after all, among the world's experts on the relatively new and sophisticated electroencephalogram. No bedside, cultivated doctors these. Instead, the Gibbses argued that, *above and beyond* the important results the electroencephalogram provided, the qualified neurologist could learn the requisite expertise to arrive quickly, accurately, and repeatedly at a proper diagnosis, via the trained eye.[34]

For scientists like those we are considering here — across a wide sphere of domains — trained judgment was not the purview only of the ascendant or declining elites who rejected the rule-governed. The supplementing of automatic procedures by trained judgment, as well as the increasing reliance on the pattern-recognition capabilities of a trained, educated audience, extended deep into domains as diverse as geology, particle physics, and astronomy, despite their very different social structures and status. These experts did not reject "objective" instruments in favor of gentlemanly tact or pronouncements by graduates of the *grandes écoles*; on the contrary, they embraced instruments, along with shareable data and images, as the infrastructure on which judgment would rest.

The pervasiveness of the trained use and assessment of images is visible not only in the open-ended audience for geological works such as Oelsner's or electroencephalography atlases like those of the Gibbses and their successors. It is also quite strikingly present in what was the most highly instrumented particle physics laboratory in the world during the 1960s — the one where Luis Alvarez presided over a vast team of senior and junior physicists, engineers, programmers, and scanners. In all sectors, personnel — down to the lowliest scanner moving a trackball across the projected image of a bubble-chamber track — were taught to see their scientific images as matters requiring computer-assisted quantification *and* trained judgment. Here is the 1968 training guide that all scanners studied in depth: "As you have seen, ionization, or track density, can help you to identify particles. As with the other scanning techniques, it is approximate and can only be relied upon as such. Experienced scanners will rarely, if ever, say 'I *know* that track was made by a [pion].' What they will more likely say is 'I bet it is a [pion],' or 'it is most likely a [pion].' One should always use track density information with the awareness that it is not foolproof." Scanners were taught that "eyeballing" was a necessary part of track analysis — alongside the vast quantification apparatus that turned wispy tracks into meson masses, momenta, and energies.[35] Relying *only* on the objective was the problem — "not foolproof" — and the Alvarez group was decidedly against the fool with blind sight alone.

   Not all particle physics groups agreed with Alvarez's group when it came to the use of trained judgment; for example, several key groups at the European particle physics laboratory, the European Organization for Nuclear Research (CERN), battled hard for a less judgment-bound approach to the great tide of bubble- and spark-chamber images washing over the physics community. But Alvarez was adamant, as in these comments from 1966: "More important than [my] negative reaction to the versatile pattern recognition abilities of digital computers is my strong positive feeling that human beings have remarkable inherent scanning abilities. I believe these abilities should be used because they are better than anything that can be built into a computer."[36] The role of judgment and "eye-balling" was emphasized again and again, from Alvarez's training guides, through popular particle-physics atlases such as C.F. Powell and G.P.S. Occhialini's *Nuclear Physics in Photographs* (1947) that aimed to train amateurs to use the new particle-physics technique, to the bible of particle physics experts, the massive *Study of Elementary Particles by the Photographic Method* (1959).[37] Skill — hard-won, trained skill — mattered when it came to making, interpreting, and classifying images.

Judgment as an act of cultivated perception and cognition was associated with a picture of reading that was both anti-algorithmic and antimechanistic. Trained judgment for an Alvarez or a Powell stood opposed to — or perhaps on top of — the fragmented building-up, the mechanically calculated, automated, protocol-driven set of procedures. Scientific image judgment had to be acquired through a sophisticated apprenticeship, but it was a labor of a very different sort from the rehearsed moves of the nineteenth-century mechanical objectivist. Interpreted images got their force not from the labor behind automation, self-registration, or absolute self-restraint, but from the expert training of the eye, which drew on a historically specific way of seeing. Scientific sight had become an "empirical art." This was made vivid in the striking, disturbing analogy deployed by the Gibbses in 1941: reading scientific images was, for them, very close to the judgment-based distinction of the face of "an Eskimo from an Indian." Here was allegedly an un-self-conscious, indeed unconscious act of holistic recognition.

This racial-facial simile was quite widely distributed, not only through the Gestalt psychologists' concerns with holistic cognition, but also via the wider (and not unrelated) preoccupation with matters of race in the 1930s and 1940s.[38] Consider an atlas whose subject was located (literally) light-years from the human brain, W.W. Morgan, Philip C. Keenan, and Edith Kellman's *Atlas of Stellar Spectra*, from 1943. (See figure 6.4.) Here the authors set out a classification of stars in the 8 to 12 magnitude range, based on their spectra. The work was carried out with a one-prism spectrograph attached to a forty-inch refracting telescope. Plates were then sorted according to a two-dimensional system. On one axis stood the spectrum (based, for example, on the intensity of the hydrogen lines), yielding the star type (O, B, A, F, G, K, M, R, N, or S). On the other axis stood the luminosity (ranked by class I–V, progressing from the dimmest to the brightest). In practical terms, the astronomers first determined a rough type, an "eyeball" estimate of the category of a given spectrum

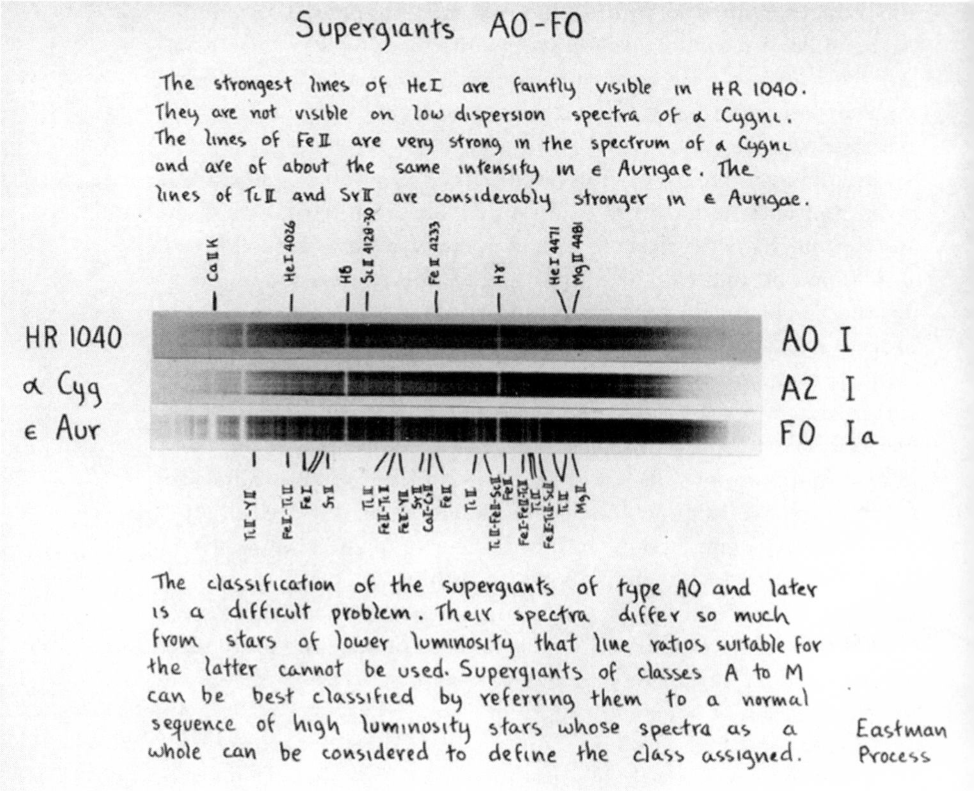

Supergiants A0-F0

The strongest lines of HeI are faintly visible in HR 1040. They are not visible on low dispersion spectra of α Cygni. The lines of FeII are very strong in the spectrum of α Cygni and are of about the same intensity in ε Aurigae. The lines of TiII and SrII are considerably stronger in ε Aurigae.

HR 1040   A0 I
α Cyg   A2 I
ε Aur   F0 Ia

The classification of the supergiants of type A0 and later is a difficult problem. Their spectra differ so much from stars of lower luminosity that line ratios suitable for the latter cannot be used. Supergiants of classes A to M can be best classified by referring them to a normal sequence of high luminosity stars whose spectra as a whole can be considered to define the class assigned.

Eastman Process

Fig. 6.4. "Supergiants A0-F0." W.W. Morgan, Philip C. Keenan, and Edith Kellman, *An Atlas of Stellar Spectra, with an Outline of Spectral Classification* (Chicago: University of Chicago Press, 1943), pl. 20. The authors of this spectral atlas explicitly invoked the need for a sorting strategy that, while *not* amenable to a purely rule-driven mechanism, was nonetheless definite. "Good judgment" was needed, even if the criteria used could not always be conscious. In this plate, the authors indicated to the user that "spectra as a whole" must be used to identify this kind of supergiant. As they add in the surrounding text, this sort of identification is very much like the identification of a human face or the establishment "of the race to which it belongs."

— say B2, a variant of the B-type. Second, using parallax measurements to fix the distance to the star, they found the star luminosity. With the luminosity in hand, they could then compare the candidate star spectrum with previously established spectra of similar luminosity. Matching the candidate spectrum against previously sorted B1, B2, and B3 spectra fixed the precise classification, which might not be B2 after all, but rather B1 or B3 (the final classification rarely differed from the rough estimate more widely than that).

It might be thought that the process of identifying a star as, say, B2 class V was purely routine, the kind of sorting that could just as well be effected by an automatic system. Not so, said Morgan, Keenan, and Kellman: "There appears to be, in a sense, a sort of indefiniteness connected with the determination of spectral type and luminosity from a simple inspection of a spectrogram. Nothing is measured; no quantitative value is put on any spectral feature. This indefiniteness is, however, only apparent."[39] Here is an interesting and important claim: the qualitative is not, by dint of being qualitative, indefinite. Again and again, one sees this cluster of terms now in the ascendant: what was needed is the subjective, the "trained eye," an "empirical art," an "intellectual" approach, the identification of "patterns," the apperception of links "at a glance," the extraction of a "typical" subsequence within a wider variation. Reflections like these point to the complexity of judgment, to the variously intertwined criteria that group entities into larger categories that defy simplistic algorithms. But for Morgan, Keenan, and Kellman, the complexity and nonmechanical nature of this identificatory process does not block the possibility of arriving at an appropriate and replicable set of discriminations. It may take judgment to sort a B1 from a B2, but such judgments can be nonmechanical and perfectly valid: there is not a whiff of the arbitrary in the trained scientific judgments that Morgan, Keenan, and Kellman had in view.

What the trained observer does, according to these authors, is combine a variety of considerations: the relative intensity of particular pairs of lines, the extension of the "wings" of the hydrogen lines, the intensity of a band, "even a characteristic irregularity of a number of blended features in a certain spectral region." None of these characteristics could be usefully quantified ("a difficult and unnecessary undertaking"). The root problem is one that has long vexed philosophers:

"In essence the process of classification is in recognizing similarities in the spectrogram being classified to certain standard spectra."[40] Of what do these "similarities" consist?

Recognition cannot be grounded in the application of algorithmically fixed procedures; any such attempt would at best be cumbersome and at worst would ultimately fail. The stellar spectroscopists continued with the familiar appeal to the physiognomic Gestalt:

> It is not necessary to make cephalic measures to identify a human face with certainty or to establish the race to which it belongs; a careful inspection integrates all features in a manner difficult to analyze by measures. The observer himself is not always conscious of all the bases for his conclusion. The operation of spectral classification is similar. The observer must use good judgment as to the definiteness with which the identification can be made from the features available; but good judgment is necessary in any case, whether the decision is made from the general appearance or from more objective measures.[41]

Note that, like the Gibbses, these star-atlas authors contrast judgment with objectivity, using the word quite clearly in the sense of mechanical objectivity: fixed, specifiable criteria of evaluation. But for both sets of twentieth-century image classifiers, "mere" objectivity was insufficient.

Classifying (judging) by luminosity, which was by no means simple, illustrates the complex way judgment had to be deployed. Certain lines or blends of lines may serve as a basis for calibrating stars relative to a standard in one spectral group; in another it may be useless — the lines may hardly vary at all. Dispersion in the spectrogram — the spreading of spectral lines on the plates — also varies for different spectral types. So long as one uses plates of low spectrographic dispersion, hydrogen lines vary with absolute magnitude in stars of type B2 and B3. In high-dispersion plates that separate off the "wings" (outlying portions of the broadened spectral line) from the central line, these wings are frequently no longer visible. And since it is the wings that vary with the absolute magnitude, when they cannot be seen the remaining line looks much the same whether the star it issues from is a dwarf or a giant. Conversely, some lines visible in the high-dispersion plates are invisible at lower dispersion. According to the stargazing spectroscopists, "These considerations show that it is impossible to give definite numerical values for line ratios to define luminosity classes. It is not possible even to adopt certain criteria as standard, since different criteria may have to be used with different dispersion." Variations like these made it impossible to specify a one-size-fits-all-rule by which to classify: "The investigator must find the features which suit his own dispersion best."[42]

One has here a subtle and interesting confluence of phenomena. On the side of the spectra themselves, there is variation that precludes naive rule-following. On the side of the observer, there is a celebration (not denigration) of the human (rather than mechanical) ability to seize patterns (metaphysically neutral, in contrast to the types of truth-to-nature) and therefore to classify even when algorithmic forms of reasoning fail. Subjectivity became an important feature of classification because the objects did not demonstrate universal essential properties *and* because in the mid-twentieth century a growing number of scientists across many fields began to take it as a good thing that people could be trained to classify objects univalently even in the absence of strict protocols. Physiognomic sight could be *taught*.

In sum, Morgan, Keenan, and Kellman draw attention to four features of judgment. First, they emphasize that classification involves the establishment of similarity relations, and that these similarity relations (such as those of luminosity) cannot be specified in terms of a fixed set of standard criteria (for example, line-intensity ratios for all spectral types). Second, the evaluative process of studying stellar spectra (like the evaluation of "race") is not necessarily a conscious one. At a glance, in a flash of recognition, one sees that a star is "racially" a B-class rather than an F-class entity. Third, the cognitive process at work in interpreted images is represented as holistic, and it is precisely this holism ("decision made from . . . general appearance") that stands in contrast to the "objective measures" of mechanical images (which were piecemeal as well as mechanical). Fourth and finally, nothing in the process of judgment is necessarily vague or indefinite — it is an error, these authors argued, to suppose that quantitative measures (even were they applicable) are the only way to a determinate classification. All four of these distinguishable features of judgment seem to be captured by the authors' racial-facial simile and its contrast to quantitative, algorithmic assessment.

335

This privileging of physiognomic sight over protocol-driven definitions or exhaustive criteria in classification evokes, once again, the philosophy of Ludwig Wittgenstein. In a 1929 lecture on ethics, Wittgenstein likened the way in which he proposed to get at the subject matter of ethics to Sir Francis Galton's composite photograph in a facial-racial metaphor:

> And to make you see as clearly as possible what I take to be the subject matter of Ethics I will put before you a number of more or less synonymous expression ... and by enumerating them I want to produce the same sort of effect which Galton produced when he took a number of photos of different faces on the same photographic plate in order to get the picture of the typical features they all had in common. And as by showing to you such a collective photo I could make you see what is the typical—say—Chinese face; so if you look through the row of synonyms which I will put before you, you will, I hope be able to see the characteristic features they all have in common.[43]

Composite photographs exercised a certain fascination for Wittgenstein; he is said to have carried with him always such a photograph made of himself and his siblings: a literal Wittgensteinian family resemblance. Wittgenstein says that his ethical expressions and Galton's images both aim to elicit the "the typical features they all had in common." Yet by choosing "to produce the same sort of effect" as the Galtonian image, he already was moving beyond the idea of a simple Venn diagram overlap. The composite picture produces a weighted average of all the features. If nine out of ten cousins have a mole, the composite picture will too—even if the tenth cousin does not.

In a later, famous passage on family resemblance in the posthumously published *Philosophical Investigations* (§67) Wittgenstein developed a more subtle notion of what unites the members of a concept like "game": not what all members had in common, but rather a weave of partially overlapping traits. Some exhibit the same nose, others the same gait. The analogy with Galton's composite photograph is gone, but something of the same physiognomic metaphor lingers—now purged of Galton's essentialism.

Wittgenstein's hunt for a way to generalize beyond simple mechanical extraction of characteristics in common had thus used the still partially mechanical form of Galton's facial-racial form of physiognomic sight, and then, by the time of the Investigations shifted to a physiognomic sight. Throughout, Wittgenstein emphasized the importance of knowledge at a glance, the human ability to fill in (conceptual) "intermediate terms" between related forms.[44]

This emphasis on the ability of the practiced eye to seize with a glance goes back a very long way — certainly it is emphasized in the early nineteenth century by the German naturalist Alexander von Humboldt, in his work the "physiognomy" of plant landscapes.[45] But it is found in a new and intense form, riding on and against highly sophisticated scientific instruments in the atlases of the twentieth century. The Gibbses likened the detection of patterns in the electroencephalogram to distinguishing Indians from Eskimos; Morgan, Keenan, and Kellman sorted out stellar spectra by a kind of "racial" classification. In different — and, to later readers, often disturbing — scientifically engaged ways, all these authors deployed the complexity of grouped facial recognition and classification to oppose what they took to be the inadequate classificatory power, the simplistic proceduralism of mechanical objectivity.

Galton, it should be said up front, was a hard-line eugenicist. For Gibbs and Gibbs, Morgan, Keenan, and Kellman, and Wittgenstein, the allusions to physiognomic classification built metaphorically on the classification of individuals into groups by race. There is no reason to think that these scientists (or Wittgenstein) shared Galton's particular eugenicist ambitions. But the timing of this kind of group reference was not incidental, and not only "in the air" — racial stereotyping was *in print* thanks to the most prolific atlas publisher in the world, Julius F. Lehmann. Lehmann had begun his publishing empire in Munich in 1886 with establishment of *Münchener medizinische Wochenschrift* (*Munich Medical Weekly*), which became the most widely circulated of all the German medical journals.[46] From that base, he began his enormously successful series of medical atlases — some forty-one small-format "hand-atlases" and seventeen full-size atlases translated into some fourteen languages. His successes included many of the atlases discussed here, including Rudolf Grashey's and Johannes Sobotta's.[47] Among Lehmann's best-sellers were not only medical tomes but also a long string of race atlases.

Lehmann declared himself actively on the political far right. Of "*volkisch*" persuasion, he subscribed to social Darwinism and worked to broaden his list of medical and biological publications to include genetics, eugenics, and hygiene. In 1922, Lehmann Verlag took over the journal *Archiv für Rassen- und Gesellschaftsbiologie* (*Archive for Racial and Societal Biology*) at a financial loss, and Lehmann encouraged Hans F.K. Günther to publish his *Rassenkunde des deutschen Volkes* (roughly, *Racial Science of the German People*, 1922), which was reprinted sixteen times between 1922 and 1933; it sold about 50,000 copies during that period and some 272,000 copies (including a shorter version, first published in 1929) by 1943. Lehmann funded lectures on racial hygiene and a prize for the best collection of "pure German portraits." As Lehmann wrote Günther in October 1920, the publisher wanted a "human field guide to the flora (*Excursions-flora*) of Germany that, first of all, would lay out the general racial markings in an exemplary fashion."[48] Günther was happy to oblige and produced, in addition to specifically German atlases, a 1925 one that extended to the "flora" of all Europe, which he divided into five main groups and their mixtures. (Lehmann clearly saw the race guides as of a piece with medico-scientific ones.) Criteria such as height, limb length, skull measurements, and skin and hair color were all useful — but racial identification was always more than this, Günther argued. In pursuit of this extra element, mental comportment (*seelisches Verhalten*), the author sought systematically to portray a great number of examples, covering page after page with exemplars of each racial type.[49] (See figure 6.5.) Only these could train the eye to see people as belonging to races, as particular flowers could be seen in their taxonomic place, or star spectra in theirs.

Opposed dangers face any discussion of the pervasive scientific

Fig. 6.5. Racial-Facial Atlas. Hans F.K. Günther, *Kleine Rassenkunde Europas* (Munich: Lehmann, 1925), p. 33. Günther divided up Europe's people into five "pure" races and their various combinations: Nordic, East-Baltic, Western, Eastern, and Dinaric. The atlas was a kind of guide to recognition — providing examples not only of pure but also of mixed races. For example, members of the Dinaric race (illustrated here) were supposed, *inter alia*, to be tall with brown or black hair, have large noses, deep-set brown eyes, and a characteristic skull shape — but pictures aimed to capture what words and measurement criteria could not. In 1932, Günther joined the Nazi party, which celebrated and used his work.

Abb. 77 a, b. Wien. K: 85,50, G: 91.04 (bei Zahnverluſt), 75 jährig.
A: braun mit trübblauem äußerem Ring

Abb. 78 a, b. St. Johann (Tir.). (Aufn.: Frl. Zuber, Anthrop. Inſt., Wien)

Abb. 79. Rauris (Tirol)　　　　　Abb. 80. Hotzenwald (südl. Baden.)
(Aufn.: Sammlung Hofrat Toldt, Wien)　　(Aufn.: Gersbach, Säckingen)

Dinariſch oder vorwiegend dinariſch

conceit of racial-facial recognition. On the one side, there is a risk that all such talk, from the metaphorical to the eugenic, will be assimilated to the war and holocaust. On the other side, it would be wrong to portray these metaphors as entirely incidental to the spread of group stereotypes by race classification during the first half of the twentieth century. Avoiding both simplifications, one can nonetheless discern a narrowing of physiognomic sight from the 1920s through the early 1940s, when it became increasingly described in terms of metaphors of racial recognition (used not just by the far right[50]), in contrast to earlier applications of facial metaphors to everything from global plant distribution to meteorological trends. For many atlas makers before the Second World War, the atlas genre itself put such group stereotypes directly at hand, providing a way of seeing that addressed the vexed and more general problem of classification and similarity. It is a mark of how loaded, how *un*-neutral, these metaphors were that after the Second World War such race-distinguishing conceits were caught before pen met the page and rarely made it to print.

Given the pervasiveness of atlases that relied on trained judgment, one could ask, Did the atlases foregrounding prepared judgment and the piecewise estimation of similarity differ simply in *subject matter* from earlier ones grounded in mechanical objectivity? Perhaps (it may be thought) the twentieth-century material in some way demanded trained judgment by its very nature, whereas the subject matter of the nineteenth century required no more than the objectivity of machines. Yet there are nineteenth-century x-ray atlases that aspire to mechanical objectivity and twentieth-century x-ray atlases that rely on judgment while referring back to their forebears; there are nineteenth-century anatomical atlases espousing mechanical objectivity and altogether comparable twentieth-century anatomical atlases predicated on judgment and critical of their predecessors. Stellar-spectra atlases provide a perfect instance of this continuity of topic, despite a sharp break in the mode of categorical classification. As we have seen, the Morgan, Keenan, and Kellman atlas argued for judgment over objectivity, root and branch. Strikingly, however, the atlas that the three explicitly identified as their direct forerunner was the *Henry Draper Catalogue* of 1918, which quintessentially advocated the image-making goals of mechanical objectivity. To make the contrast as sharp as possible, it is worth pausing to consider that predecessor volume.

The stunning *Henry Draper Catalogue* included the classification of some 242,093 spectra from 222,000 stars. Labor history is not irrelevant, even — especially — in the observatory: routinizing and managing an enterprise of this scale linked scientific and industrial work. Mechanical proceduralism joined the laboratory to the factory.[51] The *Henry Draper Catalogue* was an opus designed from the outset to last forever: the preface even assured the reader that "various authorities" expected the paper itself to be "practically permanent." Edward Pickering (the director of the Harvard College Observatory) began that preface by saying, "In the development of any department of Astronomy, the first step is to accumulate the facts on which its progress will depend." Nowhere did he expound on judgment as necessary to classify the spectra, on the absence of universal criteria of selection, or on the role of preconscious cognition. Quite the contrary; Pickering's preface to the *Henry Draper Catalogue* celebrated the use of scientific management and mechanical objectivity. These were so "automatic" that they could be suitably executed by a replaceable set of hardworking (female) assistants, of whom an average of five were at work at any given time over four years.[52]

The practice of employing women to do astronomical calculation and classification can be, and has been, read as a chapter in workplace labor history.[53] But it is more than that. First, in the nineteenth century, the very possibility of employing "unskilled" workers served as a tacit guarantee that data thus gathered were not the figment of a scientist's imagination or preexisting philosophical commitment — as we saw in the case of Claude Bernard in Chapter Two. In this respect, the workers were identified with the machines, and, like the machines, in their "emptiness" they offered a transparency through which nature could speak.[54] Second, beyond their supposed "lack of skill," women workers were presumed to offer a "natural" predilection away from the grand speculative tradition. Occasionally, in the context of mechanical objectivity, this presumption conveyed the highest praise. Annie Jump Cannon, who co-authored the great *Henry Draper Catalogue* with Edward Pickering, was hardly a "mere" computer — it was she who modified and rearranged the older star

341

spectrum classification (A, B, C, and so on) into the long-lived Harvard system of spectral classification. It was also Annie Cannon who showed how these species could be rearranged to display the spectra in a continuous fashion. But it was precisely for her deliberate abstinence from theory that she was esteemed by her contemporaries, as is clear from the characterization of her written in the year of her death, 1941: "Miss Cannon was not given to theorizing; it is probable that she never published a controversial word or a speculative thought. That was the strength of her scientific work — her classification was dispassionate and unbiased."[55] (See figure 6.6.)

Both the *Henry Draper Catalogue* of 1918 and Morgan, Keenan, and Kellman's 1943 atlas handled stellar spectra. But where the later authors saw the irreducible need for trained judgment, Pickering, Cannon, and their nineteenth-century staff viewed their ideal atlas as planted in the firm ground of scientific management and mechanical objectivity. So despite Morgan, Keenan, and Kellman's use of the Draper catalog — despite their similarity of subject — the framing of the two projects was quite different. Here and elsewhere, mechanical objectivity and scientific management yielded to a new practice of sorting nature in which trained judgment, subjectivity, artisanal practice, and unconscious intuition all were heralded as vital to the scientific project of visual classification. The blind sight of mechanical objectivity was confronted with the physiognomic sight of trained judgment.

Atlases of the mid- to late twentieth century, unlike those of the mid-nineteenth, began to be explicit about the need for subjectivity, as in the updated version (*Normal Roentgen Variants that May Simulate Disease*, 1973) of Grashey's atlas, with which this chapter began. In his update, the author insisted on the *subjectivity* now needed for this kind of work: "The proof of the validity of the material presented is largely subjective, based on personal experience and on the published work of others. It consists largely of having seen the entity many times and of being secure in the knowledge that time has proved the innocence of the lesions."[56] Identifying the bounds of the normal spectrum required exquisite judgment and extensive clinical training. The new work built on Grashey's famous atlas, *Atlas typische Röntgenbilder vom normalen Menschen* (*Atlas of Typical X-Rays of Normal People*) which had been an early call for interpreted images,

Fig. 6.6. Spectral Workers. Helen Leah Reed, "Women's Work at the Harvard Observatory," *New England Magazine* 6 (1892), p. 166. This photograph, taken at the Harvard College Observatory shows Annie Jump Cannon (far right) with colleagues in the room devoted to Draper Memorial work. *Inter alia*, these women astronomers and astronomical workers contributed fundamentally to the *Henry Draper Catalogue*, which classified almost a quarter of a million stellar spectra.

by means of which the author sought to impart to his readers a sense of the limits of the normal. To Grashey, as we saw, his radiograms were "wanted posters" that told the radiologist where the territory of the pathological began.[57] Again, one sees interpreted, exemplary images analogized to the recognition of the face.

As the star atlases indicate, there is nothing specifically medical about the strategy of trained judgment. Indeed, in particle physics one finds the same kind of argument as that advocated by the x-ray master Grashey: atlases exist to teach the range of what is known in order to highlight the unusual. In physics, however, the "pathological" is equivalent to the rare and unknown species of particles, and the "normal" becomes the known instances of particle production and decay. P.M.S. Blackett, one of the great British cloud-chamber physicists, wrote the foreword to George Rochester's 1952 cloud chamber atlas (see figure 6.7), in which he put it this way: "An important step in any investigation using [the visual techniques] is the interpretation of a photograph, often of a complex photograph, and this involves the ability to recognize quickly many different types of sub-atomic events. To acquire skill in interpretation, a preliminary study must be made of many examples of photographs of the different kinds of known events. Only when all known types of event can be recognized will the hitherto unknown be detected."[58]

Learning to recognize the scientifically novel was a matter of training the eye, whether to pick malignant lesions from normal variations or to extract a kaon from a background of pions. Key concepts included acquired skill, interpretation, recognition. Whether one was dealing with pions, skulls, stellar spectra, heartbeats, or brain waves, the problem was the same. Scientists, whether they were analyzing stellar spectra, x-rayed skulls, or cloud-chamber images had no faith that pictures could be sorted automatically: the edict of mechanical objectivity to abstain from all interpretation turned out to be sterile. According to an increasing number of mid-twentieth-century atlas makers, more than the mechanical production and use of images would be needed. Only images interpreted through creative assessment — often intuitive (but trained) pattern recognition, guided experience, or holistic perception — could be made to signify. Only through individual, subjective, often unconscious judgment could pictures transcend the silent obscurity of

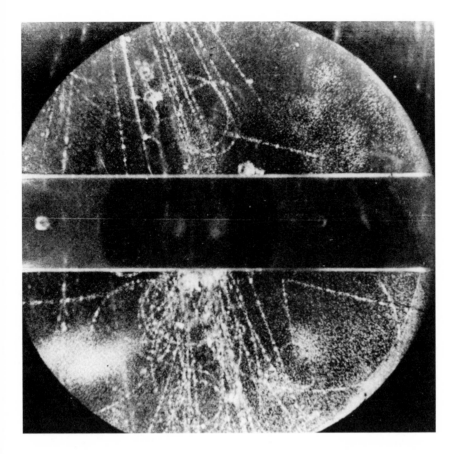

Fig. 6.7. V-Particle Decay. G.D. Rochester and J.G. Wilson, *Cloud Chamber Photographs of the Cosmic Radiation* (New York: Academic Press, 1952), pl. 103, p. 102: Cloud chamber image by George Rochester and C.C. Butler (originally published in *Nature* in 1947). This particle, known as the V^0, is neutral, so only its decay particles are visible — as an opening "V"-shaped track a few millimeters below the horizontal plate, to the right of the main shower. The authors argued that this was the spontaneous decay of a neutral particle for three reasons: first, the opening angle is too wide (67 degrees) to be an electron-positron pair, and moreover, if the track was due to an ordinary collision, other events like this one should have been seen by the hundreds originating in the (much denser) plate; second, an interaction in the gas should have produced a recoiling particle; third, energy and momentum conservation exclude the possibility of the by-then-well-known pion and muon decays. Consequently, the authors concluded that this was, in fact, a new particle, the first of what came to be known as "strange" particles.

their mechanical form. Only the judging eye could pluck the patho-logical lesion or the previously ambiguous particle track from the tangled pictorial world of "normal variations." Mechanical objectiv-ity fell short.

The Art of Judgment

Bearing in mind the twentieth-century demand for judgment of images — from skulls and electroencephalograms to stellar spectra and cloud-chamber images — we can now return, with surprisingly different conclusions, to the relation of scientist or research physi-cians to their illustrator-artists. Take surgery. In the mid- to late nineteenth century, as we have seen, a snowballing number of scien-tists — pathologists, microscopists, snowflake hunters, and splash physicists — swore that they policed every line, every dab of color for accuracy, or sought the photographic as an explicit means of avoiding the need for such surveillance. The contrast with new, judgment-invoking procedures of the mid-twentieth century could not be starker. In his 1968 *Atlas of Precautionary Measures in General Surgery*, the thoracic and general surgeon Ivan D. Baronofsky reported, with-out apology, on the active measures taken by "his" illustrator, Daisy Stilwell, "one of the finest artists in the medical field." He added: "Miss Stilwell is a superb interpreter. It would have been simple for her merely to act as a camera, but instead she brought out the fea-tures that justified the picture."[59] In the nineteenth century, for a sci-entific illustrator to be likened to a camera was compliment of the highest sort. The artist's autonomy and interpretive moves were powerful threats to the representational endeavor, threats the cam-era and vigilant "policing" alone could quell. For Baronofsky, to be a "mere" camera now carried only opprobrium. To be able to inter-pret was the key; judgment made it possible for Stilwell to sort the significant from the background, which "justified the picture." Mere camera-enabled naturalism was too blunt to reveal what the atlas makers and readers wanted to see.[60]

Baronofsky was not alone. John L. Madden's 1958 *Atlas of Tech-nics in Surgery* did not hesitate to underline just how far representa-tion stood from the surgical theater: "In illustrations, the incisions never bleed and the clamps and ligatures on the cystic and superior thyroid arteries never unlock or slip off. Furthermore, postoperative complications do not occur and there are no fatalities." Bloody incisions and slipping ligatures were the human side of the operating room, and Madden sought to join hospital-floor pragmatic realism to a representational realism founded on judgment. In the preparation of Madden's atlas, the importance of having the medical artist present at each operation was stressed. Only in this way could the illustrations include both anatomic realism and the informed interpretation of the artist. Therefore, only those operations that were witnessed by the medical artist were depicted.[61] In pursuit of this "anatomic realism," the artist would sometimes observe three or four surgical procedures, with the goal of obtaining a logical visual exposition with no "jumps." To secure that realism, Madden (like Baronofsky) was perfectly willing to eschew the mechanical objectivity of the camera, and he was enthusiastic about the adoption of the "medical artist" whose *interpretation* offered an accuracy that more automatic (camera-like) procedures could not match.

No rigid "policing" of the artist, it seemed, was desirable in these various twentieth-century atlases. (Contrast Madden and Baronofsky with Johannes Sobotta, whose famous turn-of-the-century anatomy atlas denounced woodcuts as not "true to life" precisely because they left "entirely too much to the discretion of the wood engraver" — a discretion that photomechanical reproduction would stop cold.[62]) As Madden and Baronofsky insisted, it was exactly the artist's ability to extract the salient that rendered a depiction useful.

It must be kept in view that the identification of the salient by the trained anatomist, surgeon, or scientific illustrator is far from the metaphysical "truth-to-nature" image extracted by the sage observer. Goethe, Jean Cruveilhier, Albinus, and Samuel von Soemmerring did not use exaggeration or highlighting to facilitate recognition, classification, or diagnosis — nor were they struggling to eliminate an instrument-produced artifact. They were after a truth *obscured* by the infinitely varied imperfections of individual appearance. Emphasis in the interest of operational success is a long way from perfection in the interest of metaphysical truth. The exercise of a highly trained judgment after objectivity — in response to its perceived shortcomings — is quite a different matter from drawing to unearth an ideal in the years before protocol-driven mechanical objectivity reared its head.

347

One 1954 atlas explicitly celebrated the choice to maintain drawings over actual x-ray photographs in pursuit of this operational and diagnostic utility: "The publisher has done well to retain the original illustrative sketches. A drawing can show so much better the features one is trying to emphasize than the best chosen original roentgenogram. And of course it is such ideal abstractions of sought-for morbid changes that one carries in one's mind as one searches the fluoroscopic screen for diagnostic signs."[63] Interpolation, highlighting, abstraction — all were subtle interventions needed to elicit meaning from the object or process, and to convey that meaning — to teach expertise — through the representation. Images shaped by experienced judgment are neither those of truth-to-nature nor those of mechanical objectivity.

Even when the object itself is as unchanging as the visible face of the moon, accurate representation was a task of monumental difficulty for these postmechanical atlas makers. Astrophotography, which by 1960 was far more sophisticated than Percival Lowell could have imagined, in no way ended the problem. In 1961, V.A. Firsoff published his *Moon Atlas* (see figures 6.8 and 6.9), and the difficulties of extracting realism from the vagaries of moment-to-moment astronomical appearances were all too apparent. Expert judgment could not be eliminated, even when it came to depicting something as self-evidently "out there" as the moon. (Firsoff, an older member of the British Astronomical Association, later had to backpedal in high gear when photographs taken from Apollo spacecraft canceled some of the "volcanic" peaks he had drawn in the middle of craters.) Firsoff was blunt about the limits of any purely mechanical procedure getting right the surface of the moon: "Nobody who has not himself attempted to map the Moon can appreciate the difficulties involved in such a programme. The lights and shadows shift with the phase and libration and can alter the appearance, even of a clear-cut formation, almost beyond recognition. Thus every region has to be studied under different illuminations and a true picture of the surface relief built up step by step. To some extent the result must needs be one of individual judgment."[64]

Representation need not be homomorphic.[65] That is, the pictures constructed from the world need not correspond in form to something one has actually seen — or even could see, were one to be somewhere else (or to be much bigger or smaller than our human scale). Population-density maps, for example, use the visual to express a phenomenon that may otherwise be presented in tabular form. For the physical sciences, such nonmimetic representations as tables serve frequently in all branches of theoretical and experimental work: these illustrations are often the highly processed output of a computer that has not only stored reams of data but also manipulated them in controllable ways. When Robert Howard, Václav Bumba, and Sara F. Smith composed their *Atlas of Solar Magnetic Fields*, published in 1967 (see figure 6.10), they had to choose how much to "smooth" the data as they grappled with different observations. Even here in this heartland of astrophysics, the role of objectivity was frankly contested by a subjectivism tied to the twentieth-century emphasis on judgment and interpretation:

> Considerable experience in the handling of the magnetograms has made us cautious in our approach to their interpretation, but for those unfamiliar with the instrument the variation in the quality of the observations can be a great handicap. For this reason we decided that the best way to make the information available was in the form of synoptic charts, which represent a somewhat smoothed form of the data.
>
> Inevitably many decisions had to be made concerning what were or were not real features on the magnetograms. Naturally there is a certain subjective quality to these charts.[66]

These "subjective" decisions about what was real were explicitly active; they were just the sort of intervention that had no place within the nineteenth-century scientific self, with its obsession with the self-discipline needed to create the possibility of objective depiction. In this atlas, unlike those of the Gibbses and Morgan and his colleagues, it was not just a question of learning to classify the image — it was a matter of *modifying* the image itself. Trained judgment was needed to make the image useful at all.

Gerhart S. Schwarz (from the Chronic Disease Center of New York Medical College), collaborating with Charles R. Golthamer (of Van Nuys, California), also had an active, artistic conception of pictorial production. As an eighteen-year-old, Golthamer (then called Karl Goldhamer) had served in the Austrian army and had been at

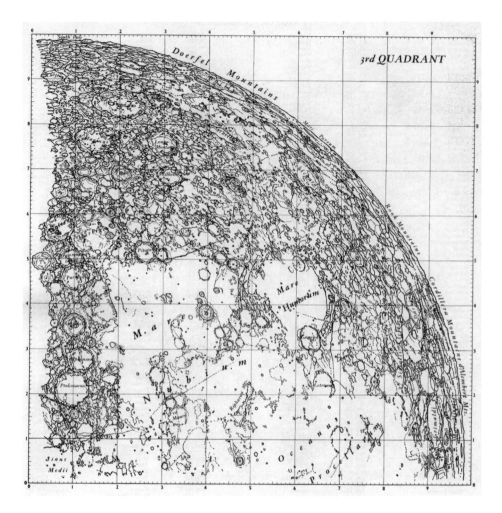

Fig. 6.8. Judging the Moon. V.A. Firsoff, *Moon Atlas* (London: Hutchinson, 1961), third quadrant map (note south at top, west on left). What could be more "objective" than a photograph of the moon — and what more timely than an accurate map when planning was getting under way for astronauts to walk there? Yet for Firsoff, it was as plain as a lunar day that the varying light conditions on the moon made photographs highly problematic and an *interpreted* drawing a better, more faithful representation. (Please see Color Plates.)

Fig. 6.9. The Moon of a Practiced Eye. V.A. Firsoff, *Moon Atlas* (London: Hutchinson, 1961), pl. 7. This photograph, reproduced by Firsoff, was taken, with the Palomar Observatory's 200-inch telescope, of the Clavius region. Even with a photograph, the author made it clear that only a "practiced geological eye" could detect the "swarm of parallel faults" that lay between craters Gruemberger and Klaproth above the main plain of Clavius.

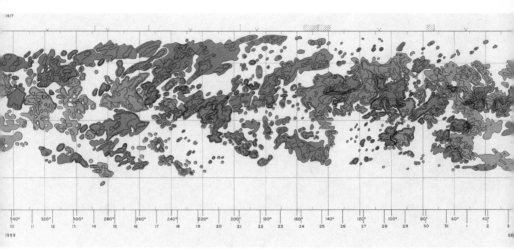

Fig. 6.10. Sun, Corrected. Robert Howard, Václav Bumba, and Sara F. Smith, *Atlas of Solar Magnetic Fields*, August 1959 – June 1966 (Washington, DC: Carnegie Institute, 1967) (courtesy of the Observatories of the Carnegie Institution of Washington, DC). In this magnetogram, the authors actively modified the image itself – to remove artifacts (that which was not "real"). But this "smoothing," as the authors dubbed it, was not in any way meant to claim for the atlas images an ideal (metaphysical) status – the authors still wanted their chart to be an image not of the sun's fields in the abstract but of the particular rotation of the sun measured in the late summer of 1959 – minus instrumental artifacts. (Please see Color Plates.)

the front during the opening salvos of the First World War, and it was not long before his leg was smashed by shrapnel. In part because of his wartime experiences, he began studying medicine; he rose quickly through the ranks of the Department of Anatomy at the University of Vienna. In 1930–1931, he published a two-volume atlas, *Normal Anatomy of the Head as Seen by X-ray*, which appeared in four languages. By the mid-1930s, he was in charge of all pediatric radiology in all the municipal hospitals of Vienna and had some fifty articles to his credit. None of this protected him. After the 1938 German *Anschluss* of Austria, Golthamer was thrown into the Dachau concentration camp, and only by dint of his war service and wounds was he "conditionally" released – and given just days to get out of the country. Just before his time ran out, putting him at risk of instant rearrest, he obtained an exit visa. For his part, Schwarz had studied with Golthamer in Vienna in the 1930s and had gathered, modified, and published an updated version of a radiographic wall chart first put out by Rudolf Grashey in the 1930s.[67]

Schwarz and Golthamer teamed up to produce a 1965 Röntgen atlas of the human skull. By this time, the authors argued, the discipline had advanced to the point where familiarity with normal skull radiology could be simply assumed as background knowledge: now radiologist, orthopedic surgeon, dental surgeon, neurologist, neurosurgeon, otolaryngologist, and forensic specialist needed exposure to the normal variants and pseudolesions that could "vex" even the expert. Several simultaneous demands made the task complex. First, the radiologists wanted to reproduce radiographs such that they actually looked like the originals, with prints of actual size or even larger than life. Back in 1930, Golthamer had solved the difficulty of reproducing the image so the copy resembled the individual by brute force: he had printed each image with a photographic contact print on bromide paper and had them stitched into the atlas by hand. Even if this craft procedure had been economically feasible in the United States of the 1960s (which it was not), the goal of the atlas had shifted. Second, Golthamer and Schwarz wanted more than a mere reproduction of a "normal" radiograph in facsimile. They were after more — "a theoretical composite of many different skulls, containing more than one hundred variants and pseudo-lesions on each printed plate." These two constraints — the necessity of resemblance and theoretical compositeness — threatened to overwhelm any possible text.[68]

Schwarz writes, "It was then that Dr. Golthamer suggested that we might reproduce all radiographs by hand." Even though the x-rays already existed, drawings, deliberately altered from the original, would be created. It was a move that was unimaginable seventy-five years earlier. After the struggle to extract a photograph of Mars, could Lowell conceivably have reverted to a hand-produced image if he had had a sharp photograph available? Realism (in this mid-twentieth-century context) aimed not at the reflexive correspondence of nature with reproduction, but at the half-tone drawing that interpreted particular radiographs.[69] Golthamer, although he was (by his own account) "an expert painter with many awards to his credit," could not produce a "sufficiently realistic" rendering, nor could

353

Schwarz. Finally, with the aid of the director of the art department of the Columbia College of Physicians and Surgeons, they met with success; the volume represented the combined efforts of two other artists (Helen Speiden and Harriet E. Phillips). Once the artistic technique had been perfected, a more subtle set of concerns arose, issues that get at the very heart of the problem of objectivity as atlas makers came to celebrate intervention on the basis of a trained and training eye:

> The question as to how true to nature the image should be arose for more than one reason. Our initial intention was to make the plates look as "natural" as possible, depicting the normal variant, or pseudo-lesion, as true to its appearance on an actual radiograph as the artist's skill could achieve it. However, after our first plate had been drawn in this manner, we came to realize that painstaking copying of nature was not the purpose of drawings in an anatomic atlas. In many instances, a normal variant, depicted "naturally," remained invisible except to the trained eye of a specialist who was familiar with the lesion to begin with. Reading the completely "natural" plates turned out to be an exercise in "rediscovering" lesions, rather than viewing them. Since a laborious search for lesions in an atlas was surely neither desirable nor practicable, this "natural" manner of graphic presentation would have missed the point altogether. We became convinced that our atlas would gain proportionately in usefulness the more each lesion could be made to look so obvious that a reader would recognize it instantly and without effort.[70]

To bring out the pseudolesions, the authors depicted the basic structures of the skull, such as the foramen lacerum, "naturally," but subdued them. The practice of judgment went like this. Schwarz and Golthamer received the hand-drawn facsimile radiographs, then inserted the lesions that interested them on an acetate overlay superimposed on the picture. The artist then "reinterpreted" the drawings and produced a new acetate overlay that "blended with her original art work." Over and over, radiographers and artists iterated the cycle until "all lesions seemed to possess the desired appearance."[71]

Had the image been produced "as it appears in a skull" (that is, had the original x-rays been copied objectively), the images would have obscured and overlapped lesions that were precisely the point

of interest. Had the images departed unrecognizably from the radiographs, they would have had no significance. So using "slight optical distortion," the authors "overemphasized" normal variants and
pseudolesions — only in this way could the radiologists be sure the
important elements would be evident against a "de-emphasized" but
recognizable background. "The lesson we learned in preparing the
plates for the atlas was that nature may be depicted realistically only
by setting off the uncommon and unusual against the background of
the 'natural' and common."[72] If one needed evidence that mechanical
objectivity no longer could simply be assumed to be the first and
only epistemic virtue, the virtue trumping all others, here it is: the
"realistic," which these authors wanted, had become the *enemy* of the
"natural," which they subordinated (see figure 6.11). As Golthamer
and Schwarz said, "We came to realize that painstaking copying of
nature was not the purpose of drawings in an anatomic atlas." At
these words, many mechanical objectivists would have revolted.

The real emerged from the exercise of trained judgment. So
while the mechanical transfer of object to representation may well
be "natural," the natural was no longer the sole object of scientific
desire. Differing both from the genial improvement of the "natural"
object *and* from the objectivist's mechanical reproduction of the
working object, the interpreted image — used in this way — is something new. Manipulated to build on the natural, but structured to
bring out specific features by means of expert understanding, the
twentieth-century image embodies professional experience; it is pictorial presentation by (and for) the trained eye. True, the older form
of self-restrained mechanical objectivity lives on — as we saw in
Henry Alsop Riley's 1960 polemic against the "artist's interpretation" that stacked so poorly against the photograph, which was "the
actual section." But throughout the mid-twentieth century, a new
form of scientific visualization came to be photographed, painted,
and written across sagas like these of magnetograms and x-rayed
lesions. More and more scientists wanted an interpreting, physiognomic vision, not the blind sight of mechanical objectivity.

Here, in the already interpreted image of figure 6.11, realism has
been redefined. It has become a realism that forcefully takes already-
existing photographs and replaces them with a photographically
inflected artwork; this is a realism explicitly positioned against the

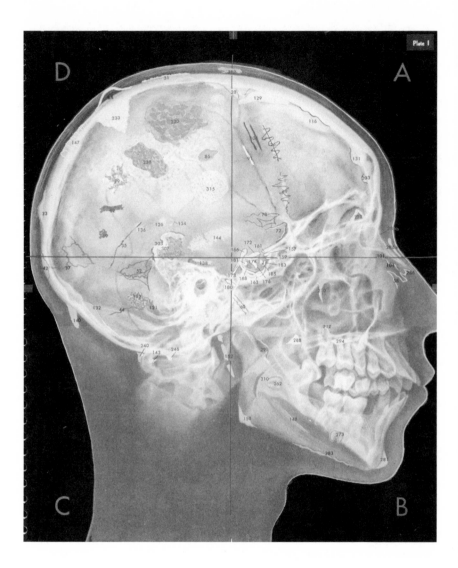

Fig. 6.11. Realism Versus Naturalism. Gerhart S. Schwarz and Charles R. Golthamer, *Radiographic Atlas of the Human Skull: Normal Variants and Pseudo-Lesions* (New York: Hafner, 1965), pl. 1 (reproduced by permission of Harriet E. Phillips). Unlike a 1930s atlas by Golthamer that had original photographs stitched into each copy, this atlas used hand-painted prints and transparent overlays that were (as the authors put it) a "theoretical composite." More than one hundred variants and pseudo-lesions could be found on each printed plate. Judgment was necessary not only in the radiographer-authors' choice of pseudo-lesions but also in creating the artwork — which, in this case, was done by Harriet E. Phillips, director of the art department of the College of Physicians and Surgeons, who did the line drawings, and Helen Erlik Speiden, who had "the manual skill" to execute the half-tone drawings so that in reproduction they would resemble original radiographs. (Please see Color Plates.)

automaticity of unvarnished photographic naturalism, against mechanical objectivity. In making their claim, Schwarz and Golthamer resituated the nature of depiction; the whole project of nineteenth-century mechanically underwritten naturalism suddenly seemed deeply inadequate. For the image to be purely "natural" was for it to become, *ipso facto*, as obscure as the nature it was supposed to depict: a nightmare reminiscent of Borges's too-lifelike map. Only by highlighting the oddities against a visual background of the normal could anyone learn anything from the sum of Schwarz and Golthamer's vast labor of compilation.

Golthamer and Schwarz wrote, disarmingly, that it was only after painstaking efforts to depict nature as it was that they "discovered" that the "purpose" of their atlas was to achieve realism, not naturalism. Their discovery was qualitatively unlike the unearthing of a new fossil or the recognition of a never-seen star. Yet it was just as surely a discovery, one that turned inward to reconstruct not only the kind of evidence they would allow but also the kind of persona that they as scientists would need to have. Instead of wanting to create transparent vehicles for the transport of forms from nature to the reader, the scientist now aspired to another ideal, one in which an expert, trained eye counted for more than a mechanical hand. To understand the "discovery" Golthamer and Schwarz had made — to see it repeated over and again, as judgment supplemented objectivity — is to realize just how impossible the interpreted image would have been in the blind sight of mechanical objectivity.

Practices and the Scientific Self

Sage to worker to trained expert; reasoned image to mechanical image to interpreted image. This epigram, albeit too schematic, joins the epistemological history of the image to the ethical epistemology of the author-scientist. More enters with the interpreted image than what stands on the page. At the beginning of the twentieth century, a new kind of opportunity appeared for scientists to cultivate a different kind of scientific self. Poincaré, in his *Valeur de la science* (*The Value of Science*, 1905), put enormous emphasis on the role of intuition as a tool of discovery in science. Some mathematicians, he wrote, work through logic, through analysis, through a kind of extended arithmetic. The other group — *not* separated by field

357

of work or even by education (according to Poincaré) — was all for physical reasoning, visual depiction, immediate grasp. These "sensual" intuitionists manifested the difference in writing, in teaching, "in their very look." For Poincaré, this contrast was never forgotten by anyone who had witnessed it — as he had in the contrast between an Ecole Polytechnique colleague, the mathematician Joseph Bertrand, who specialized in analytical mechanics, probability, and thermodynamics, and Charles Hermite, the much more formal algebraist from the Collège de France and the Sorbonne: "While speaking, M. Bertrand is always in motion; now he seems in combat with some outside enemy, now he outlines with a gesture of the hand the figures he studies. Plainly he sees and he is eager to paint, this is why he calls gesture to his aid. With M. Hermite, it is just the opposite; his eyes seem to shun contact with the world; it is not without, it is within he seeks the vision of truth."[73]

"Shunning" the world (as Hermite did) risked losing it altogether, as Poincaré warned. "'What you gain in rigour,' [philosophers say,] 'you lose in objectivity.'" Infallible science would come, or so it could be argued, only by isolating mathematics from the world it purportedly described. Pure spatial, physical intuition (of Bertrand's kind) had much to offer mathematics — but it could also be fooled by its weaker attachment to strict rigor. Only a logic inflected by the mathematical analogue of "seeing," "painting," and "gesturing" could lead forward. "The two kinds of intuition [logical and sensual] have not the same object and seem to call into play two different faculties of our soul; one would think of two search-lights."[74] (Frege would no doubt have detested such a psychology of invention.)

But many among Poincaré's contemporaries increasingly took the nonprocedural, the intuitive, the immediate grasp as a crucial part of science — not just in the empirical world, but even, perhaps especially, on the icy heights of mathematics. If processes were *unconscious*, that did not constitute a hindrance. On the contrary, the bright light of deliberate, logical, procedural work was insufficient, as Poincaré emphasized when he recalled the hidden trials of the mind. The French mathematician Jacques Hadamard built on Poincaré's reflections when he wrote, to widespread acclaim, on the psychology of mathematical invention. He, too, stressed the unconscious as an inevitable part of the productive mathematical self. This was not the detailed, articulated unconscious of Freudian theory — there was no talk of drives, instincts, or the ego as the boundary between id and reality principle. No Oedipal complexes here. The scientific unconscious was instead closer to the unconscious suggestibility that Pierre Janet found in his patients (the French psychiatrist could induce them to see or not see crosses marked on cards), or to the unconscious criteria for pattern recognition invoked by the Gestalt psychologists.

Like his many predecessors who had invoked facial-racial recognition, Hadamard's central example was the unconscious pattern assessment employed in the recognition of a human visage. Here the mathematician's judgment joined that of the astronomer and the electroencephalographer. The scientific unconscious tries different combinations, invokes a myriad of hidden factors, joins them together, and then seizes the right array. Approvingly, Hadamard quoted Poincaré: "The unconscious self 'is not purely automatic; it is capable of discernment; it has tact, delicacy; it knows how to choose, to divine. What do I say? It knows better how to divine than the conscious self, since it succeeds where that has failed.'"[75]

This judging, unconscious-intuitive scientific self is a long way from a self built around the imperious will. Nor was it a return to the fragmented self of the eighteenth-century savants. Though expert trained judgment, like truth-to-nature, stood in opposition to mechanical objectivity, trained judgment and truth-to-nature are far from identical. The atlas author of the twentieth century is a more adept version of the reader — a trained expert — not a debased echo of the sage. To the reader-apprentice of the twentieth century, there was no need to rely on the guiding genius's qualitatively different sensibility. The Gibbses may have been more familiar with the erratic markings of an electroencephalogram than the advanced medical student or up-to-date physician, but the aspiring electroencephalogram reader is promised 98 percent reading accuracy in twelve short weeks. No part of the self-confidence displayed here is grounded in genius. The trained expert (doctor, physicist, astronomer) grounds his or her knowledge in guided experience, not special access to reality. (Imagine Goethe promising his readers the ability to construct the ur-forms of nature after a Gibbs-like high-intensity training course.) Nor are the interpreted images that

359

judgment produces to be likened to the metaphysical images of an earlier age. Explicitly "theoretical," the new depictions not only invited interpretation once they were in place but also built interpretation into the very fabric of the image — but they did so as an epistemic matter. Theirs were exaggerations meant to teach, to communicate, to summarize knowledge, for only through exaggeration (advocates of the interpreted image argued) could the salient be extracted from the otherwise obscuring "naturalized" representation. The extremism of iconography generated by expert judgment exists not to display the ideal world behind the real one but to allow the initiate to learn how to see and to know.

Along with this conjoint history of scientific self and image comes a reshaping of the presupposed audience for the scientific work itself. For different reasons, both the reasoned and the objective images took for granted an epistemic passivity on the part of those who viewed them. The reasoned image is authoritative because it depicts an otherwise hidden truth, and the objective image is authoritative because it "speaks for itself" (or for nature). But the interpreted image demands more from its recipient, explicitly so. The oft-repeated refrain that one needs to learn to read the image actively (with all the complexity that reading implies) also transforms an assumed spectator into an assumed reader. Both the maker and the reader of images have become more active, more dynamic, drawing on unconscious as well as conscious faculties to effect something far more complex than a simple Manichaean struggle of the will between (good) receptivity and (dangerous) intervention.

If the objective image is all nature, nothing of us, then oughtn't there to be (by antisymmetry) an image that is all us, no nature? There is. Hermann Rorschach produced his plates more or less exactly in the time period that trained judgment emerged as an epistemic virtue. Nowhere could the active, unconscious self be more evident in image making and using than in Rorschach's eponymous test. Having designed the test in the 1910s and early 1920s to explore the very nature of perception, Rorschach systematically "scored" his patients' responses to standardized plates as a way of exploring their subjectivity. The contrast of the Rorschach test with the objective image is illuminating from all angles. His plates were standardized, "working objects"; he had a strict protocol for interrogating his subjects on their associations to the ink-blots and grading their responses to them. Yet Rorschach designed his plates, at least ostensibly, to be "random" — that is, without any direct reference to the world — precisely so they would serve as the screen onto which the subject would make visible (objective) his or her pure subjectivity.[76]

Rorschach's cards — indeed, the whole test — presupposed a certain kind of self, precisely one marked by the presence of a characterizable and quasi-stable unconscious that could be defined by the particular ways the subject "read" the images: How much color? How much form? How much implied motion? And then, more specifically: What kinds of associations, which content, what role for the blank spaces, and much more. Like Poincaré and Hadamard, Rorschach emphatically rejected the older idea that the cognitive and the affective were natural enemies; like them, he was committed to an unconscious, broadly and narrowly construed, that was a necessary and fundamental part of scientific work. Also like them, he was commited to the unconscious in a broad-church rather than sectarian manner — he drew importantly on recent work but was rigidly attached to no particular psychological system.

Poincaré had insisted that productive scientific work *demanded* that the two "searchlights" (conscious logic and unconscious intuition) function together. Rorschach, equally committed to the idea of joining affect and cognition, put forward a related thought in his magnum opus, *Psychodiagnostik* (*Psychodiagnostics*, 1921): "Coartivity [constriction of affect] is necessary if there is to be talent in the field of systematic scientific endeavor [but] maximum coartation leads to empty formalism and schematization."[77]

Whether they were astronomers sorting spectrographs or physicians examining x-rays, whether they saw themselves as philosophers peering into science or mathematicians judging the roots of their inventions, early twentieth-century scientists reframed the scientific self. Increasingly, they made room in their exacting depictions for an unconscious, subjective element. Psychologists, meanwhile, were busy finding ways of measuring the deepest aspects of subjectivity against the grid of procedure and protocol. By the mid-twentieth century, objectivity and subjectivity no longer appeared like opposite poles; rather, like strands of DNA, they executed the complementary pairing that underlay understanding of the working objects of science.

CHAPTER SEVEN

Representation to Presentation

Seeing Is Being:Truth, Objectivity, and Judgment
Making a scientific image is part of making a scientific self (see figures 7.1, 7.2, 7.3, 7.4 and 7.5): Through each of these atlas images of natural objects shimmers an image of an ideal atlas maker. True, none of these epistemic ambitions could be completely realized. Mechanical objectivists could never completely remove themselves from the process of image making, any more than seekers of truth-to-nature ever revealed the one and only ur-form of a plant, animal, or crystal. Nonetheless, regulative ideals were never mere Sunday sermons. They were assiduously practiced, as techniques of shaping the self as well as of picturing nature. The sage who sought truth-to-nature cultivated memory and synthetic perception; the hardworking hero of objectivity steeled the will to resist wishful thinking and even mental images; the self-confident expert trusted to judgment informed by well-schooled intuitions. Atlas images—whether reasoned, mechanical, or interpreted—bear the marks of both epistemology and ethos.

This book has traced how epistemology and ethos emerged and merged over time and in context, one epistemic virtue often in point-counterpoint opposition to the others. But although they may sometimes collide, epistemic virtues do not annihilate one another like rival armies. Rather, they accumulate: truth-to-nature, objectivity, and trained judgment are all still available as ways of image making and ways of life in the sciences today. All of these images are taken from mid-twentieth-century atlases. There is nothing intrinsically surprising about this accumulation; after all, political virtues such as

363

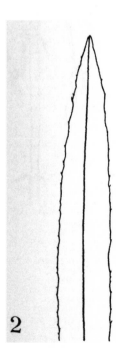

Figs. 7.1, 7.2. Truth-to-Nature. *Potamogeton gramineus*
L., Olaf Hagerup and Vagn Petersson, *Botanisk Atlas*
(Copenhagen: Munksgaard, 1956), vol. 1, p. 15.
The "L." in this plant's official Latin name stands
for Linnaeus, and the mid-twentieth-century image
remains faithful to Linnaeus's principles of botanical
description: sharply outlined forms, clear rendering
of proportions and characteristic features, and no color.
Although the atlas is printed on high-quality glossy
paper, color photography is eschewed. Note the stylized
leaf detail (fig. 7.2), the direct descendant of
Linnaeus's "Genera foliorum" (fig. 2.3).

2

Fig. 7.3. Mechanical Objectivity. *Cirrostratus fibratus*, World Meteorological Organization, *International Cloud Atlas*, rev. ed. (Geneva: World Meteorological Organization, 1987), p. 114 (© Howard B. Bluestein). The highly particular circumstances — photographer, place, date, time of day, part of sky — under which this color photograph was taken are recorded with the image itself. But like all atlas images, this one is meant to be emblematic of a whole "genus" of clouds. The classificatory language (along with the binomial Latin nomenclature) of botany and zoology was self-consciously adopted by the late nineteenth-century meteorologists who assembled the first cloud atlases. However, they repeatedly remarked on the distinctiveness and mutability of every individual cloud formation and turned to photography to record it. (Please see Color Plates.)

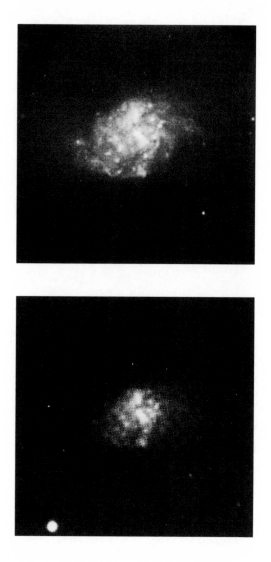

Figs. 7.4, 7.5. Trained Judgment. NGC (New General Catalogue of Nebulae and Clusters of Stars) 1087, James D. Wray, *The Color Atlas of Galaxies* (Cambridge: Cambridge University Press, 1988), p. 13 (reprinted with the permission of Cambridge University Press). These two images of the same galaxy are presented with the explicit aim of schooling the reader's judgment, "to provide a further basis for judging the repeatability of colors, not only from one telescope to another, but from one night to another, for different zenith distances and different air masses, different image tubes and any other parameters that could enter in to produce the final results. You will find that the agreement is on the whole reasonably good, with occasional obvious differences which you should consider in your own interpretation of the information conveyed in these color images." (Please see Color Plates.)

freedom and solidarity come to be endorsed in different historical contexts and yet eventually coexist in a society that is heir to these several traditions. In both the epistemic and the ethical realm, coexistence is sometimes peaceful and sometimes not. Epistemic virtues that exist side by side implicitly modify one another by the very possibility of choice among them, however dimly the facts of diversity and choice are recognized.

The same point can be made for scientific selves. It is a familiar observation that there are more scientists at work today than in the entire previous history of humanity. In this multitude coexist not only many individual personalities but also distinct collective traditions of schooling and sustaining scientific selves, perpetuated by much the same mechanisms as research traditions are. As we have seen (literally seen, in the images from scientific atlases over three centuries), to learn to observe and depict in a science is to acquire at once an ethos and a way of seeing. The same cultivated patterns of attention that single out certain objects in a certain way — in the way of a Bernhard Albinus as opposed to a Rudolf Grashey anatomical atlas, a Wilson Bentley rather than a Gustav Hellmann atlas of snowflakes, the Henry Draper versus the W.W. Morgan, Philip C. Keenan, and Edith Kellman atlas of stellar spectra — also pattern a self. Perceptions, judgments, and, above all, values are calibrated and cemented by the incessant repetition of minute acts of seeing and paying heed.

Indignation bears vehement witness to the fact that values, not just habits, are instilled by these and other practices. When epistemic virtues confront one another, so do scientific selves — as in the case of Santiago Ramón y Cajal squaring off against Camillo Golgi or Wilhelm His upbraiding Ernst Haeckel, but also in more recent cases of alleged scientific fraud. Where one side sees a breach of scientific integrity, another may see loyalty to the discipline's highest standards.[1] The differences that provoke mutual outrage may split along the lines of generation, discipline, or research group. But they are never merely idiosyncratic, one personal style clashing with another. There are no purely private values, any more than there are purely private languages: the ethical, even the narrowly scientific ethical, is always a matter of collectives, and historical ones at that.

The ways of seeing we have explored are the achievements of no individual, not even of any particular laboratory or discipline. No

Nobel prizes honored the introduction of trained judgment into making and classifying images. There is no single domain of phenomena that monopolized the impulse to find the idea in the observation. Neither crystallographers nor anatomists nor astrophysicists can take credit for developing the regulative ideal of mechanical objectivity, of transferring images from objects to the page without human interference. Instead, these kinds of scientific sight, in their rise and fall, constitute the development of a truly *collective* empiricism.

Scientific sight as described in this book is epistemologically saturated. Making and reading of atlas images crystallize what is meant by truth-to-nature or mechanical objectivity or trained judgment. The four-eyed sight required to depict the idea in the observation, the blind sight needed to forestall interpretation, the physiognomic sight cultivated to detect family resemblances — these visual habits were also expressions of epistemological loyalties. The collaboration of René-Antoine Ferchault de Réaumur and Hélène Dumoustier de Marsilly brought truth-to-nature to the page in the form of rigidly symmetrical insects. Arthur Worthington abandoned his exquisitely etched splashes for the "objective view" captured by the much messier split-second photographs. Frederic A. Gibbs and Erna L. Gibbs in turn threw mechanical objectivity overboard, embracing the trained judgment that permitted them to sort out electroencephalograms as confidently as they would distinguish "Eskimos from Indians." Ways of seeing become ways of knowing.

But close consideration of these practices seldom enters into the ancient and still continuing philosophical debate over the epistemological status of vision *per se*. Whether vision is repudiated as a false guide, leading the unwary astray with the gleam of mere appearances, or defended as the noblest and most intellectualized of the senses, it is conceived abstractly in this debate, as the same faculty for Plato and George Berkeley, René Descartes and Arthur Schopenhauer. Proponents and opponents treat theories and valorizations of vision historically and with discerning attention to nuance, but they rarely address the actual activity of seeing.[2] In this book, we have focused on practices of seeing, rather than theories of vision.[3] We nonetheless hold these practices as well as theories to be of philosophical import. They dictate not just how the world looks but also what it is — what scientific objects are and how they should be known.

Ways of scientific seeing are where body and mind, pedagogy and research, knower and known intersect. To weaken these oppositions is also to weaken the conventional philosophical understanding of epistemology. Yet historicized, collective ways of seeing undeniably produce knowledge and therefore qualify as the stuff of epistemology. The four-eyed sight that reveals the universal in the particular, the blind sight that blocks projection, the physiognomic sight that puts a face to the data — these were all corporeal skills to be learned as well as cognitive stances to be mastered.

Once internalized by a scientific collective, these various ways of seeing were lodged deeper than evidence; they defined what evidence was. They were therefore seldom a matter of explicit argument, for they drew the boundaries within which arguments could take place. Atlases provide a rare and precious glimpse of ways of seeing in the making, as a place where established practices are transmitted and innovations explicitly advanced. For centuries, atlas images have taught scientists what to look for and how to see it. At crucial junctures, when new epistemic virtues clash with old, ways of seeing were revised — and atlases along with them. The subjects and objects of inquiry, knower and known, were thereby transformed: different ways of seeing picked out different working objects and shaped different scientific selves.

We can use these three opening images to sharpen the distinctions among the epistemic virtues described in this book. Truth-to-nature seeks to reveal a type of a class that may correspond to no individual member of that class and yet stands for all of them. Even if the metaphysics of immutable natural kinds is replaced by a Darwinian notion of evolving species or by statistical reference classes, the class crystallized in a true-to-nature image still performs scientific work, often of a taxonomic or correlational sort. The plant in figure 7.1 may not be a pure Goethean archetype, but it still stands for an entire species. Truth-to-nature counters an epistemological worry that is as much about nature as it is about would-be knowers of nature: what if the variability of nature is so great as to swamp the infirm human senses and intellect? The cloud in figure 7.3 answers the corresponding worry about variable observers: what if your subjectively construed cloud differs from mine, equally subjective? Because the evil is believed to lie in intervention, the remedy is

369

sought in automatism. The variability of knowers is suppressed, even at the expense of readmitting the variability of nature: *this* cloud, formed at *this* place, at *this* time, in all its accidental uniqueness.

Trained judgment differs from truth-to-nature in discerning patterns rather than types. The galaxy shown in figure 7.4 and figure 7.5 has been "interpreted"—the original atlas caption advises readers to practice with various photographic filters to hone their judgment—on the basis of family resemblances rather than species types. Within a family (or, as some atlas makers would have had it, a race), variability is taken for granted. Whether the patterns indicate natural kinds or not is a matter of indifference for most practitioners of judgment; for them, pattern detection is the preface to action, not just to classification. Their paramount fear is of paralysis; hence their impatience with the scruples of objective atlas makers who abdicate their primary responsibility to supply working objects for their sciences. The understanding of patterns may be roughly statistical, in the sense of corresponding to the distribution of cases around a mean (as in Grashey's collection of deviant x-rays), or it may appeal to Wittgensteinian family resemblances. But in neither case is it essentialist, in the sense of compressing an entire class into a type. Both trained judgment and truth-to-nature trust to long experience, but whereas truth-to-nature makes prodigious demands on memory, both natural and written, trained judgment relies on unconscious processes that cannot even be introspected, much less recorded.

The mere fact of plural possibilities among epistemic virtues in science provokes comparisons, justifications, even defensiveness. Atlas makers who embrace trained judgment are pugnaciously frank about the intrusion of the subjective into their images. Those who defend true-to-nature images are impatient with the particularities and peculiarities of objective ones; proponents of objective images insist that only mechanical procedures can ward off distortions. Moreover, mutual modifications occur: the judgment exercised, for example, by the self-assured twentieth-century expert differs fundamentally from that cultivated by an eighteenth-century savant. For the latter, judgments were universal, a realization of universal reason in interaction with universal nature; for the former, they were personal, an expression of the unconscious harnessed by training. The historical divide that separates them is the distinction between objective and subjective, which requires that *all* judgments be per-sonal, clearly located on the subjective side, even if they are in the service of a more faithful depiction of nature.

By a kind of ratchet effect, the epistemic virtue of objectivity, once established, makes it impossible simply to replicate earlier virtues and practices. Judgment before and judgment after the emer-gence of objectivity in the mid-nineteenth century both stand op-posed to it, but they are also opposed to each other, by dint of the interposition of objectivity between them. This is a history not of the oscillations of a pendulum between two fixed extremes in a two-dimensional plane, but of orthogonal innovation into the third dimension. Historical sequence matters: mechanical objectivity was a reaction to truth-to-nature, and trained judgment was a reaction to — and different from — both.

One of the aims of this book has been to point out the bare exis-tence of a plurality of epistemic virtues, as well as to trace the history of some of them. Moral philosophers have argued for an irreducible plurality of visions of the good, which can be reasonably debated in specific cases but never eliminated in principle by reason alone.[4] Analogously, we believe that a plurality of visions of knowledge, understood in the most capacious sense of fidelity to nature, is likely to be a permanent aspect of science.

Objectivity stands at the center of this book. We have flanked it with accounts of truth-to-nature and trained judgment to show that its emergence is recent and contingent: there can be, there has been, there is science without mechanical objectivity. This table offers a simplified overview of the covariance of scientific self, image, proce-dure, and object.

Layered, Pressured Periodization (*time* →)	
Trained Judgment	scientific self: trained expert; interpreted image; pattern depiction; families of objects
Mechanical Objectivity	scientific self: will-abnegating worker; mechanical image; procedural depiction; particular objects
Truth-to-Nature	scientific self: sage; reasoned image; idealized depiction; universal objects

With this framework in mind, return to the images that opened this chapter: figure 7.3 from the cloud atlas could have been a type or a pattern (see figures 7.6 and 7.7). Objectivity is one epistemic virtue among several, not the alpha and omega of all epistemology. Objectivity is not synonymous with truth or certainty, precision or accuracy. Sometimes, as we have seen in concrete instances, objectivity can even be at odds with these: an objective image is not always an accurate one, even in the view of its proponents. Objectivity is neither inevitable nor uncontested. Indeed, juxtaposed to alternatives, it can even seem bizarre. Why knowingly prefer a blurred image marred by artifacts to a crisp, clear, uncluttered one?

Why, then, is objectivity so powerful as both ideal and practice? How did it come to eclipse or swallow up other epistemic virtues, so that "objective" is often used as a synonym for "scientific"? In order to answer these questions, we must first of all insist that there *are* other epistemic virtues besides objectivity. One reason we have focused on scientific atlas images is that only at the level of specific practices do the distinctions among epistemic virtues such as truth-to-nature, objectivity, and trained judgment sharpen. At the more abstract level of epistemological analysis, objectivity tends to be used as shorthand for all epistemic virtues — the whole of epistemology. The history of epistemology (and of science) is often narrated as if it were identical to the history of objectivity. Francis Bacon and Descartes, even Plato and Aristotle, are recruited into a lineage that has allegedly always battled subjectivity, as if the Kantian terms merely rechristened a distinction present since the beginnings of Western philosophy.[5] We have argued that this homogenized view of the history of epistemology and of science is false. But if the view is in error, why is the error so widespread, so irresistible?

All epistemology begins in fear — fear that the world is too labyrinthine to be threaded by reason; fear that the senses are too feeble and the intellect too frail; fear that memory fades, even between adjacent steps of a mathematical demonstration; fear that authority and convention blind; fear that God may keep secrets or demons deceive. Objectivity is a chapter in this history of intellectual fear, of errors anxiously anticipated and precautions taken. But the fear objectivity addresses is different from and deeper than the others. The threat is not external — a complex world, a mysterious God, a devious demon.

372

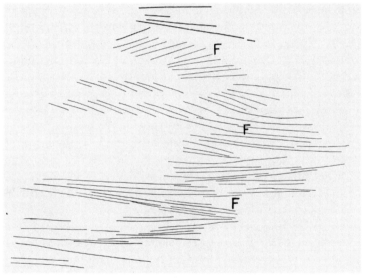

Figs. 7.6, 7.7. Training Observers. *Cirrostratus filosus,* Internationales Meteoro-
logisches Komitee, *Internationaler Atlas der Wolken und Himmelsansichten* (Paris:
Office National Météorologique, 1932), pl. 21. A new cloud atlas was issued in part
because "it was urgent to provide observers with a new atlas [to replace the 1896
edition], because the quality of observations steadily declined and differences in
identification emerged" (*ibid.,* p. v). As in fig. 7.3, the image is labeled with the
particular circumstances under which the photograph of the cloud was taken. The
schematic diagram (drawn exactly to the scale of the photograph) directly below,
however, directs the reader's attention to the "essential details" of this species of
cloud: "distinct threadlike structure" (*ibid.,* p. xi). (Please see Color Plates.)

Nor is it the corrigible fear of senses that can be strengthened by a telescope or microscope or memory that can be buttressed by written aids. Individual steadfastness against prevailing opinion is no help against it, because it is the individual who is suspect.

Objectivity fears subjectivity, the core self. Descartes could discount the testimony of the senses because sensation did not belong to the core self as he conceived it, *res cogitans*. Bacon believed that the idols of the cave, those intellectual failings that stemmed from individual upbringing and predilection, could be corrected by the proper countermeasures, as a tree bent the wrong way could be straightened. But there is no getting rid of, no counterbalancing post-Kantian subjectivity. Subjectivity is the precondition for knowledge: the self who knows.

This is the reason for the ferociously reflexive character of objectivity, the will pitted against the will, the self against the self. This explains the power of objectivity, an epistemological therapy more radical than any other because the malady it treats is literally radical, the root of both knowledge and error. The paradoxical aspirations of objectivity explain both its strangeness and its stranglehold on the epistemological imagination. It is epistemology taken to the limit. Objectivity is to epistemology what extreme asceticism is to morality. Other epistemological therapies were rigorous: Plato's rejection of the senses, for example, or Descartes's radical doubt. But objectivity goes beyond rigor. The demands it makes on the knower outstrip even the most strenuous forms of self-cultivation, to the brink of self-destruction. Objectivity is not just one intellectual discipline among many. It is a sacrifice — and was often so described by its practitioners: Worthington surrendered symmetry, Robert Koch gave up three-dimensional corrections, Erwin Christeller lived with artifacts.

Whether they took the form of spiritual exercises as taught in the ancient philosophical schools or of regimens of observation followed by Enlightenment naturalists, lives of the mind have long aimed to shape the self as a recipient of wisdom and knowledge. The suppression of subjectivity attempted by scientists striving for objectivity went much further. Subjectivity is not a weakness of the self to be corrected or controlled, like bad eyesight or a florid imagination. It *is* the self.

Or rather, it is the self in a particular mental universe in which all that exists is divided into the opposed and symmetrical provinces of the objective and subjective. This mental universe in which we moderns are now so at home had its Big Bang a scant two hundred years ago. Just as there was epistemology before (and after) the advent of objectivity, there were selves before and after the emergence of subjectivity. Mechanical objectivity was called into being in the mid-nineteenth-century sciences to rein in the excesses of a dynamic, will-centered self that threatened to remake the world in its own reflection. Neither hard facts nor clear images could, it was feared, block the projections of the self of subjectivity. In contrast, the friable, fractious self of eighteenth-century sensationalist psychology prompted quite different worries about being overwhelmed by the tumult of experience or seduced by the imagination. And neither model of the self as knower made a place for the unconscious processes of perception and intuition invoked in early twentieth-century accounts of trained scientific judgment. The personas of the sage, the indefatigable laborer, and the expert exemplified the idealized knower who had successfully overcome the characteristic frailties of each sort of self in the service of truth-to-nature, objectivity, and trained judgment, respectively.

A history of knowledge that links epistemic virtues with distinctive selves of the knower traces a trajectory of a different shape from familiar histories of philosophy and science. Instead of a jagged break in the seventeenth century, in which knowledge is once and for all divorced from the person of the knower — the rupture that allegedly announces modernity — the curve is at once smoother and more erratic: smoother, because knowledge and knower never became completely decoupled; more erratic, because new selves and epistemic virtues, new ways of being and ways of knowing, appear at irregular intervals. It is a story of sporadic collective creativity, still ongoing, rather than one of a single explosive revolution after which history froze. The changes are nonetheless dramatic, even in the few centuries covered by this book, and are heightened still more by a longer historical baseline.

Contrast, for example, the vision of knower and knowledge embodied by Socrates in the *Symposium* and that embraced by Albert Einstein in his autobiography: Socrates pursued knowledge through Eros, as a beautiful soul in an ugly body who seduces others to seek the truth;[6] Einstein yearns for a "paradise beyond the personal," in which knowledge is the eternal and collective possession of a community of thinking beings dispersed over time and space.[7] These are very different models of knowers and knowing, but both take a certain kind of knower to be the precondition for knowledge. Perhaps the most dramatic change of all in this long history of knowers and knowing was the emergence of objectivity: a novelty so blinding as to become invisible, it came to be perceived as an inevitability rather than as an innovation.

It is a misconception, albeit an entrenched one, that historicism and relativism stride hand in hand, that to reveal that an idea or value has a history is *ipso facto* to debunk it. But to show that objectivity is neither an inevitable nor an eternal part of science passes no verdict on its validity, desirability, or utility — any more than to document that the prohibition against cruel and unusual punishment first emerged at a particular time and place would *per se* subvert that judicial principle. Conversely, to point out that certain beliefs and practices have enjoyed widespread acceptance in various cultures and epochs is not necessarily an endorsement: no one thinks better of slavery or geocentrism on learning that many people in many places at many times have subscribed to them.

All history can do is to demonstrate the possibility of alternatives, thereby turning an apparent axiom — things could never have been otherwise than as we know them — into a matter for reasoned argument. Between dogmatism and relativism stretches a wide plain of debate. To claim that there are multiple virtues, be they epistemic or moral, is very different from the claims that all virtues (or none) are equally well- (or ill-) grounded and that whim may decide among them. It is a commonplace in ethics and politics that hard choices must sometimes be made, but this idea is something of a novelty in epistemology. One of the aims of this book is to open such a debate about epistemic virtues, by using history to clarify what they are, how they work, and how much hangs in the balance if one is obliged to choose among them.

The implications of a history of epistemic virtues reach further. Far from relativizing these virtues, history exhibits their rationale, if not their transcendent rationality. Truth-to-nature, mechanical objectivity, and trained judgment all combat genuine dangers to knowledge: the dangers of drowning in details, of burking a fact to support a theory, of being straitjacketed by mechanical procedures. The atlas makers who embraced one or another of these virtues were not just tilting at windmills, even if (in the opinion of their colleagues who espoused other virtues) they exaggerated the risk they dreaded most. The reality of the risks explains the persistence of the countermeasures. Truth-to-nature, for example, endures, despite the existence of alternatives, because, at least in some sciences, the danger of being overwhelmed by particulars is still paramount.

Yet some (philosophers especially) may still be uneasy about the corrosive power of history to dissolve whatever it touches. They will persist: Doesn't the very existence of multiple epistemic virtues undermine the unity of truth? And doesn't the fact that they emerge historically threaten the permanence of truth? If epistemology marks out the most reliable route to truth, how can it repeatedly change course without losing its way? Once again, the answer must be framed in terms of fear. To continue the ancient metaphor of the stony way to truth, epistemology is less about trailblazing than about path clearing. Epistemology seeks first and foremost to identify and remove sources of error, rather than to define the nature of truth. Errors notoriously proliferate; so do the strategies for blocking them. That epistemic virtues should be multiple and historical is the unsurprising consequence of the largely negative mission of epistemology: they were called into being to counter equally multiple and historical epistemic vices. Truth itself may indeed have a history, but whether it does or not cannot be concluded from the fact that the means devised to attain it vary over time.

To grant objectivity a history is also to historicize the framework within which much philosophy, sociology, and history of science has been cast in recent decades. The opposition between science as a set of rules and algorithms rigidly followed versus science as tacit knowledge (Michael Polanyi with a heavy dose of the later Ludwig Wittgenstein) no longer looks like the confrontation between an official ideology of scientists as supported by logical positivist philosophers versus the facts about how science is actually done as discovered by sociologists and historians.[8] Instead, both sides of the opposition emerge as ideals and practices with their own histories — what we have called mechanical objectivity and trained judgment.

Neither epistemic virtue is ever realized fully, any more than any other virtue, but both objectivity and judgment are efficacious and consequential in shaping how workaday science is done. To contend that mechanical objectivity (or, for that matter, trained judgment) is a fraud and a delusion because it is never realized in purest form is a bit like making the same claim for equality or solidarity. These ethical values can change society without ever being perfectly fulfilled, and the same is true for epistemic virtues in science.[9] It is a case not of ideology versus reality but of two distinct and sometimes rival regulative visions of science, each as real as the images it makes and both products of specific historical circumstances.

For students of science, to recognize that objectivity (and truth-to-nature, as well as trained judgment) has a history is to reflect upon our own terms of analysis, be they objectivity and subjectivity or Wittgensteinian family resemblances. Wittgenstein, Frege, and Henri Poincaré — to name only a few patron saints of current historical and philosophical analyses of science — don't float above this history; they are part of it, perusing anthropological atlases, responding to the latest psychophysiological experiment, sorting out electrodynamical theories. To historicize their analyses does not *ipso facto* invalidate them. It does, however, unsettle their self-evidence. They have not existed everywhere and always, and none of them reigns supreme even now.

Once the hidden history of objectivity is revealed, what new light is cast upon debates about objectivity in the here and now? Objectivity is still a fighting word, and not just among scientific atlas makers. Critics have attacked it as a fraud, an impersonal mask that veils the very personal and ideological interests it purports to suppress, or as a crime, an arrogant attempt to play God by pretending to a view from everywhere and nowhere. Like other keywords in our conceptual vocabulary — such as "culture" — "objectivity" has more layers of meaning than a mille-feuille.[10] Historians use it as a rough synonym for impartiality or disinterestedness.[11] Philosophers variously define it as "standing in an immediate relation to a nonhuman reality,"[12] as being "cut loose from the idiosyncratic peculiarities of individuals by being of such a nature that any normal person whatsoever could reasonably be expected to have the same experience (or the same feeling) in the circumstances at issue,"[13] as "formed by the kind of critical discussion that is possible among a plurality of individuals about a commonly accessible phenomenon,"[14] as that which "is invariant under all (admissible) transformations,"[15] or as "what constitutes correct use of an expression in particular circumstances... settled somehow independently of anyone's actual dispositions of response to those circumstances."[16] Sometimes objectivity refers to ontology: "an objective world of particulars independent of experience." Sometimes it refers to epistemology: "beliefs, judgments, propositions or products of thought about what is really the case." And sometimes it refers to character: "impartiality, detachment, disinterestedness and a willingness to submit to evidence."[17] Among scientists, objectivity slides between mechanical and structural senses, as we saw in Chapters Three and Five, and each sense implies different metaphysical, methodological, and moral commitments.

What process of historical fusion soldered the metaphysical, the methodological, and the moral into the amalgamated concept of scientific objectivity? How was each distinct component of the amalgam formed, and what affinities among components made their bonding first thinkable and then apparently inevitable? It is not enough to say simply that history has united what logic would have put asunder. History's unions may be less constrained than logic's, but even history cannot arbitrarily recombine elements — otherwise we would have chimeras instead of concepts. A history of objectivity must explain why some ideas and practices melded with one another and others slid away.

All the multiple senses of objectivity intersect in their opposition to subjectivity. The multiplicity of the one is simply the photographic negative of the multiplicity of the other. And in contrast to many other historical views of the self, subjectivity is intrinsically multiple, both among and within individuals. Objectivity and subjectivity are expressions of a particular historical predicament, not merely a rephrasing of some eternal complementarity between a mind and the world. The self captured by subjectivity is highly individualized, in contrast to the self of the rational soul, whose most salient feature was the faculty of reason shared with all other rational souls. Therefore, one sense of objectivity, which we have called structural, strips away all individual peculiarities: the marks of this place and that time, of creed and nationality, of sensory apparatus and species. These are the "thinking beings" of Peirce's (and Einstein's) dreams and Moritz Schlick's nightmares.

Subjective selves also tend to overflow their boundaries, to project themselves into the world, in contrast to the Enlightenment self under siege from the bombardment of sensation. Another sense of objectivity, which we have called mechanical, checks willful self-assertion by enforced passivity and rigid procedures. Each facet of the subjective self, like the forms of objectivity that countered them, had its own distinctive practices — whether it be the Bohemian exaggeration of individuality praised by Charles Baudelaire or the resolute exercise of the unfettered will hammered into the head of French lycée pupils in the mid-nineteenth century by the Cousinians. We have focused on the practices of scientific objectivity, but those of artistic subjectivity were no less concrete and specific and — our chief point here — in reversed-mirror-image relationship to one another.

We are now in a better position to understand the odd associations of bedrock reality with emotional distance, or mechanical procedures with the escape from perspective, that objectivity makes possible. What they all have in common is the repudiation of one or another aspect of the subjective self, but not always the same one. Take the case of emotion. Many intellectual traditions have considered reason and the passions immiscible but have nonetheless deemed certain personal characteristics — being able to split a double star with the naked eye or to remember the names and forms of thousands of plant species — a positive advantage in probing reality.

What makes objectivity different is the conviction that *all* such individuating features interfere with knowledge. As we saw in Chapter Five, Hermann von Helmholtz did not question Jan Purkinje's exceptional ability to register certain visual phenomena that other researchers (including Helmholtz himself) could not, but the very rarity of that ability made it a dubious basis for the psychophysiology of vision. Emotion *per se* was no disqualification; a fiery temper and a passionate commitment to research were, as in the cases of Michael Faraday and Cajal, regarded as perfectly compatible with scientific objectivity. But passionate preferences for one's *own* theories and speculations (Cajal's reproach to Golgi) or even for one's own sensations and intuitions (Frege's reproach to psychologizing mathematicians) count as dangerous expressions of subjectivity. We can also ascertain which forms of quantification intersect with objectivity and which do not: mathematical models may be as idealizing as images from an eighteenth-century atlas; precision measurements often enlist trained judgment to separate signal from noise. Only when quantification is invoked to suppress some aspect of the self — for example, its judgments by means of inference statistics — does the appeal to numbers become a call to objectivity. Similarly, there is no direct link between mechanical procedures and the escape from perspective, except that each seeks to neutralize an aspect of subjectivity, although not the same one. There is a coherence to the concept of objectivity, after all, but it is a coherence that can be detected only against the background of its history.

It is therefore not hard to understand why objectivity has been equated with the complete elimination of the self and consequently dismissed as impossible, whether as an illusion ("a noble dream") or as a deception ("the God trick"). But scientific objectivity never undertook to erase the self, even the self of subjectivity, completely. Rather, its practices, like all techniques of the self, cultivated certain aspects of the self at the expense of others. The will was at once the citadel of the subjective self and the sword and buckler of objectivity. It was the will straining against the will that gave objectivity its peculiar pathos, its tension between personal sacrifice and liberation from the personal, between active intervention in and passive registration of nature. However far scientists' *comprehension* of objectivity and subjectivity may have diverged from their Kantian origins, their *practices* retained a faint echo of the Kantian injunction that the truly free will expresses itself in binding laws, not caprice. Objectivity is at once the enemy of the arbitrary and the highest expression of *liberum voluntatis arbitrium*, the will's free choice.

The story of epistemic virtues in science is one of novelty and transformation: truth-to-nature, objectivity, and trained judgment all have birth dates and biographies; each remade science and self — and scientific images — in its own image. Yet these three virtues all served, each in its way, a common goal: what we have called a faithful representation of nature. This book has documented how various the understanding and, above all, the practices of fidelity could be: nature's types plumbed, nature's appearances registered, nature's patterns intuited. But nature was always in the picture, literally so. The images with which this chapter began, different as they are, are all attempts at *re*presentation. Whether drawing or photograph or digital image, type or individual or pattern, they assume a distinction between nature and image — of this species of Danish plant, that cirrostratus cloud in the skies over Dillon, Colorado, on the afternoon of January 5, 1978, this remote galaxy made visible by processing faint electromagnetic radiation into shapes and colors. Each aimed to be faithful to nature, in its fashion, yet none of them pretended to be, much less to transform nature. Representation is always an exercise in portraiture, albeit not necessarily one in mimesis. The prefix *re-* is essential: images that strive for representation present again what already is. Representative images may purify, perfect, and smooth to get at being, at "what is." But they may not create out of whole cloth, crossing over from nature into art.

Focusing on one or another form of scientific sight keeps two questions front and center: What kinds of practices are needed to produce this kind of image? And what kinds of practices are needed to cultivate the scientific self such that this sight is possible? The history of scientific sight always demands this double motion, toward the unfolding of an epistemology of images, on the one side, and toward the cultivated ethics of the scientific self, on the other. Fidelity to nature was always a triple obligation: visual, epistemological, ethical.

What happens when fidelity itself is abandoned and nature merges with artifact? We close with a peek at scientific atlases right now: images in which the making is the seeing.

Seeing Is Making: Nanofacture
However much atlas images have changed their form in the last three hundred years, however dramatically the persona of the atlas maker has altered, one feature of image making has remained constant. Atlas makers aimed to fix nature on the pages of books, to represent stones, skulls, and snowflakes as faithfully as possible. Toward the end of the twentieth century, however, that seemingly self-evident aspiration began to be edged aside. For many scientists pursuing nanotechnology, the aim was not simply to get the images right but also to manipulate the images as one aspect of producing new kinds of atom-sized devices. This shift from image-as-representation to

image-as-process wrenched the image out of a long historical track. No longer were images traced *either* by the mind's eye *or* by "the pencil of nature." Images began to function at least as much as a tweezer, hammer, or anvil of nature: a tool to make and change things.[18]

In this necessarily tentative section about what is happening as we write, we want to look at a type of atlas — or successor to the atlas — that still aims to organize scientific images systematically for many kinds of uses, but in which images are, to a certain degree, interactive, not fixed. With clicks and keystrokes, these digital images are meant to be *used*, cut, correlated, rotated, colored. Their subjects are as diverse as ever: there are e-atlases of flora, fauna, and fluid-flow, but also of microbiological, chemical, physical, and astrophysical structures. In exploring the novel uses of these interactive atlases-in-the-making, we will attend to examples of two sorts. On the one hand, there are atlases that are based on digital archives — these range from studies of simulated turbulent flow to the Visible Human Project. An increasing number of these archives allow the user to zoom, excise, rotate, or fly through the images. On the other hand, there are images that depart even further from the traditional bound volume: images that are used to alter the physical world. This new tool-like role for images in the expanding field of nanotechnology has come to be known as *nanomanipulation*. For our purposes here, it is worth distinguishing these two kinds of manipulable, interactive images. We will call navigation through given data sets *virtual images* and navigation through the image to modify physical objects in real time *haptic images.*

In the context of the more engineering-inspired, device-oriented work that surrounds much of nanotechnology, images function less for representation than for *presentation*. We use the term presentation in a triple sense. First, because nanomanipulation is no longer necessarily focused on copying what already exists — and instead becomes part of a coming-into-existence — we find it makes more sense to drop the prefix *re-*, with its meaning of repetition. Second, the objects really are being presented like wares in a shop window. By the early twenty-first century, images from nanotechnology and related areas were being produced to entice — scientifically *and* entrepreneurially. Their makers were often ostentatiously uninterested in faithful coloration or spatial fidelity. Instead, atlas-like image collections sought to highlight chosen features, promising things to come by displaying devices that so far existed only in fragmentary, prototype, or imaginary form. Finally, freed from the asceticism of mechanical objectivity or even the interpretation of trained judgment, the nano-image and other interactive images slid more easily into an artistic presentation. It became routine, not just in nanotechnology but in many scientific domains (from fluid dynamics to particle physics and astronomy), to see the virtual scientific image not as competing with art or even employing art but positioned as art itself.

Turning to the nanomanipulated images as an introduction to presentational pictures, consider the following sequence (see figure 7.8). Already it is remarkable that the scientists could manipulate polymer spheres just 120 billionths of a meter across. But it is the picture sequence itself that arrests our attention. Produced by an atomic force microscope that measures the force between a tiny probe and a surface over which the probe scans, figure 7.8 is not intended to depict a "natural" phenomenon. Instead, this and similar haptic images are part and parcel of the fabrication process itself. A second example will clarify the technology.

Normally, the atomic force microscope consists, schematically, of a cantilever that is used to measure the force between its probe tip and the surface over which it is passing. In the case illustrated in figures 7.9 and 7.10, the probe, charged negatively, is hovering above a surface that contains a two-dimensional (flat) gas of electrons, and the charge on the probe "pushes" electrons to flow along the surface. But the probe does not just disturb this flat electron gas, it also scans it, producing an image (figure 7.10) by measuring the variable current produced in the probe itself (rather than the force between tip and surface) — a particular adaptation of the atomic force microscope. The probe acts as both manipulator of the electron gas and as its image-maker.

In such haptic images, seeing and making entered together — unlike the more familiar image making that marked so many generations of science, holding fast to a two-step sequence. The older method meant *first* smashing a proton against an antiproton in an accelerator, *then* imaging the detritus for analysis in a bubble-chamber photograph or a digital display. Or, in a very different domain of science, *first* preparing a tissue sample, *then* imaging it in the electron microscope. For early twenty-first-century nanoscientists, such after-the-fact representations were often entirely beside the point.

Frequently, the nanographers want images to engineer things. In the first instance, these were *images-as-tools*, entirely enmeshed in making, much more than *images-as-evidence* to be marshaled for a later demonstration. In Chapter Three, our interest was in images that aimed to show the actual (rather than ideal, as in Chapter Two) configuration of snowflakes, liquid-drop impacts, or physiological crystals. In Chapter Six, we examined interpreted images — images produced to highlight important features of a lesion or to smooth out the artifacts of production of a solar magnetogram. Here, in this concluding glimpse at image collections of working objects of science, we want to highlight images-as-tools, images that were themselves manipulated.

Some interactive images — virtual ones — can be manipulated to learn something about a configuration of a molecular structure, an anatomical detail, or a structure of a galaxy. Other interactive images — haptic ones — were to be manipulated as part of the modification or construction of a physical object, as in nanomanipulation. Our first aim is to explore the ways these virtual and haptic images have shifted the status of image compendiums — and at the same time to ask how haptic images seem to mesh with a new kind of engineering self. Our second goal will be to point — all too briefly — to the ways new, more presentational images have begun to circulate at the blurred edge of science and art.

Both interactive virtual and haptic (nanomanipulated) images often find their atlaslike homes under the ever-widening rubric of the "image gallery," which, as will become clear shortly, often embraces the older remit of the classical atlases. Though image galleries can — and often do — carry functions far beyond anything in atlases, there is no doubt that this superordinate category is a central place to look if one wants to track images of record in the early twenty-first century.

For a glimpse at the emergent atlases (or atlas successors) of the virtual sort, take the Visible Human Project, begun in 1989 and sponsored by the National Library of Medicine. Designed to make a complete three-dimensional anatomy of both the male and the female, its goal was to offer a widely shared digital resource accurate to

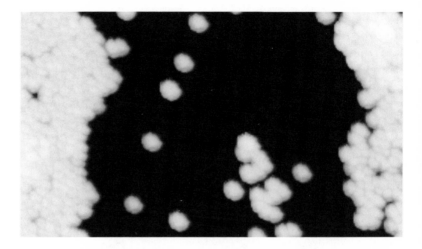

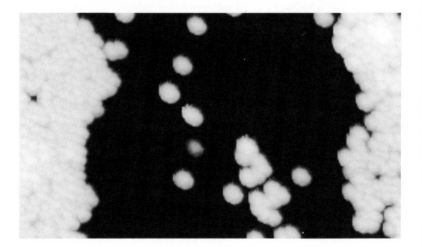

Fig. 7.8. Rolling Nanospheres. JPK Instruments, http://www.jpk.com/spm/spheres-manipulation1.htm, accessed 20 June 2005 (reproduced courtesy of JPK Instruments AG). Here an atomic force microscope is used to manipulate polymer spheres (with a diameter of 120 nanometers) into a cluster. The probe's motion is tracked by yellow lines. (Please see Color Plates.)

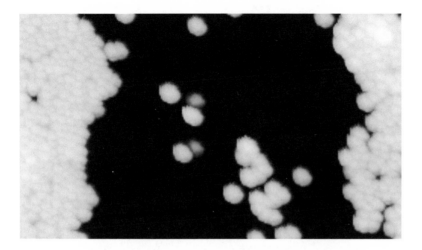

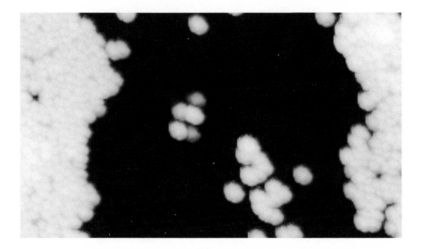

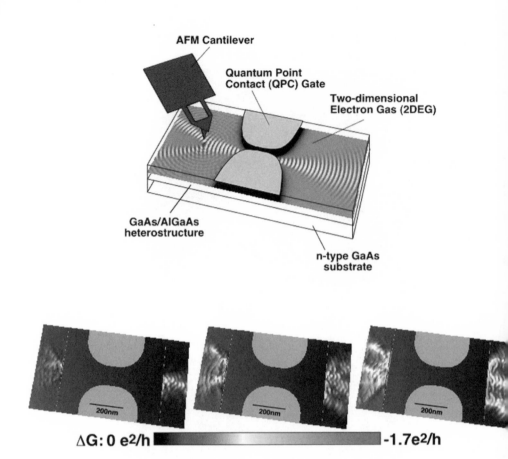

Figs. 7.9, 7.10. Building with the Brush. Robert Westervelt, Schematic of device (*top*); map of charge density that is both created and measured by the probe (*bottom*). http://meso.deas.harvard.edu/spm.html, accessed 8 June 05 (courtesy of Robert Westervelt). The negatively charged tip of the cantilever pushes electrons away from the region directly under it; because there are fewer electrons there, this "depletion zone" is more positively charged than the surrounding areas. This causes the electron gas to scatter from it. Quantum mechanics predicted that electrons flowing through a very narrow passage would only be able to pass with a quantized current — only certain wavelengths pass easily, because any other wavelength causes destructive interference. The three bottom images show the first three modes, that is, three wavelengths such that constructive interference facilitated passage. But in addition to altering the flow of electrons, the probe scans it: a tool and a brush.

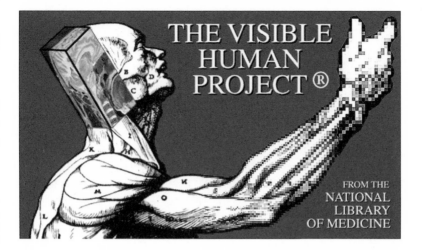

Fig. 7.11. Visible Human. http://www.nlm.nih.gov/research/visible/vhpconf98/MAIN.HTM.
Quotations from the on-line fact sheet, http://www.nlm.nih.gov/pubs/factsheets/visible_
human.html, both sites accessed 9 September 2006 (reproduced by permission of the
National Library of Medicine). The Visible Human Project consists of two digital data sets,
one based on a male cadaver, with slices taken at 1-millimeter intervals (15 gigabytes)
and one taken at 0.33 millimeters from a female cadaver (40 gigabytes). These data sets
are "to serve as a set of common public domain data for testing medical imaging algorithms, and to serve as a test bed and model for the construction of network accessible
image libraries." By 2006, these data sets were in use for "educational, diagnostic,
treatment planning, virtual reality, artistic, mathematical, and industrial uses by nearly
2,000 licensees in 48 countries." As an interactive, shared source, the Visible Human
Project signaled a new form of the atlas, one that nonetheless clearly links, as is clear
from the project's iconic image (Vesalius, half-digitized), to the older, paper forms.

1 millimeter. Ambitious in scope, the project aimed to link physiological functions to a vast, online image library — one that would be
used to develop imaging technologies and diagnostic and prognostic
techniques, alongside mathematical techniques and artistic applications. Its creators built many-gigabyte data sets to allow a myriad of
different uses, from a visual "fly-through" of the body in transverse
section to static, precise displays of particular tissues. (See figure 7.11.)

From this vast collective project involving hundreds of participants scattered across many countries, atlases could be made. For
example, one group took the data and, with the contribution of
several radiologists, constructed a properly labeled "interactive

musculoskeletal anatomical atlas" that anyone with a personal computer could use. It permitted the user not only to view sections of the body but also to alter them, highlight in color, excise elements, render elements transparent, rotate the images, or produce two-dimensional images from a variety of angles. The labels were designed, using sophisticated software, to track with their referents even as the user piloted in, around, and through the volume.[19] The Visible Human Project, in other words, is a form of meta-atlas imaging that subsumed — and could be used to produce — interactive *atlases*, understood as such, in the early twenty-first century. Epistemically, these virtual atlases differed from the older medical atlases such as those of Jean Cruveilhier, Albinus, and William Hunter: using the navigational and image-modifying capability of the program, they could produce in a moment an image that no one had ever seen — this was not, *sensu stricto*, a matter of *re*-presentation.

But anatomy is just one example of the proliferation of atlas sets of images on the Internet. Search the digital world for just about any classical atlas form (for example, electron microscope atlases or botanical atlases or mineral atlases) and results rush in. From the vast Sloan Digital Sky Survey, another meta-atlas, one group of astronomers created, in 2005, a new galaxy atlas.[20] Many of these sites are entirely recognizable as digitized and more widely distributed analogues of the nineteenth-century atlas. Some make use of animation, simulation, and other forms of interactive depiction — along the lines of, though not so sophisticated as, the Visible Human Project. Certain examples, now in color, are very clearly of the mechanical objective type so characteristic of the nineteenth century. There exists an atlas-style collection of meteorite photographs, each with its date and time recorded.[21] Others, like the Sloan Digital Sky Survey, explicitly embraced the combination of interpretive and algorithmic procedures, as the project indicated: "Combining computer-aided data analysis with manual, subjective human visual classification, the new galaxy compendium is based on ages, masses, and other physical properties. In time, this may become the largest and most useful visual atlas of galaxies ever produced."[22] Plant atlases, virus atlases, fluid-flow atlases — these and many others greeted the twenty-first century in full digital bloom.

But the digital update and proliferation of the nineteenth and twentieth centuries by the twenty-first was only one piece of an even larger picture. Anyone using images within the sciences soon discovered that the superordinate category including atlases had come to be referred to as the "image gallery." (The expanded category of atlases was accompanied by many other types of image collections from informal pictures of the research group through formal conference proceedings to exemplary images taken by a particular device.) Clicking into Iowa State University's Entomology Index of Internet Resources, for example, leads immediately to hundreds of image galleries addressing myriad insect species. There one can find the many more specific atlases — for example, "An Illustrated Atlas of the Laemophloeidae Genera of the World (Coleoptera)." By the early twenty-first century, image galleries came in a multitude of forms, including scholarly metasites like Iowa State's.

Within this heterogeneous genre, the nanotechnological "image gallery" stands out. Some galleries depict a collection of "working objects" familiar from earlier atlases, although now the images are produced not by a light or electron microscope but by the family of scanning probe microscopes. In this genre, we find what earlier would have been classed as atlases: electronic compendiums of endothelial cells, liquid effects, particular surfaces, all imaged using a particular device, such as the atomic force microscope, or produced with the aid of a given software program. Such instrument-specific image galleries are precisely the analogue of instrument-based atlases that handled x-rays or opthalmoscopes — or, for that matter, the photomicrographic compendium of Richard Neuhauss. Where Neuhauss's atlas carried advertisements for microscopes on the inside cover and in the closing pages, these new image galleries were often posted on the Web by particular instrument or software manufacturers — as we will see in more detail in a moment.

What is *not* familiar, however, is the new category of "nano-manipulation." True, we have here collections of working objects — examples of what a tool can accomplish. Nanomanipulative atlases, however, aim not so much at depicting accurately that which "naturally" exists, but rather at showing how nano-scale entities can be made, remade, cut, crossed, or activated. In the realm of nanomanipulation, images are examples of right depiction — but of objects that are being made, not found.

Nanomanipulable images have other goals. In this corner of science, the representation of the real — the use of images to finally get nature straight — may be coming to a close. In the battles over how we gain knowledge through the senses, one side traditionally espoused observation as the key to understanding. According to this view, scientists pursued a *vita contemplativa*, watching the distant objects of the sky through telescopes, peering through microscopes, scrutinizing flora and fauna. Opposed to this strategy (as the philosopher Ian Hacking has noted) was a proudly active, Baconian stance: intervene in the world as a way of establishing what we actually understand and therefore what really is out there. This was the *vita activa* of science. As Hacking put it, "Maybe there are two quite distinct mythical origins of the idea of 'reality.' One is the reality of representation, the other, the idea of what affects us and what we can affect. Scientific realism is commonly discussed under the heading of representation. Let us now discuss it under the heading of intervention."[23] The Baconian goal, according to Hacking, was to count as real that which can be used in the world, through experimental intervention, to affect something else. If you can spray positrons to do something, Hacking remarked, how can positrons not count as real? Scientists establish the reality of entities not by displaying them but by using them, for example, to achieve their goals in particle physics.

According to the interventionist ideal, seeing (pure receptivity) was not enough. *Action* produced knowledge; action showed what did and did not exist in realms too small or too large to grasp with our unaided senses. As Hacking saw it, in the early 1980s, the long history of scientific depiction — tracing, drawing, sketching, even photographing — was doomed to fail. It would always be possible to invent a plausible reason to treat the reality of objects as merely a useful assumption, a helpful fiction. Hacking, seconding Bacon, contended that only *use* could provide a robust realism. It was a strong salvo in a long-standing debate over whether and under what conditions scientific objects may be taken as real. On the side of representation: we should take as real that which offers the best explanations. On the side of intervention: we should accept as real that which is efficacious.

By the early twenty-first century, nanomanipulation, suspended between science and engineering, sidestepped the long-standing struggle between representing and intervening. Atomic physicists, surface chemists, and cellular biologists began making common cause with electrical engineers. Their goal in this hybrid venture was not to prove the existence or nonexistence of particular entities. In this sense, the work was *not* like that of the elementary particle physicists out to establish the reality of neutral currents, positrons, the omega meson, or the Higgs boson. It would be to miss the point of these efforts to characterize their fundamental concerns as being like those of the atlas makers of the eighteenth *or* the nineteenth century *or* (for that matter) the twentieth. Having rolled a carbon nanotube and deployed it in a circuit (see figure 7.12), these nanoscientists were not worried that they were being fooled into thinking that a nanotube existed when it did not — or deceived into thinking their image was true-to-nature when it wasn't. Nanoscientists making image galleries were not concerned in the first (or second) instance that their own theoretical presuppositions were clouding their vision. They were not even casually offering indirect proof of the existence of nanotubes by employing them to other effects. Instead, they were after haptic images *as tools.*

Ontology is not of much interest to engineers. They want to know what will work: what will function reliably under harsh conditions, what can be mass-produced — whether they are building airplanes, magnetic memories, or, increasingly, things in the nanodomain. Nanoscientists in the early twenty-first century were after the fabrication of devices at the atomic scale. They wanted to know how reliably the billionth-of-a-meter-long transistor would work. Here an engineer's traditional way of working is at least as important as the scientist's. Back in the 1870s and 1880s, as the Roebling team was building the Brooklyn Bridge, when they had completed their *in situ* facility for fabricating steel rope and were stringing cables across the 15,000-ton structure, their abiding question was not whether the 1,600-foot suspension bridge *existed.* Questions of existence might well keep astronomers awake at noon when they should be sleeping: Was this nebula *really* nothing but a collection of stars? Roebling's worries were different: he wanted to know if his bridge would stand against tides, currents, traffic, and hurricanes. Ontological problems as such fade from interest for engineers, the way evil demons faded from the anxieties of early-modern natural philosophers.

During the 1990s and early 2000s, many scientific institutions

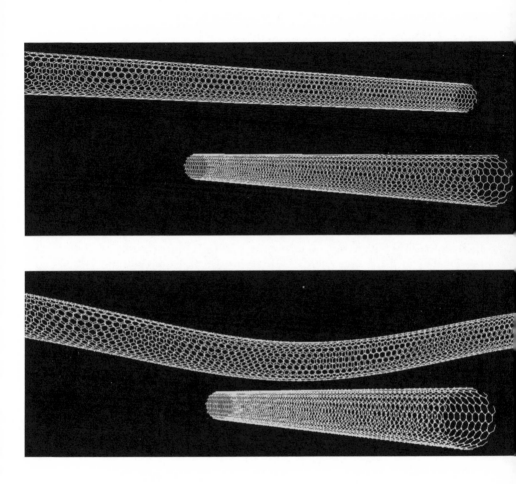

Fig. 7.12. Switchable Nanotubes. When the nanotubes are not touching (*top*), they are in the "off" position; when they draw together (*bottom*), they are in the "on" position (courtesy of Charles Lieber, Lieber Research Group, Harvard University). (Please see Color Plates.)

around the world worked to combine the "purer" sciences (atomic physics, surface chemistry, microbiology); further, there was a scramble to join these sciences to more specifically applied ends through engineering, especially electrical engineering. In one of the leading and early reports, released in 1999, a worldwide survey of nanoscience by a high-level joint committee of academics, industrialists, and government officials pushed for "an integrative science and engineering approach" that would weld quantum effects, on the one hand, with manufacturing, on the other. This "new educational paradigm" (as the authors put it) aimed to alter more than courses. It was meant to foster "regional coalitions of industry and technology" and "ease intellectual property restrictions" that the authors took to be obstacles. Universities were obliging: scientists were now more keen to join industrial and academic laboratories and to create interdisciplinary graduate and postdoctoral positions. Similar pleas and plans reverberated across many individual countries — from France, Britain, and Germany to Japan — not to speak of multinational collaborations bolstered, for example, by the European Union.[24] Again and again, the mantra was that training had to shift, as the codirector of the program at the University of Massachusetts made clear: "Our conversations with industry leaders have convinced us that a strong scientific education alone is not sufficient to make headway in a rapidly developing field like nanotechnology. We're purposefully pushing this program toward technology applications and asking the students to understand the business angles behind getting new technologies to market, so their value to society will be all the greater." Among the tasks the students were asked to tackle were group projects specifically directed toward the design of devices that could, in principle, be taken all the way to the sales-presentation stage.[25]

In science, this engineering-style *presentational* approach to the real was, early in the 2000s, still relatively new. But, one might counter, weren't there, famously, older struggles to bring the newly "purified" sciences together with the more applied arts of engineering, such as Berlin's Physikalisch-Technische Reichsanstalt (German National Institute for Science and Technology) *circa* 1900? True, the institute housed some remarkable alliances and produced very important work — but listen to the tone and tenor of the incoming head of the institution, Friedrich Kohlrausch, as he took over from Helmholtz in 1895. Somewhat apologetic about the applied character of the institution's research, he insisted in his inaugural address that the goal was pure science (*"reine Wissenschaft"*). Science need not only be pure, he allowed. But he hoped his own thirty years of pure scientific work would not be put aside, because without the life of pure research he could not go on living.[26] The joining of the pure and applied in the early twentieth century kept the identity of each quite distinct — it was a mixture, so to speak, not a compound.

A century later, more than allying mixed specialties was at stake. The self-definition of what it meant to be a scientist was in flux. Among traditionally trained scientists in physics, biology, and chemistry, device making often registered as an extraordinary, even disturbing alteration in their research practice — with consequences for their understanding of the scientific self. Colleagues uneasily asked about the making of nanopores or nanocircuits: "This device you are working on: Is that kind of activity really physics or chemistry — or is it in fact engineering?" The truth of the matter is that it often was both — or, rather, all three: surface chemistry, atomic physics, electrical engineering. And along with this activity in the trading zone between the scientific and the engineered, a new role came into existence for the visual, one that is only awkwardly and irrelevantly reducible to faithful depiction — direct or indirect — of what can exist.

For example, in 2005, Veeco Instruments ran a Web-based "Nano-Theatre" that displayed a gamut of image galleries: a gallery of materials and surface science, a gallery of semiconductors, a gallery of nanolithography and nanomanipulation. Figure 7.13, for example, shows a nanoscale rectifier. A rectifier is a familiar piece of electronics that allows current to pass in one direction but not the other, but at the nanoscale it is dramatically different. When constructed atom by atom, a rectifier resembles a child's marble game: electrons coming up the main "street" from the lower left hit the triangle and scatter right or left. Electrons coming from the upper right simply bounce back.

Like the contents of the more generic galleries that are to a certain degree generalizations of the older atlas forms, many of these pictures illustrate the capacities of the depicting instruments themselves. This is not that different from Neuhauss's showing, one after the other, his microphotographic captures of a snowflake, a typhus bacillus, and a cortical organ (lamina reticularis). For example, in one figure, the atomic force microscope is used to image the hardness of a particular cell, bit by bit, by tracking the force measured by the probe as it moves and the response of the cell.

But the most striking difference between the older atlases and their successors lies in the image collections that demonstrate nanomanipulation. The original of this genre was the famous picture, published in 1990 in which scientists at IBM managed to write their company logo at the scale of atoms (see figure 7.14). Asylum's Web site advertised its atomic force microscope controller, which would let one compose just about anything one wanted at the nanoscale: "Remember the 1990 famous IBM image of Xenon atoms.... With MicroAngelo, you can manipulate and modify samples and surfaces on the nanometer and picoNewton scale — even down to the level of single molecules."[27]

The camera and the tweezer had merged, so to speak, and with that fusion the whole point of image making had shifted. Moving and making nanoscopic entities became the order of the day; this was an active, *haptic sight* that sought neither to re-present nature through idealization nor to re-present natural objects by a ferociously policed copying. Now scientists wanted to be able to move nanotubes, to shape them as they pleased. In figure 7.15, this is precisely what is happening: the probe, entering from the lower right, is bending the carbon nanotube that stretches from lower left to upper right. If, instead of a single-walled carbon nanotube, the object to be manipulated is biological, the discipline is swapped, but not the technique (see figure 7.16). Again, the Asylum image gallery depicts the results of an atomic force microscope with the proprietary software Micro-Angelo. Scientists use these image-gallery sites to learn about the possible use of the visualization tools and to compare the images provided with their own actual and planned research.

Two purposes of images like these — showing the cutting of flagella or the rolling of carbon nanotubes — is to demonstrate what the technology can do and to sell machines. But the sites themselves also rapidly became something else: digital crossroads where nanoscientists working in different arenas encountered each other. By 2005, for example, Pacific Nanotechnology had long been posting a "picture of

397

Fig. 7.13. Nanorectifier. Device by Aimin Song, image from http://www.veeco.com/
library/nanotheater_detail.php?type=application&id=459&app_id=21; discussion on
http://personalpages.manchester.ac.uk/staff/A.Song/ research/BallisticRectifier.htm, both
accessed 9 September 2006 (courtesy of Veeco Instruments). Nanolithography pattern
of a rectifier (a device that produces direct current from alternating current) created by an
atomic force microscope tip. Unlike conventional macroscopic rectifiers, this "ballistic"
one acts on the electrons one-by-one as if they were billiard balls. Electrons scatter from
the side channels down to the lower left from the triangle — whether they come from the
upper left or from the lower right.

the month" on its site and offering a reward for its proper identification; scientists from all over the world contributed images and
guessed at their distant colleagues' prize specimens.

Relatively quickly, nanoscientist engineers began to live in different kinds of spaces, buildings more suited by their bleached wood,
indirect lighting, and high-end furniture to the comings and goings
of corporate planners, venture capitalists, and visiting politicians.
These nanoresearchers had to learn to move easily in the marketing
world, to think in terms of patents, to dress differently as they met
with their corporate-world homologues, and to retool their work to
meet a business standard. Images hand-scrawled with marker pens
on overhead transparencies may have been good enough for a scientific meeting in 1980; in 2000, before a mixed group of investors and
scientist-entrepreneurs, they certainly were not. Digitized slide
shows gave way to elaborate, often moving simulations.

Fig. 7.14. Atomic IBM (1990). http://www.almaden.ibm.com/vis/stm/images/stm10.jpg, accessed 9 September 2006 (courtesy of International Business Machines Corporation © 1990 IBM). One of the first dramatic examples of using a scanning probe microscope (a device related to the atomic force microscope) both to image and to manipulate individual xenon atoms — here spelling out the company name.

As the standard of production values and pictorial presentation clicked upward, another element may have come into play — and here once again we must speak speculatively. Scientists, who were used to a rather elaborate economy of name-based credit, began to encounter the more anonymous ethos of industry-oriented engineers. (However unfair it may be, who outside the aerospace engineering community remembers the name of the lead engineer on the Boeing 747?) In this environment, scientist-engineers increasingly began to present their images as artistic as well as technical accomplishments. (Art, whose practitioners are highly conscious of intellectual property rights, is currently among the most "authored" of all practices.)

Visual presentation was becoming part and parcel of the making of new kinds of things, from quantum dots to switchable nanotubes. It is no accident that even the first generation of university nanolaboratories integrated visualization facilities architecturally within the

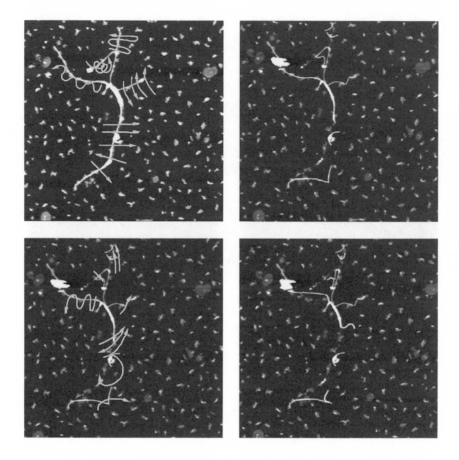

Fig 7.15. Cutting and Pushing Nanowires. http://www.asylumresearch.com/Applications/
MicroAngelo/ MicroAngelo.pdf, accessed 9 September 2006. The arrow-headed yellow
lines in the image of the upper left and lower left indicate the motion of the operator-
controlled cantilever tip. Images on the upper right and lower right indicate the resulting
state of the nanowires after this manipulation. The images have a scan size of 7.4
micrometers. Atomic Force Microscopy image taken with the Asylum Research MFP-3D
AFM. (Please see Color Plates.)

Fig. 7.16. Cutting Bacterial Flagella. http://www.asylumresearch.com/ImageGallery/
Litho/Litho.shtml#4, accessed 9 September 2006 (sample courtesy of Dr. Jim Cooper,
University of California, Santa Barbara). Nanomanipulation here is applied to the cutting
of flagella. The uncut sample is on the left; in the center image the yellow lines indicate
the areas to be cut, and on the right is the sample after the cuts have been made.
The scan size is 5 micrometers. Atomic Force Microscopy image taken with the Asylum
Research MFP-3D AFM. (Please see Color Plates.)

fabrication facility. It is frequently not possible to make things without depicting them visually — and, quite often, it is not possible to represent them without the procedure of making. The atomic force microscope and the scanning tunneling microscope were perfect examples of this compound: the same device was used at one and the same time to image and to alter.

Within the domain of the nanopictorial, some visual effects were, or aimed to be, aesthetic interventions — concatenations of scanned microscopic data, simulations, and artifactual modifications of color, scale, and presentation created striking images. Other researchers made broader claims (sometimes interestingly, sometimes less so) to be straddling art and science. This in itself is a noteworthy phenomenon. For centuries, atlases, especially anatomical atlases, counted as both objects of art and objects of science. Leonardo da Vinci's explorations of water motion were at once art and science in ways that largely obviated the need for a distinction — and, as we saw in Chapter Two, so were the works of Carolus Linnaeus and Bernhard Albinus.[28] But with the proliferation of mechanical objectivity, art and science were self-consciously pitted against each other; Cajal, like many of his contemporaries, saw the deliberate aestheticization of the scientific image as one of the worst crimes against right depiction.

During the decades of the mid-twentieth century, the overwhelming preference for an unvarnished, automatic image diminished. Trained interpretation became not a vice to be suppressed but a supplement to mechanical objectivity to be celebrated. Gerhard S. Schwarz and Charles R. Golthamer, far from apologizing for the interpretive, noncameralike work of "their" illustrators, found the medical artists' ability to reveal salient aspects of the atlas images essential to the project. While Schwarz and Golthamer made no claim for the fine-art value of the painted images, they explicitly rejected the ambition of providing "only" an automatic registration of that which stood on the x-ray plate.

Toward the end of the twentieth century, the balance between art and artlessness began to tip again, in still-unstabilized ways. In his *Album of Fluid Motion* (1982), the physicist Milton Van Dyke assembled images of projectiles, turbulence, shock waves, and instabilities — all carefully photographed in black and white (see figure 7.17). Bullets, water, tubes of liquid: "Scattered through this century's literature of fluid mechanics," Van Dyke asserted, "is a treasure of beautiful and revealing photographs, which represent a valuable resource for our research and teaching."[29] Van Dyke's atlas of fluid-flow images became a standard tool of fluid-dynamics training. One sees it in syllabus upon syllabus, across the myriad disciplines that make use of fluid dynamics.[30] Those who teach the subject argue, in course after course, that it is one thing to know how to calculate an instability and quite another to gain the qualitative understanding of the phenomena that these photographs permit.

Propelled by Van Dyke's atlas, the American Physical Society launched a photo contest starting in 1983. Each year, researchers submitted images of fluids in motion to be judged under two announced criteria: "the artistic beauty and novelty of the visualizations" and "the contribution to a better understanding of fluid flow phenomena." The field's journal of record, *The Physics of Fluids*, published the winning picture with an article, and the editors made sure that every participant at the annual meeting of the Division of Fluid Dynamics' yearly meeting received one. In 2000, *The Physics of Fluids* took the publication of the article with the best image online; a few years later, it issued a print version of the best of the best in *A Gallery of Fluid Motion*. Readers, in the first instance the research community, were enjoined both to "enjoy the beauty of the images" and to "ponder more deeply the physical significance of the flow visualizations" as a prelude to further research.[31]

The self-conscious aestheticization of scientific depictions was not restricted to choosing striking images, or even touching up images to make particular phenomena evident. Using simulations, scientific gallery makers could just as easily produce images *outside* real space — that is, in mathematical spaces — as within it. Phase space (with axes of position and momentum) provided one arena of nonmimetic display; others exploited curves of constant energy or entropy. Some of the more sophisticated computation-based images used a technique (the wavelets representation) that, since the 1980s, had become a powerful aid in visually expressing the dynamics of turbulent flow.

Marie Farge, a computational fluid dynamicist working at the Ecole Normale Supérieure in Paris, used these various methods not only to take snapshots of complex, time-dependent turbulence, but

Fig. 7.17. Turbulent Streets. Sadatoshi Taneda, "Kármán Vortex Street Behind a Circular Cylinder at R = 140," in Milton Van Dyke (ed.), *An Album of Fluid Motion* (Stanford, CA: Parabolic Press, 1982), p. 56. This image was produced by a flow of water passing at 1.4 centimeters per second past a 1-centimeter cylinder, barely visible at left. The streaks are produced by the electrolytic precipitation of a white colloidal smoke, illuminated by a sheet of light. As the turbulence moves to the right, it grows in diameter. Van Dyke's book is widely used in fluid-dynamics courses as an "intuitive" supplement to more formal treatment of fluid dynamics.

also to formulate simulations of them. From the 1980s into the 2000s, she was a critic of what she considered sloppy, "subjective" uses of color and simulation; at the same time, she enthusiastically backed simulation in science. Borrowing from the Bauhaus artist Johannes Itten and other color theorists, Farge attended to the problem of "simultaneous contrast" (which was not a new concept): that is, the psychophysiological tendency, upon seeing a particular color, to produce its complement. For Goethe, the dependence of our perception on the surround had been a good thing — it made color useful aesthetically. For Farge, since she was trying to standardize color use, the context-dependence of color perception was a disaster: it practically guaranteed the production of unintended information. To cut down on simultaneous contrast, Farge designed software to insert gray between color fields. Another Ittenian contrast is that there is a great physiological difference between the bodily response to blue-green (which induces a feeling of coldness) and the response to red-orange (which produces the sensation of warmth). To make use of that felt difference, Farge designed her displays — as in figure 7.18 — to use red and blue to capture opposite values of certain parameters (say, the intensity of vorticity, the rate of rotational spin in two-dimensional virtual moving images — "movies" — or turbulent fluid flow). Blue values of the vorticity are much less than zero; yellow indicates zero; red is much greater than zero.

Building on Itten's contrasts (choosing a palette that she judged to be "as objective as possible"), Farge restructured Itten's twelve-part color wheel using the 593 standard colors widely distributed by Pantone and used by graphic artists, printers, and designers. For Farge, this structural objectivity meant that the standard palette was transmissible and shared — in a nightmarish context in which every researcher chose a different palette and then worsened the situation by making the colors signify differently. "Faced with the development of [computer] graphical methods that are more and more sophisticated," Farge continued, "we run the risk of letting ourselves be carried away by a tool that we do not master and of being deceived by a seductive aestheticism stripped of information content — if the choice of palettes is left haphazardly to subjective and changing appearances."[32] Imagine, she insisted, that all road maps had completely different choices of the colors by which they depicted their different basic elements — or, worse, picture a set of maps, each with a different color scheme and no legend anywhere in sight. That, Farge lamented, is precisely where computer simulations all too often left us.[33] Color standardization might prove a tool to block subjectivity — to halt the person-to-person variability in the interpretation of the data. Figures 7.18 and 7.19 present turbulent flow in a way that, in no nineteenth-century sense, simply draws itself from a real-world fluid to the page. Artificially colored, virtual, moving on demand — we are a long way from the black-and-white photographic images of Van Dyke's atlas.

For years, Farge and the Ecole Polytechnique computer engineer Jean-François Colonna struggled against the "subjective" and toward an objective use of color. Color was for them a tool to express properties of the fluid flow; it was a construction, in the sense that simulated liquids do not come naturally in color the way an amethyst crystal does. But the objectivity they were after was not the mechanical form characterized by a hands-off stance toward the visual material. What status did the color have? Farge considered that a proper approach to the choice of color was neither "scientific" strictly speaking (the palette choice was a means to encode the properties of the simulation, not a contribution to the understanding of color vision) nor purely "artistic" (it was not her goal to use colors to create an aesthetic, spiritual, or sensory response). Instead, she labeled her efforts "pragmatic," for they were part of a project to use systematization and empiricism to transmit graphic information effectively and clearly.[34]

Just this pragmatic approach made possible a new stance toward the relationship between science and art. It certainly was not one in which the artist had to be "policed" or "repressed." Nor was it one in which the scientist (like Schwarz and Golthamer) gave the artist a free hand to interpret. Instead, by working through the color theory of "objective" artists such as Itten, Farge came to treat the field of visual simulation as indisputedly *constructed*, but constructed under immense and articulated constraints, arising not only from physics and computational structure but from color theory as well. Farge and Colonna produced a film, *Science pour l'art*, at the Ecole Polytechnique.[35] Indeed, they collaborated on a variety of projects at the edge of science and art; in 1991, Farge, assisted by Colonna, produced a winning image for *The Physics of Fluids'* gallery of fluid motion — the online atlas that, as we have seen, aimed to extol both "artistic beauty and novelty of... visualizations" *and* "contribution[s] to a better understanding of fluid flow" (see figure 7.19). Here Farge and Colonna used computer-generated data to depict the vortex field (the height of the peaks is proportional to the intensity of vorticity) but chose the light-scheme more for aesthetic emphasis than for specific scientific depiction.[36] In one sense, figure 7.19 is the direct descendant of Van Dyke's black-and-white atlas photograph. But in another sense, it takes us a long way into a domain that is neither experiment nor theory, neither mechanical-objectively mimetic nor subjectively artistic.

There are now conferences of science and art organized by fluid dynamicists — and, in the domain of the nanotechnological, hundreds of sites (real and virtual) that explore the boundary between art and science. The Harvard University condensed-matter theorist Eric J. Heller has displayed his work on the flow of two-dimensional electron gases not only in scientific contexts (including the March 8, 2001 cover of the scientific journal *Nature*) but also in a wide variety of museums and virtual and real-world art galleries (see figures 7.20 and 7.21). Originally, these studies of electron flow were performed by his colleague the experimental physicist Robert Westervelt (see figures 7.7 and 7.8). In an effort to understand the coursing of electrons over a flat surface in which positive ions were present, Heller ran computer simulations, which both matched the experiments in important ways and yielded new information about electron flow. "Transport II" shows the channeled, branched flow of these (simulated) electrons — the scientifically surprising element is that the branching continues farther from the electron source (located at the center of the image). This distant but correlated flow may even have consequences for future device designs.[37]

In his simulation images, Heller kept the data intact but added coloring and some shading, displaying the image as a work of art. He put his aim this way: "Digital artists need no longer emulate traditional media only! The computer allows us to create new media, with new rules, more naturally suited to the new tool. But such rules are best when they too follow physical phenomena, instead of arbitrary mathematical constructs. I have learned to paint with electrons

Figs. 7.18, 7.19. Digital Liquid; Organized Turbulence. Fig. 7.18: Two-dimensional vorticity field from simulation, http://wavelets.ens.fr/; accessed 28 April 2006 (courtesy of Marie Farge, CNRS, France and Jean-François Colonna, CMAP, Ecole Polytechnique, France); fig. 7.19: Marie Farge, "Wavelet Analysis of Coherent Structures in Two-Dimensional Turbulent Flows," *Physics Fluids* A 3 (1991), p. 2029, fig. 1, chosen as a winning entry for the Eighth Annual Picture Gallery of Fluid Motion, in 1991: http://pof.aip.org/pof/gallery/ 1991toc.jsp, accessed 28 April 2006 (courtesy of Marie Farge, CNRS, France and Jean-François Colonna, CMAP, Ecole Polytechnique, France © 1991 American Institute of Physics). Marie Farge (with Jean-François Colonna) used numerical simulations to depict the flow of turbulence with organized elements, coherent aleatory elements, and residual incoherent flow produced by nonlinear interactions between vortices. In her design of the palette, Farge wanted an explicit standard that would grade the amount of vorticity by color and would make the map intersubjective — even correctly interpretable, because of the choice of luminance, hue, and saturation, by a color-blind person. (Please see Color Plates.)

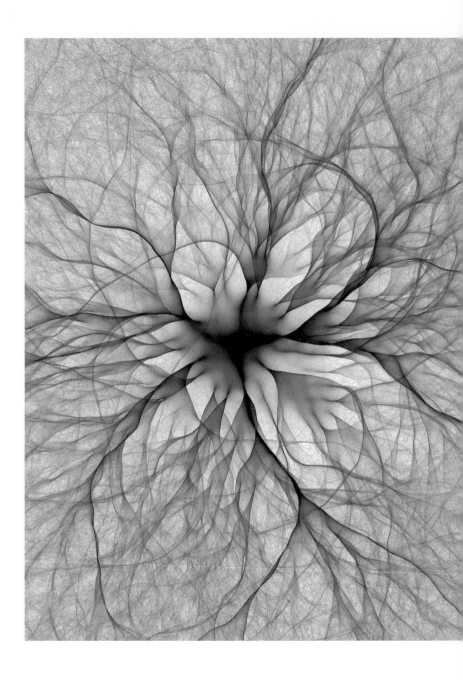

Figs. 7.20, 7.21. Transport II (Image, *Nature*). Eric J. Heller, "Transport II," http://www.ericjhellergallery. com/index.pl?pageimage;iid=8, accessed 16 July 2005 (Eric J. Heller, Transport II, 2000). This simulation tracks virtual electrons as they are sent from the center, fan out, and form branches as indirect effects of traveling over bumps (positive ions). This image is the theoretical correlate of Robert Westervelt's experimental work on the electron flow in the thin plane between semiconductors discussed earlier. It appears on the cover of *Nature* (March 8, 2001, see fig. 7.21) as a scientific image *and* circulates in the art world of galleries and exhibitions (in 2006, it sold for a substantial amount of money and was exhibited at 50 inches by 36 inches, recorded as having been produced using a LightJet-Lumniange process printer on archival color photographic paper.) (Please see Color Plates.)

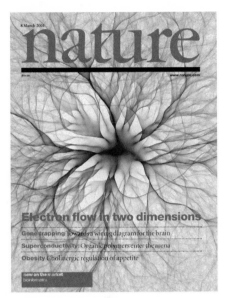

moving over a potential landscape, quantum waves trapped between walls, chaotic dynamics, and with colliding molecules. Nature often mimics herself, and so these new media, exposing the beauty and mystery of the atomic world, yield a variety of effects that recall familiar aspects of our macroscopic experience."[38] Heller chooses to restrict his computationally generated data to that which emerged from the science, and then to experiment aesthetically with the results, using lighting, shadowing, contrasting, and tinting. Most important, he positions the work in a space that is at once scientific and artistic.

At this point, the relationship of science to aesthetics has departed from all our earlier models. Art and science are not self-evidently a single enterprise (few today assume that the True and the Beautiful must necessarily converge), nor do they stand in stalwart opposition to each other. Instead, they uneasily but productively reinforce each other in a few borderline areas.

Right Depiction

How do the scientific image galleries relate to traditional atlases? Perhaps they relate in that the overarching goal in both cases is right depiction, but right depiction itself splits in two. On the one side are the older atlases that aimed, through *representation*, at fidelity to nature. Getting nature correctly on the page might mean following the eighteenth-century idea of truth-to-nature, but it also might be beholden to the nineteenth century's mechanical objectivity or the twentieth century's trained judgment. On the other side are the newer forms of image gallery that are *presentations*, where the presentational strategy can refer either to new kinds of things (rearranged nanotubes, DNA strands, or diodes) or to the presentations' proud espousal of deliberate enhancements to clarify, persuade, please — and, sometimes, sell.

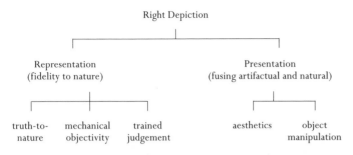

Right Depiction

Representation
(fidelity to nature)

Presentation
(fusing artifactual and natural)

truth-to-nature mechanical objectivity trained judgement

aesthetics object manipulation

The image-as-tool seems to enter the scene inseparably from the creation of a new kind of scientific self—a hybrid figure, who very often works toward scientific goals, but with an attitude to the work that borrows a great deal from engineering, industrial application, and even artistic-aesthetic ambition. By all means, make rectifiers and switchable carbon nanotubes. But always ask, Is the device robust, is it reliable, can it be scaled up to mass production? Can it be presented to a wide audience beyond the research specialty? To researchers in general? To the public?

We can capture the linked aspects of scientific self, image, practice, and ontology in a form parallel with the earlier moments schematized in the first chart of this chapter. Images have frankly and explicitly surrendered any residual claim to being a version of "seeing," in a classical sense — the four-eyed sight of truth-to-nature, blind sight of mechanical objectivity, and the physiognomic sight of trained judgment have given way to something much more manipulable, something more like haptic sight. Simulations, artificial color, rescaling, virtual cutting — in all these and other ways, the image itself no longer is held to be a copy. Procedures, too, have altered. The intervention of the nanoscientist is not that of a Goethian idealizer. The nanoimage in no way pretends to reveal a truer reality lying behind mere appearances. But at the same time, the nanoimage is not merely altered by a "trained expert," confident in a honed ability to extract the real from the machine-generated artifact. Robert Howard, Václav Bumba, and Sara F. Smith may have removed a machine artifact from their solar magnetogram. Schwarz and Golthamer laid in lesions in a way that made them visible to the acolyte. But none of them deliberately and self-consciously altered the diagram in aspect, hue, or scale to make it artistically pleasing. To extend our earlier chart of the representational, the presentational might be schematized like this:

Persona	Combines ethos of late twentieth-century scientist with device orientation of industrial engineer and authorial ambition of artist
Image	Hybrid of simulation, mimesis, manipulation
Practice	Simultaneity of making and seeing
Ontology	"Nanofactured" goods straddling the divide between natural and artifactual

Nanotechnology is manipulation of quite a different kind — intervention by the scientist, through the image, to make things, to cut, move, combine, weld, or set in operation. In some respects, the deepest change is at the level of the scientific self — or, should one now say, the engineering-scientific self. In a myriad of ways, in this hybrid field at least, the scientist and the engineer as distinct personas have begun to lose their distinctness. Traditionally, the scientist would tend to eschew device making for its own sake. The physicist might build devices, but their importance lay in what they would reveal about something else — a galaxy, a superconductor, an elementary particle. Correspondingly, the engineer wanted more efficient, powerful, flexible tools.

Once the scientific-engineering self begins to stabilize, however, it does so in concert with a new stance toward the images. Images become tools like other tools, part of the apparatus — more like the computer screen that shows the workings of a distantly controlled robotic manipulation in remote surgery, the alteration of a satellite in space, the mixing of toxic chemicals, or the defusing of a bomb.

Our schema cannot possibly capture all the image making in the early twenty-first-century sciences, or even all the atlaslike collections of images. But perhaps in the image-as-tool we can recognize a new form of image collection, this time one that has discarded the ideal of fidelity in favor of right manufacture.

It is too early to know how this form of hybridized science and engineering will look in the long run — how far it will go and the changes it will carry with it, not only in the institutional structure of research but also in the ethos of being a researcher. At a more abstract level, it raises questions about the fate of the epistemology of images. For a very long time, scientific images of record have served to ward off particular threats to knowledge acquisition: they have combated the fears of individual variation, willful, individual intervention, and instrument-produced artifacts. Through this study, we have been able to follow a powerful practical side of the history of scientific epistemology and a lab-bench view of how scientific objects come to qualify as real. But with the haptic image, fear of being in error is not really the issue — the classically conceived struggle to see in images secure knowledge and the trace of the real seems beside the point.

Is there a shift to other kinds of anxieties, anxieties not about whether we have seized the real right but about whether we are instead making the right real? Perhaps fearful discussion about cloning, genetically modified organisms, and sentient nanobots is a harbinger of a turn in how we will need to study the development of scientific virtues.

In the era of truth-to-nature, images were inspired passages to an idealized world; later, they became very much of this world, their automaticity aiming to make them, in their vaunted objectivity, all nature and none of us. In the exercise of trained judgment, images stood as bridges, part us, part not-us. Now, as images become part toolkit and part art, what are they? Nanofacturers use them as aesthetic objects, as marketing tags, all the while reaching through them to create and manipulate a brave new world of atom-sized objects. The scientific image begins to shed its representational aspect altogether as it takes on the power to build. Once again, images are in flux. Once again, so is the scientific self.

Acknowledgments

This book has been an unconscionably long time in the making, and we are all the more in debt to the institutions and people who graciously and patiently saw us through the project. We are deeply grateful for the help of many generous students, colleagues, and friends. Over the years, student research assistants (several of whom by now have students of their own) ransacked libraries from Stanford to Göttingen, Cambridge to Cambridge, on our behalf: we warmly thank Naomi Oreskes, Thomas Sturm, Michael Gordin, André Wakefield, Kathrin Willkommen, Katja Günther, Stefanie Klamm, and Jeanne Haffner. Colleagues near and far answered our queries with unfailing erudition and good cheer; above all, they gave us the benefit of critical comments as we presented our work in various stages of becoming. We are particularly grateful to the audiences of the Isaiah Berlin Lectures at Oxford, in 1999, the American Council of Learned Societies, in 1999, the Leibniz Lectures at the University of Hannover, in 2000, and the Tarner Lectures at Trinity College, Cambridge, in 2006, for their attentive and acute responses.

We owe a great deal to those colleagues who read and commented on our manuscript in whole or in part, saving us from innumerable lapses of logic and language, not to mention outright mistakes: Nancy Cartwright, Wendy Doniger, Gerd Gigerenzer, Hannah Ginsborg, Jan Goldstein, Nick Hopwood, David Kaiser, Robin Kelsey, Ursula Klein, Ramona Naddaff, Susan Neiman, Otto Sibum, Joel Snyder, Emma Spary and Norton Wise read one or more chapters with gimlet eyes; Michael Gordin, Caroline Jones, Theodore Porter, and Robert Richards heroically read and commented on the entire manuscript, and we learned enormously from their comments. At a crucial stage both our thinking and prose were clarified by Amy Johnson's excellent editing. Elio Raviola and Paolo Mazzarello had very helpful advice for us about the history of neurohistology, as did Marie Farge about computational fluid dynamics and Eric J. Heller about two-dimensional electron flow experiments and calculation. We are deeply indebted to Arnold Davidson, with whom we have thought for years about the historicity of the self. The book has been vastly improved by their suggestions, queries, and challenges, and we can only hope that it is worthy of their efforts.

The making of books does not stop with the writing of them — at least not when the images are as important as the text to the main argument. We were very fortunate to have the energetic and careful assistance of Josephine Fenger in the final preparation of the manuscript, particularly in tracking down scores of images from hither and yon. Kelley Wilder kindly photographed the largest and most unwieldy of our atlases under difficult conditions. The manuscript was much improved by the careful and thorough reading by Zone's copyeditor, Sierra Van Borst. Meighan Gale, Ramona Naddaff, and Gus Kiley at Zone Books have been generous with help and encouragement. We thank Amy Griffin, also at Zone Books, for her tenacity in securing the permissions to use the images reproduced in the book. We count ourselves very lucky to have had Julie Fry as our designer. We gratefully acknowledge the support of our home institutions, Harvard University and the Max Planck Institute for the History of Science, as well as the Max Planck Gesellschaft-Alexander von Humboldt Foundation, especially for making it possible for each of us to spend a year as guest at the other's institution.

Last and most lastingly, our families have lived with this book for as long as we have, and we thank them from the bottom of our hearts for their love and forbearance throughout.

Color Plates

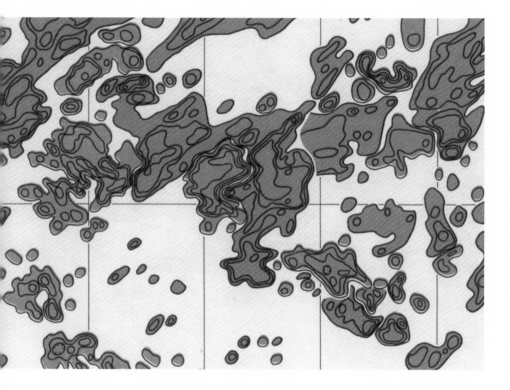

Fig. 1.3. Trained Judgment.

Fig. 1.4. Double Elephant, *Stanhopea tigrina*.

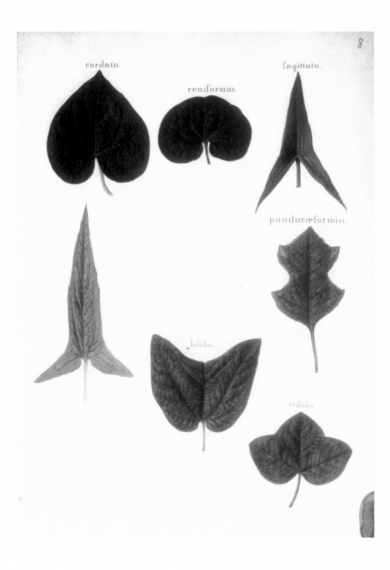

Fig. 2.4. Leaf Types Embodied.

66.

NOUVELLE-HOLLANDE : ÎLE DES KANGUROOS.

CASOAR de la Nᵉˡˡᵉ Hollande. (*Casuarius novæ Hollandiæ Lath.*)

1. *Casoar mâle.* 2. *Casoar femelle.* 3. *Jeune Casoar de 5 demaines environ. Les deux individus marqués de bandes longitudinales sont âgés de 10 à 15 jours.*

Fig. 2.6. "Cassowaries of New Zealand."

Crested Titmouse. Male 1.F.2.
PARUS BICOLOR.
Plant. Pinus Strobus.

Drawn from Nature and Published by John J.Audubon. F.R.S.E. F.L.S. M.W.S.   Engraved Printed & Coloured by R.Havell Jr London 1875

Fig. 2.12. Posed Tufted Titmouse.

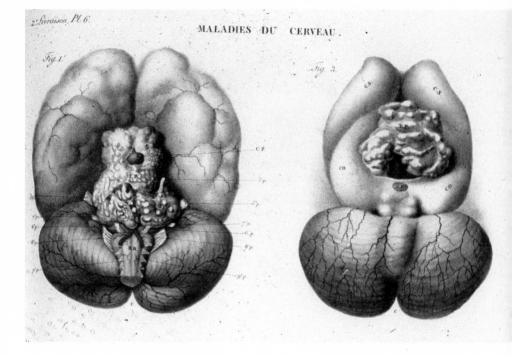

Fig. 2.13. Pathology in Color.

Strelitzia Reginæ   *Strelitzia de la Reine*.

Fig. 2.17. Flowers for the Queen.

Fig. 2.19. Visual Tug-of-War.

Fig. 2.21. Luxury Botanicals.

Naturselbstdruck.

Aus der k. k. Hof- und Staatsdrukerei zu Wien. 1855.

Fig. 2.26. Nature Prints Itself.

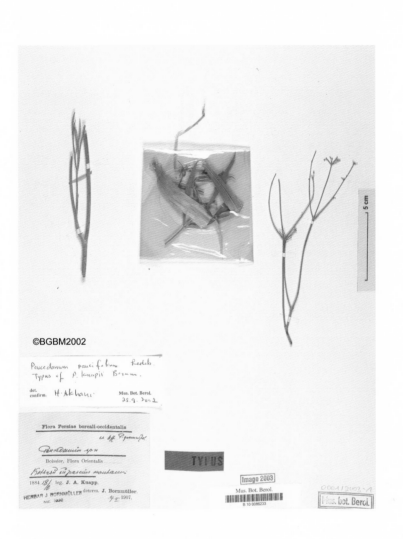

Fig. 2.27. Holotype.

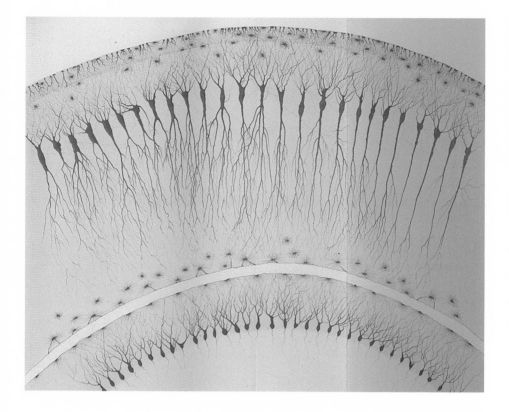

Fig. 3.1. Simpler than Nature.

Fig. 3.4. Arrangement of Fossil Shells.

Plate X.

Fig. 1.

Fig. 2.

Fig. 3.

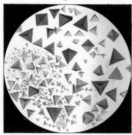

Fig. 4.

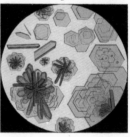

Fig. 5.

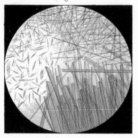

Fig. 6.

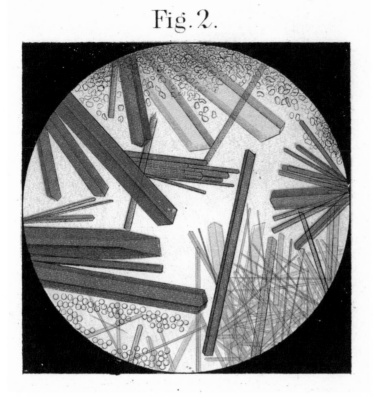

Fig. 2.

Figs. 3.15, 3.16 (detail). Blood Crystals.

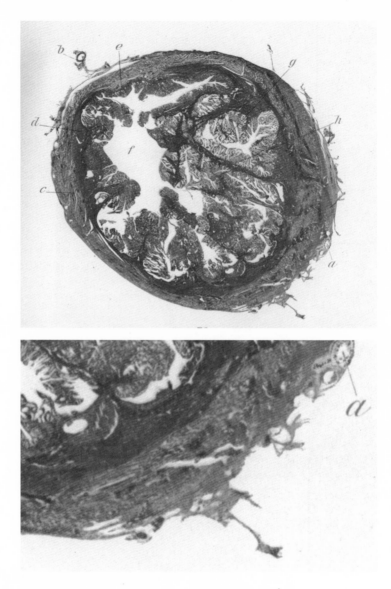

Figs. 3.32, 3.33. Tattered Objectivity, Detail.

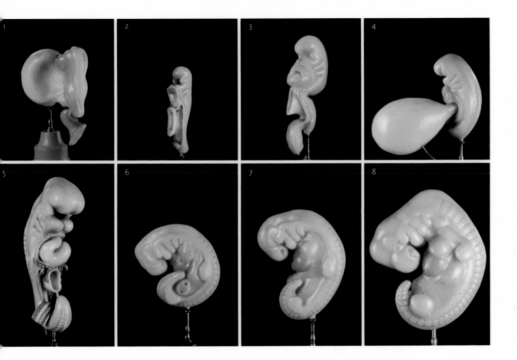

Fig. 4.2. Model Embryos.

Fig. 4.4. Newton Deified.

Fig. 4.5. Newton Domesticated.

Fig. 4.7. Claude Bernard at Work.

Fig. 4.8. Thomas Henry Huxley Plays Hamlet.

Fig. 4.12. "Nature Unveiling Herself Before Science."

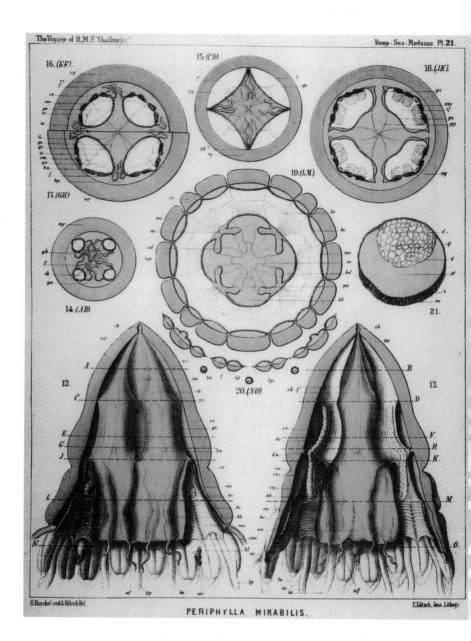

PERIPHYLLA MIRABILIS.

E.Haeckel and A.Giltsch Del.

F.Giltsch, Jena .Lithogr.

Fig. 4.13. The Science of Medusae.

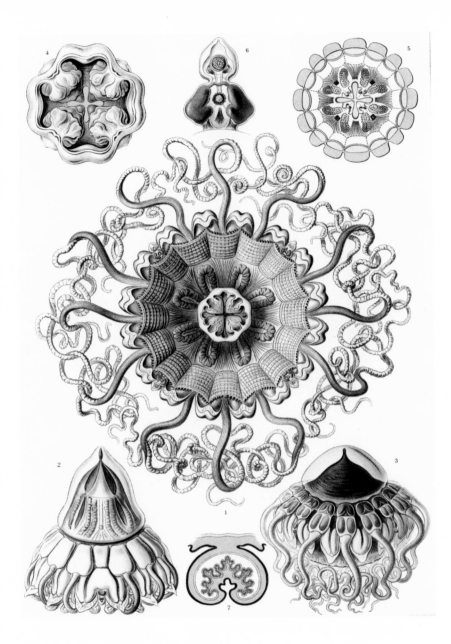

Fig. 4.14. The Art of Medusae.

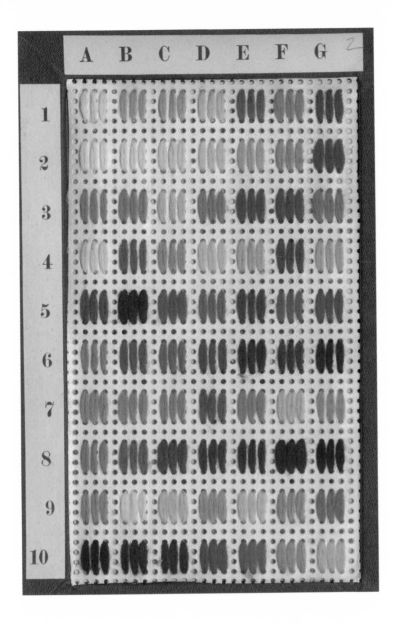

Fig. 5.5. Testing Color Sensations.

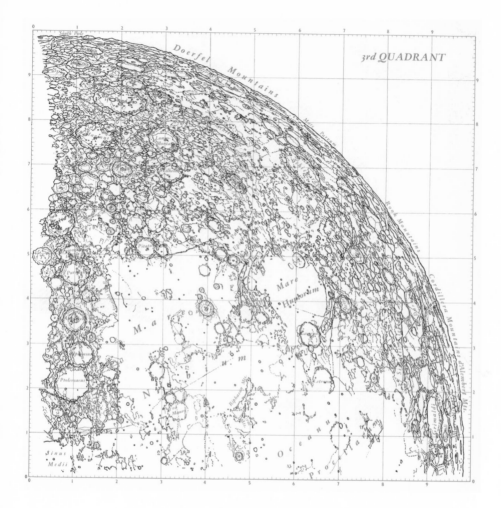

Fig. 6.8. Judging the Moon.

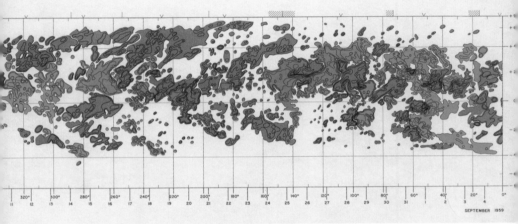

Fig. 6.10. Sun, Corrected.

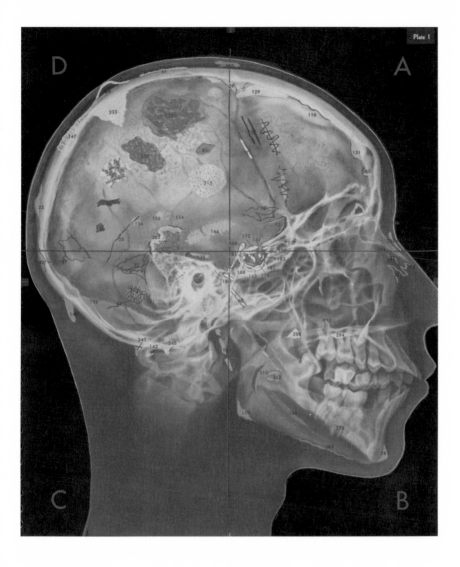

Fig. 6.11. Realism Versus Naturalism.

Fig. 7.3. Mechanical Objectivity.

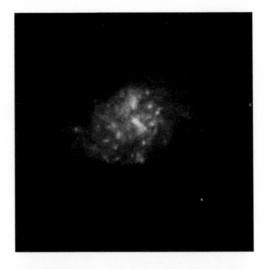

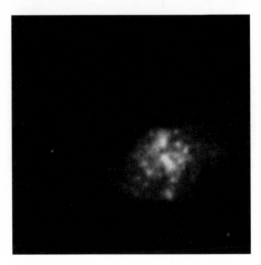

Figs. 7.4, 7.5. Trained Judgment.

Fig. 7.6. Training Observers.

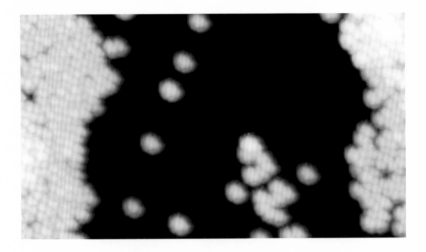

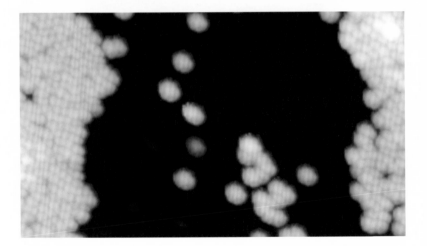

Fig. 7.8. Rolling Nanospheres.

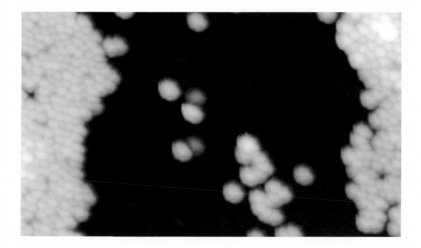

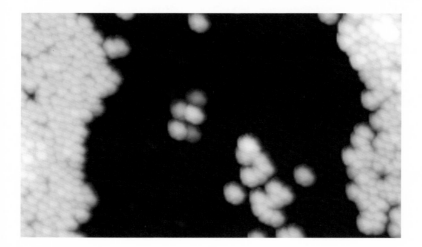

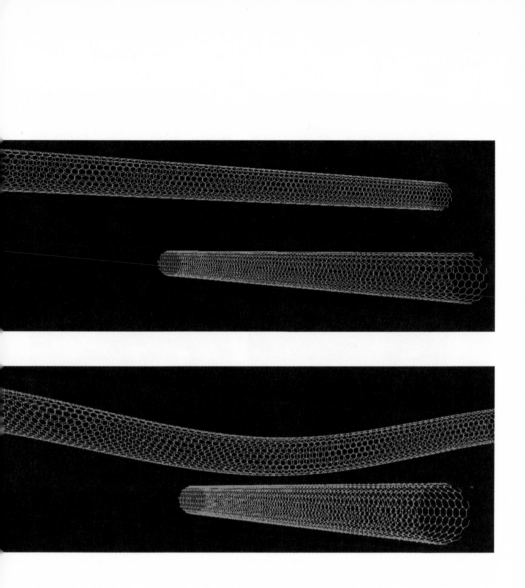

Fig. 7.12. Switchable Nanotubes.

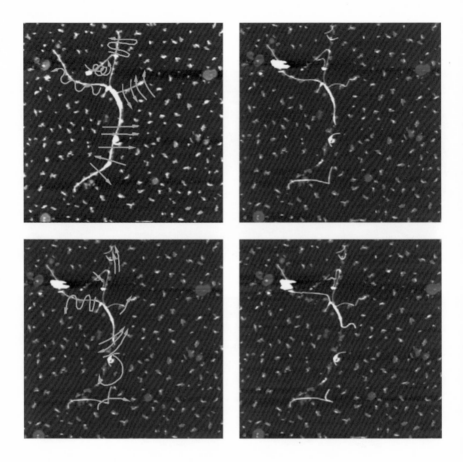

Fig 7.15. Cutting and Pushing Nanowires.

Fig. 7.16. Cutting Bacterial Flagella.

Figs. 7.18, 7.19. Digital Liquid; Organized Turbulence.

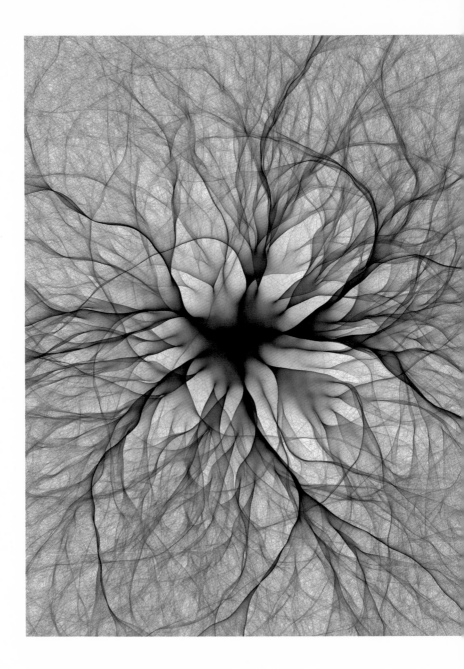

Figs. 7.20, 7.21. Transport II (Image, *Nature*).

8 March 2001

£6.00

www.nature.com

nature

Electron flow in two dimensions

Gene trapping: Towards a wiring diagram for the brain

Superconductivity: Organic polymers enter the arena

Obesity Cholinergic regulation of appetite

new on the market
bioinformatics

Notes

N.B. In lieu of a bibliography, full citations are given in every note.

PREFACE TO THE FIRST EDITION

1. Lorraine Daston and Peter Galison, "The Image of Objectivity," *Representations* 40 (1992), pp. 81–128. Earlier versions of some of the material contained in Chapters Two and Three appeared in this article, and in Chapter Six, in Peter Galison, "Judgment against Objectivity," in Caroline A. Jones and Peter Galison (eds.), *Picturing Science, Producing Art* (New York: Routledge, 1998), pp. 327–59.

PROLOGUE

1. Arthur Worthington, *The Splash of a Drop* (London: Society for Promoting Christian Knowledge, 1895), p. 64.

2. Arthur Worthington, *The Splash of a Drop* (London: Society for Promoting Christian Knowledge, 1895), p. 66.

3. Arthur Worthington, *The Splash of a Drop* (London: Society for Promoting Christian Knowledge, 1895), p. 74.

4. Arthur Worthington, *The Splash of a Drop* (London: Society for Promoting Christian Knowledge, 1895), pp. 55–58, citations on pp. 57–58.

5. All quotations from Arthur Worthington, *The Splash of a Drop* (London: Society for Promoting Christian Knowledge, 1895), pp. 74–75.

6. Arthur Worthington, *The Splash of a Drop* (London: Society for Promoting Christian Knowledge, 1895), p. 74.

7. Arthur Worthington and R.S. Cole, "Impact with a Liquid Surface, Studied by the Aid of Instantaneous Photography," *Philosophical Transactions of the Royal Society of London* 189 (1897), p. 148.

CHAPTER ONE: EPISTEMOLOGIES OF THE EYE

1. On inference statistics: Gerd Gigerenzer, *The Empire of Chance: How Probability Changed Science and Everyday Life* (Cambridge: Cambridge University Press, 1989),

pp. 70–122. On clinical trials: Anne Harrington (ed.), *The Placebo Effect: An Interdisciplinary Exploration* (Cambridge, MA: Harvard University Press, 1997); Harry M. Marks, *The Progress of Experiment: Science and Therapeutic Reform in the United States, 1900–1990* (Cambridge: Cambridge University Press, 1997). On self-registering instruments: Lorraine Daston and Peter Galison, "The Image of Objectivity," *Representations* 40 (1992), pp. 81–128; Soraya de Chadarevian, "Graphical Method and Discipline: Self-Recording Instruments in Nineteenth-Century Physiology," *Studies in the History and Philosophy of Science* 24 (1993), pp. 267–91; Robert Brain, "Standards and Semiotics," in Timothy Lenoir (ed.), *Inscribing Science: Scientific Texts and the Materiality of Communication* (Stanford: Stanford University Press, 1998), pp. 249–84.

2. The literature on the role of the visual in science is vast. Especially relevant are Martin Rudwick, "The Emergence of a Visual Language for Geological Science, 1760–1840," *History of Science* 14 (1976), pp. 149–95; Bruno Latour, "Visualization and Cognition: Thinking with Eyes and Hands," *Knowledge and Society* 6 (1986), pp. 1–40; John Law and Michael Lynch, "Lists, Field Guides, and the Descriptive Organization of Seeing: Birdwatching as an Exemplary Observational Activity," in Michael Lynch and Steve Woolgar (eds.), *Representation in Scientific Practice* (Cambridge, MA: MIT Press, 1990), pp. 267–99; Michael Lynch, "Science in the Age of Mechanical Reproduction: Moral and Epistemic Relations Between Diagrams and Photographs," *Biology and Philosophy* 6 (1991), pp. 205–26; Gordon Fyfe and John Law (eds.), *Picturing Power: Visual Depiction and Social Relations* (London and New York: Routledge, 1988); Jonathan Crary, *Techniques of the Observer: On Vision and Modernity in the Nineteenth Century* (Cambridge, MA: MIT Press, 1990); Ann Shelby Blum, *Picturing Nature: American Nineteenth-Century Zoological Illustration* (Princeton, NJ: Princeton University Press, 1993); Jennifer Tucker, "Photography as Witness, Detective, and Impostor: Visual Representation in Victorian Science," in Bernard Lightman (ed.), *Victorian Science in Context* (Chicago: University of Chicago Press, 1997), pp. 378–408; Peter Galison, *Image and Logic: A Material Culture of Microphysics* (Chicago: University of Chicago Press, 1997); Nicolas Rasmussen, *Picture Control: The Electron Microscope and the Transformation of Biology in America, 1940–1960* (Stanford, CA: Stanford University Press, 1997); Caroline A. Jones and Peter Galison (eds.), *Picturing Science, Producing Art* (New York: Routledge, 1998); Alex Soojung-Kim Pang, "Visual Representation and Post-constructivist History of Science," *Historical Studies in the Physical and Biological Sciences* 28 (1997), pp. 139–71; Klaus Hentschel, *Mapping the Spectrum: Techniques of Visual Representation in Research and Teaching* (Oxford: Oxford University Press, 2002); Soraya de Chadarevian and Nick Hopwood, (eds.), *Models: The Third Dimension of Science* (Stanford, CA: Stanford University Press, 2004); and Jennifer Tucker, *Nature Exposed: Photography as Eyewitness in Victorian Science* (Baltimore: Johns Hopkins University Press, 2005). Still classic on the training of the scientific eye is Ludwik Fleck, *Enstehung und Entwicklung einer wissenschaftlichen Tatsache: Einführung in die Lehre vom Denkstil und*

Denkkollektiv (Basel: Benno Schwabe, 1935); see also Ilana Löwy (trans. and ed.), *The Polish School of Philosophy of Medicine: From Tytus Chalubinski (1820–1889) to Ludwik Fleck (1896–1961)* (Dordrecht: Boston, 1990).

3. Antoine-Clair Thibadeau, *Rapport fait au nom du comité d'instruction et des finances, sur le Muséum national d'histoire naturelle, à la séance du 21 frimaire, l'an 3* (Paris: Imprimerie nationale, 1795), pp. 4–5. MS 2737, Muséum National d'Histoire Naturelle, Paris.

4. See, for example, Johann Gabriel Doppelmayr, *Atlas coelestis* (Nuremberg: Heredum Homannianor, 1742), or Andreas Cellarius, *Harmonia macrocosmica seu atlas universalis et novus* (Amsterdam: Joannem Janssonium, 1661).

5. See Gerhard Mercator, *Gerard Mercator's Map of the World* (Rotterdam: Maritime Museum, 1961), p. 17. The term spread to astronomical maps by the early eighteenth century: see the titles in Deborah J. Warner, *The Sky Explored: Celestial Cartography, 1500–1800* (New York: Liss, 1979). Because of the oversize format of these works, the word "atlas" came in the eighteenth century to designate a very large size (thirty-four inches by twenty-six and a half inches) of drawing paper: Emile Joseph Labarre, *Dictionary and Encyclopaedia of Paper and Paper-Making*, 2nd ed. (London: Oxford University Press, 1952), pp. 10–11. The term was apparently transferred to all illustrated scientific works in the mid-nineteenth century, when figures were printed separately from explanatory texts, in large-format supplements — hence "atlases," deriving from their size: for example, text volume in octo, accompanying atlas in folio. Especially for engraved figures, which had to be printed on higher-quality paper, usually bound separately into the back of the book, this two-volume format had the advantages that text and images could be looked at side by side. As text and figures merged into a single, often oversize, volume, "atlas" came to refer to the entire work, and "atlases" described the whole genre of such scientific picture books. We shall use the term retrospectively to refer to all such works, even those earlier ones that may not use the word "atlas" in the title.

6. So, for example, Ockham writes against the existence of universals: "Universale non est aliqud reale habens esse subjectivum nec in anima nec extra animam, sed tantum habet esse obiectivam in anima et est quoddam fictum habens tale in esse obiectivo, quale habet res extra in esse subiectivo." *Commentary on the Sentences*, Quoted in *Oxford English Dictionary*, compact ed. (New York: Oxford University Press, 1971), s.v. "Objective."

7. René Descartes, *Mediationes de prima philosophia* [1641], *Oeuvres de Descartes*, ed. Charles Adam and Paul Tannery (Paris: Cerf, 1910), vol. 7, p. 42. On Descartes's sources in medieval philosophy, see Calvin Normore, "Meaning and Objective Being: Descartes and His Sources," in Amélie Oksenberg Rorty (ed.), *Essays on Descartes' Meditations* (Berkeley: University of California Press, 1986), pp. 223–41.

8. Ephraim Chambers, "Objective/*objectivus*," *Cyclopaedia, or, An Universal Dictionary of Arts and Sciences* (London: J. and J. Knapton, 1728), vol. 2, p. 649.

9. Judging from dictionary entries in French, English, and German, the most common use of "objective" and its cognates from the late seventeenth century on was to describe microscope lenses. From 1755 onward, Samuel Johnson's *Dictionary of the English Language* gave one sense of "objective" as "[b]elonging to the object; contained in the object," a definition repeated verbatim (including the illustrative quotation from Isaac Watts's *Logick*) well into the nineteenth century: see, for example, John Ogilvie, *The Imperial Dictionary* (Glasgow: Blackie and Son, 1850), s.v. "Objective." For a parallel shift in meaning, compare Christian August Crusius's distinction between *objektivische oder metaphysische* and *subjektivische oder logikalische* truths in *Die philosophischen Hauptwerke*, vol. 3, *Weg zur Gewißheit und Zuverlässigkeit*, ed. G. Tonelli (1747; Hildesheim: Olms, 1965), p. 95. More generally for the pre-Kantian meanings of the words in philosophical texts, see Michael Karskens, "The Development of the Opposition Subjective Versus Objective in the 18th Century," *Archiv für Begriffsgeschichte* 36 (1993), pp. 214–56. See also S.K. Knebel, "Wahrheit, objektive," in Joachim Ritter and Karlfried Gründer (eds.), *Historisches Wörterbuch der Philosophie* (Basel: Schwabe, 2004), vol. 12, cols. 154–63.

10. René Wellek, *Immanuel Kant in England 1793–1838* (Princeton, NJ: Princeton University Press, 1931); Joachim Kopper, "La signification de Kant pour la philosophie française," *Archives de philosophie* 44 (1981), pp. 63–83; Frederick C. Beiser, *The Fate of Reason: German Philosophy from Kant to Fichte* (Cambridge, MA: Harvard University Press, 1987); François Azouvi and Dominique Bourel (eds.), *De Königsberg à Paris: La réception de Kant en France (1788–1804)* (Paris: Vrin, 1991); Rolf-Peter Horstmann, *Die Grenzen der Vernunft: Eine Untersuchung zu Zielen und Motiven des deutschen Idealismus* (Frankfurt am Main: Hain, 1991); Sally Sedgwick (ed.), *The Reception of Kant's Critical Philosophy: Fichte, Schelling, and Hegel* (Cambridge: Cambridge University Press, 2000). See Chapter Four for the scientific reception of the terms.

11. Samuel Taylor Coleridge, *Biographia Literaria, or, Biographical Sketches of My Literary Life and Opinions*, ed. James Engell and W. Jackson Bate (Princeton, NJ: Princeton University Press, 1983), note 3, vol. 1, pp. 172–73; Peter Galison, "Objectivity is Romantic," *American Council of Learned Societies Occasional Paper* 47 (1999).

12. Samuel Taylor Coleridge, *Biographia Literaria, or, Biographical Sketches of My Literary Life and Opinions*, ed. James Engell and W. Jackson Bate (Princeton, NJ: Princeton University Press, 1983), vol. 1, pp. 254–55; the relevant quotation from Schelling is "Wir koennen den Inbegriff alles blos *Objectiven* in unserm Wissen *Natur* nennen"; see note on *ibid.*, p. 253.

13. Thomas De Quincey, *Confessions of an English Opium Eater* [1821], *The Works of Thomas De Quincey*, 2nd ed. (Edinburgh: Adam and Charles Black, 1863), vol. 1, p. 265. Ironically, De Quincey's own use of the word hearkens back to its scholastic meaning: "These [dreams of water] haunted me so much, that I feared lest some dropsical state or tendency of the brain might thus be making itself (to use a metaphysical word) *objective*; and that the sentient organ might be projecting itself as its own object."

14. Historians of philosophy have routinely used the vocabulary of objectivity and subjectivity to analyze the work of Bacon and Descartes: see, for example, Bernard Williams, *Descartes: The Project of Pure Enquiry* (Hassocks, England: Harvester Press, 1978), and Perez Zagorin, *Francis Bacon* (Princeton, NJ: Princeton University Press, 1998).

15. René Descartes, *Principia philosophiae* [1644], 1.68-69, *Oeuvres de Descartes*, ed. Charles Adam and Paul Tannery (Paris: Vrin, 1982), vol. 8, pt. 1, pp. 33-34; cf. vol. 9, pt. 2, pp. 56-57.

16. Francis Bacon, *Novum organum* [1620], *The Works of Francis Bacon*, ed. Basil Montagu (London: Pickering, 1825-34), 1.liii-lviii, vol. 9, pp. 204-206.

17. Francis Bacon, "Of Nature in Men" [1612], *The Works of Francis Bacon*, ed. Basil Montagu (London: Pickering, 1825-34), vol. 1, p. 132.

18. Martha C. Nussbaum, *The Fragility of Goodness: Luck and Ethics in Greek Tragedy and Philosophy* (Cambridge: Cambridge University Press, 1986); Isaiah Berlin, *The Crooked Timber of Humanity: Chapters in the History of Ideas*, ed. Henry Hardy (London: John Murray, 1990); Bernard Williams, *Shame and Necessity* (Berkeley: University of California Press, 1993); J.B. Schneewind, *The Invention of Autonomy: A History of Modern Moral Philosophy* (Cambridge: Cambridge University Press, 1998); Stuart Hampshire, *Justice Is Conflict* (Princeton, NJ: Princeton University Press, 2000). For other examples of epistemic histories that take a repository rather than a rupture view, see Ian Hacking, "'Style' for Historians and Philosophers," *Studies in the History and Philosophy of Science* 23 (1992), pp. 1-20; John V. Pickstone, *Ways of Knowing: A New History of Science, Technology and Medicine* (Chicago: University of Chicago Press, 2001).

19. On historians of philosophy, see Bernard Williams, *Descartes: The Project of Pure Enquiry* (Hassocks, England: Harvester Press, 1978), p. 69, and Nancy Cartwright, *How the Laws of Physics Lie* (Oxford: Clarendon, 1983). On historians of science, see Peter Galison, *How Experiments End* (Chicago: University of Chicago Press, 1987), and M. Norton Wise (ed.), *The Values of Precision* (Princeton, NJ: Princeton University Press, 1995).

20. Henry James, "The Wallace Collection in Bethnal Green" [1873], *The Painter's Eye: Notes and Essays on the Pictorial Arts*, ed. John L. Sweeney (Madison: University of Wisconsin Press, 1989), p. 74.

21. Henry James, "Preface to the 1908 edition," *The Awkward Age*, ed. Vivien Jones (1899; Oxford: Oxford University Press, 1984), pp. xl-xli.

22. Académie des Sciences, Paris, "Orographie — Rapport relatif à des études photographiques sur les Alpes, faites au point de vue de l'orographie et de la géographie physique, par M. Aimé Civiale," *Comptes rendus hebdomadaires des séances de l'Académie des Sciences* 62 (1866), p. 873.

23. The most complete recent survey is Jerrold Seigel, *The Idea of the Self: Thought and Experience in Western Europe since the Seventeenth Century* (New York: Cambridge University Press, 2005).

24. Pierre Hadot, *Philosophy as a Way of Life*, ed. Arnold I. Davidson, trans. Michael Chase (Malden, MA: Blackwell, 1995), pp. 79–125 and 179–213.

25. Michel Foucault, "L'herméneutique du sujet," *Résumé des cours, 1970– 1982* (Paris: Julliard, 1989), pp. 145–66.

26. See, for example, William Eamon, "From the Secrets of Nature to Public Knowledge," in David C. Lindberg and Robert S. Westman (eds.), *Reappraisals of the Scientific Revolution* (Cambridge: Cambridge University Press, 1990), pp. 333–65.

27. Naomi Oreskes, "Objectivity or Heroism? On the Invisibility of Women in Science," *Osiris* 11 (1996), pp. 87–113; Stuart Strickland, "The Ideology of Self-Knowledge and the Practice of Self-Experimentation," *Eighteenth-Century Studies* 31, no. 4 (1998), pp. 453–71; Mary Terrall, *The Man Who Flattened the Earth: Maupertuis and the Sciences in the Enlightenment* (Chicago: University of Chicago Press, 2002); Londa Schiebinger, "Human Experimentation in the Eighteenth Century: Natural Boundaries and Valid Testing," in Lorraine Daston and Fernando Vidal (eds.), *The Moral Authority of Nature* (Chicago: University of Chicago Press, 2004), pp. 384–408.

28. See, for example, Robert Nozick, *Invariances: The Structure of the Objective World* (Cambridge, MA: Harvard University Press, Belknap Press, 2001).

29. Bernard Williams, "The Scientific and the Ethical," in S.C. Brown (ed.), *Objectivity and Cultural Divergence* (Cambridge: Cambridge University Press, 1984), p. 211. Those too cautious to invoke the "really real" may prefer to speak of "stability" or "reliability" rather than "truth," but in all cases "objective" is the accolade awarded to the highest grade of knowledge. Consider, for example, Richard Rorty's antifoundationalist but approving definition of objectivity as "a property of theories which, having been thoroughly discussed, are chosen by a consensus of rational discussants." *Philosophy and the Mirror of Nature* (Princeton, NJ: Princeton University Press, 1979), p. 338. For thoughtful reflections on the import of objectivity in science and scholarship, as well as its multiple meanings, see the articles in Allan Megill (ed.), *Rethinking Objectivity* (Durham, NC: Duke University Press, 1994).

30. On the association of statistical methods with objectivity, see Zeno Swijtink, "The Objectification of Observation: Measurement and Statistical Methods in the Nineteenth Century," in Lorenz Krüger, Lorraine J. Daston, and Michael Heidelberger (eds.), *The Probabilistic Revolution*, vol. 1, *Ideas in History* (Cambridge, MA: MIT Press, 1990), pp. 261–86, and Gerd Gigerenzer, "Probabilistic Thinking and the Fight Against Subjectivity," in Lorenz Krüger, Gerd Gigerenzer, and Mary S. Morgan (eds.), *The Probabilistic Revolution*, vol. 2, *Ideas in the Sciences* (Cambridge, MA: MIT Press, 1990), pp. 11–34; on the association with numerical methods more generally, see Theodore M. Porter, "Objectivity as Standardization: The Rhetoric of Impersonality in Measurement, Statistics, and Cost-Benefit Analysis," *Annals of Scholarship* 9 (1992), pp. 19–60; "Quantification and the Accounting Ideal in Science," *Social Studies of Science* 22 (1992), pp. 632–51; and *Trust in Numbers: The Pur-*

suit of Objectivity in Science and Public Life (Princeton, NJ: Princeton University Press, 1995).

31. Thomas Nagel, *The View from Nowhere* (New York: Oxford University Press, 1986), p. 5.

32. Karl Pearson, *The Grammar of Science* (London: Scott, 1892), p. 11; Donna Haraway, "Situated Knowledges: The Science Question in Feminism and the Privilege of Partial Perspective," *Feminist Studies* 14, no. 3 (1988), pp. 575–99.

33. See, for example, Vaclav Havel, "Politics and the World Itself," *Kettering Review* (Summer 1992), p. 12: "The world today is a world in which generality, objectivity, and universality are in crisis.... Original ideas and actions, unique and, therefore, always risky, often lose their human ethos, and, therefore, de facto, their human spirit after they have gone through the mill of objective analysis and prognoses. Many of the traditional mechanisms of democracy created and developed and conserved in the modern era are so linked to the cult of objectivity and statistical average that they can annul human individuality."

CHAPTER TWO: TRUTH-TO-NATURE

1. Carolus Linnaeus, *Hortus Cliffortianus* (Amsterdam: n. p., 1737).

2. Carolus Linnaeus, "The Preparation of the *Species Plantarum* and the Introduction of Binomial Nomenclature," *Species plantarum: A Facsimile of the First Edition of 1753* (London: Ray Society, 1957–59), pp. 65–74. The current *International Code of Botanical Nomenclature* (the Saint Louis code of 1999) still dates the beginning of officially accepted botanical nomenclature from Linnaeus's *Species plantarum*: ch. 4, sec. 2, art. 13.4–5.

3. See Chapter One concerning the history of the word "objective" and its cognates.

4. Joachim Ritter and Karlfried Gründer (eds.), *Historisches Wörterbuch der Philosophie* (Basel: Schwabe, 2004), s.v. "Wahrheit," vol. 12, cols. 48–123. To our knowledge, no exhaustive history of truth has yet been published, but see for the seventeenth century the classic work by Steven Shapin, *A Social History of Truth: Civility and Science in Seventeenth-Century England* (Chicago: University of Chicago Press, 1994) and the suggestive sketch in Lorenz Krüger, "Wahrheit und Zeit," in Wolf Lepenies (ed.), *Wissenschaftskolleg zu Berlin: Jahrbuch 1987–88* (Berlin: Nicolaische Buchhandlung, 1989), pp. 72–75. The English phrase "true to nature" means "conformable to nature," but the word "true" here also retains the older connotation of "faithful" (as in "a true friend"): cf. the French phrase *d'après nature*, especially in its normative, seventeenth-century sense, as an aesthetic model for painting; cf. the German *naturgetreu*, a *faux ami* that means "faithful to nature," "true" being expressed by *wahr, wahrhaftig*, and variants.

5. "Génie (*Philosophie & Littér.*)," in Jean Le Rond d'Alembert and Denis Diderot, *Encyclopédie, ou, Dictionnaire raisonné des sciences, des arts et des métiers* (Paris: Briasson, 1751–65), vol. 7, p. 583.

6. Johann Wolfgang von Goethe, "Erfahrung und Wissenschaft" [1798, pub. 1893], *Goethes Werke*, 7th ed., vol. 13, *Naturwissenschaftliche Schriften*, ed. Dorothea Kuhn and Rike Wankmüller (Munich: Beck, 1975–76), p. 25; translated by Douglas Miller as *Scientific Studies* (New York: Suhrkamp, 1988), p. 25; translation slightly emended.

7. Carolus Linnaeus, *The "Critica Botanica*," trans. Arthur Hort, rev. Mary Lauretta Green (London: Ray Society, 1938), aphorisms 256, 259, 266, 282, pp. 116, 122, 139, 161. The *Critica Botanica* was first published in 1737. For the development of these methods in Renaissance botany, see Brian W. Ogilvie, *The Science of Describing: Natural History in Renaissance Europe* (Chicago: University of Chicago Press, 2006), pp. 139–208

8. Alphonse de Candolle, *La phytographie* (Paris: Masson, 1880), pp. 238 and 242.

9. See Gunnar Eriksson, "Linnaeus the Botanist," in Tore Frangsmyr (ed.), *Linnaeus: The Man and His Work*, rev. ed. (Canton, MA: Science History Publications, 1994), pp. 63–109, esp. pp. 83–84, concerning Linnaeus's idea of the "most natural form" or "archetype" of flowers; on the context of Linnaeus's career, see Lisbet Koerner, *Linnaeus: Nature and Nation* (Cambridge, MA: Harvard University Press, 1999), and Staffan Müller-Wille, *Botanik und weltweiter Handel: Zur Begründung eines natürlichen Systems der Pflanzen durch Carl von Linné (1707–78)* (Berlin: Verlag für Wissenschaft und Bildung, 1999). More generally on botanical illustration in this period, see Kärin Nickelsen, *Wissenschaftliche Pflanzenzeichnungen — Spiegelbilder der Natur? Botanische Abbildungen aus dem 18. und 19. Jahrhundert* (Bern: Bern Studies in the History and Philosophy of Science, 2000); Gill Saunders, *Picturing Plants: An Analytical History of Botanical Illustration* (Berkeley: University of California Press in association with the Victoria and Albert Museum, 1995); Claus Nissen, *Die botanische Buchillustration: Ihre Geschichte und Bibliographie*, 2nd ed. (Stuttgart: Hiersemann, 1966); and Gavin D.R. Bridson and James J. White, *Plant, Animal and Anatomical Illustration in Art and Science: A Bibliographical Guide from the 16th Century to the Present Day* (Winchester: St. Paul's Bibliographies in association with Hunt Institute for Botanical Documentation, 1990).

10. René-Just Haüy, *Essai d'une théorie sur la structure des crystaux: Appliquée à plusieurs genres de substances crystallisées* (Paris: Chez Gogué & Née de la Rochelle, 1784), p. 3.

11. D.J. Carr (ed.), *Sydney Parkinson: Artist of Cook's Endeavour Voyage* (London: British Museum of Natural History in association with Croom Helm, 1983). On the impact of new taxidermy techniques, see Paul Lawrence Farber, "The Development of Ornithological Collections in the Late Eighteenth and Early Nineteenth Centuries and Their Relationship to the Emergence of Ornithology as a Scientific Discipline," *Journal of the Society for the Bibliography of Natural History* 9 (1980), pp. 391–94; and Hanna Rose Shell, "Skin Deep: Taxidermy, Embodiment and Extinction in W.T. Hornaday's Buffalo Group," in Alan E. Leviton and Michele Aldrich

(eds.), *Museums and Other Institutions of Natural History, Past, Present, and Future: A Symposium Held on the Occasion of the 150th Anniversary of the California Academy of Sciences* (San Francisco, CA: California Academy of Sciences, 2004), pp. 79–102.

12. Georges Cuvier, "Rapport fait au gouvernement par l'Institut Impérial, sur le Voyage de Découvertes aux Terres Australes," June 9, 1806, in François Péron, *Voyage de découvertes aux terres Australes* (Paris: Imprimerie impériale, 1807–1816), plus Atlas (Paris: Bertrand, 1824), p. vi. On the illustrations made in connection with these voyages, see Jan Altmann, "Exakte Beobachtung der Natur und des Menschen: Die Bildwerke der Entdeckungsreise zu den Terres Australes (1800–1804)," Ph.D. diss., Humboldt-Universität zu Berlin: 2005, and, more generally, on the history of zoological illustration, Claus Nissen, *Die zoologische Buchillustration: Ihre Bibliographie und Geschichte* (Stuttgart: Hiersemann, 1969–78).

13. Jean Cruveilhier, "Avant propos," *Anatomie pathologique du corps humain* (Paris: Baillière, 1829–42), pp. i–ii.

14. Francis Bacon, *Novum organum* [1620], in *The Works of Francis Bacon*, ed. Basil Montagu (London: Pickering, 1825–34), 2. xxviii–xxix, pp. 137–38.

15. Lorraine Daston and Katharine Park, *Wonders and the Order of Nature, 1150–1750* (New York: Zone Books, 1998), pp. 220–40.

16. Lorraine Daston and Katharine Park, *Wonders and the Order of Nature, 1150–1750* (New York: Zone Books, 1998), pp. 350–60.

17. Carolus Linnaeus, aphorism 310, *The "Critica Botanica,"* trans. Arthur Hort, rev. Mary Lauretta Green (London: Ray Society, 1938), p. 196.

18. Johann Wolfgang von Goethe, "Fortunate Encounter" [1794; pub. 1817], *Scientific Studies*, ed. and trans. Douglas Miller (New York: Suhrkamp, 1988), p. 20.

19. Johann Wolfgang von Goethe, "Erster Entwurf einer allgemeinen Einleitung in die vergleichende Anatomie, ausgehend von der Osteologie" [1795, pub. 1820], *Goethes Werke*, 7th ed., vol. 13, *Naturwissenschaftliche Schriften*, ed. Dorothea Kuhn and Rike Wankmüller (Munich: Beck, 1975–76), p. 172; translated by Douglas Miller in *Scientific Studies* (New York: Suhrkamp, 1988), p. 118. Translated slightly emended. On the *Urpflanze* in relation to Goethe's doctrine of plant metamorphosis, see Olaf Breidbach, *Goethes Metamorphosenlehre* (Munich: Fink, 2006), pp. 103–16.

20. Johann Wolfgang von Goethe, "Erfahrung und Wissenschaft" [1798, pub. 1893], *Goethes Werke*, 7th ed., vol. 13, *Naturwissenschaftliche Schriften*, ed. Dorothea Kuhn and Rike Wankmüller (Munich: Beck, 1975–76), p. 24; translated by Douglas Miller as *Scientific Studies* (New York: Suhrkamp, 1988), p. 24.

21. On Wandelaar and other illustrators of this period, see individual entries in Hans Vollmer (ed.), *Allgemeines Lexikon der bildenden Künstler von der Antike bis zur Gegenwart* (Leipzig: Seemann, 1907–1950). Many of the eighteenth-century atlas illustrators were Dutch or Dutch-trained; on the Dutch tradition of descriptive art, see Svetlana Alpers, *The Art of Describing: Dutch Art in the Seventeenth Century* (Chicago: University of Chicago Press, 1983).

22. Bernhard Siegfried Albinus, "Historia hujus operis," *Tabulae sceleti et musculorum corporis humani* (Leiden: J. & H. Verbeek, 1747), n.p.; translated as *Tables of the Skeleton and Muscles of the Human Body* (London: John and Paul Knapton, 1749), sig. c.r.

23. Bernhard Siegfried Albinus, "Historia hujus operis," *Tabulae sceleti et musculorum corporis humani* (Leiden: J. & H. Verbeek, 1747), n.p.; translated as *Tables of the Skeleton and Muscles of the Human Body* (London: John and Paul Knapton, 1749), sig. b.r.

24. Londa Schiebinger, "Skeletons in the Closet: The First Illustrations of the Female Skeleton in Eighteenth-Century Anatomy," *Representations* 14 (1986), pp. 42–82. On the choice of adult male animals in eighteenth-century zoology, see Kirsten Winther Jørgensen, "Between Spirit and Matter: An Ethnographic History of British Zoology and Zoologists, ca. 1660–1800," Ph.D. diss., European University Institute, 2003, p. 200.

25. Bernhard Siegfried Albinus, "Historia hujus operis," *Tabulae sceleti et musculorum corporis humani* (Leiden: J. & H. Verbeek, 1747), n.p.; translated as *Tables of the Skeleton and Muscles of the Human Body* (London: John and Paul Knapton, 1749), sig. b.r.

26. Henry Bowman Brady, *A Monograph of Carboniferous and Permian Foro- minifera (the Genus Fusulina Excepted)* (London: Paleontolographical Society, 1876), p. 7. On perfecting specimens, see, for example, James Sowerby, *The Mineral Con- chology of Great Britain* (London: Benjamin Meredith, 1812), pp. 101 and 156; Daniel Sharpe, *Description of the Fossil Remains of Mollusca Found in the Chalk of England* (London: Paleontolographical Society, 1853–56), pl. 11, figs. la and lb. For idealiz- ing-cum-theoretical tendencies in geological illustrations of the late eighteenth and early nineteenth centuries, see Martin Rudwick, "The Emergence of a Visual Lan- guage for Geological Science, 1760–1840," *History of Science* 14 (1976), p. 171.

27. Ludwig Choulant, the great nineteenth-century historian of anatomical illustration and champion of idealizers such as Albinus and Soemmerring, stated the naturalistic alternative only to reject it: "Whenever the artist alone, without the guidance and instruction of the anatomist, undertakes the drawing, a purely indi- vidual and partly arbitrary representation will be the result, even in advanced peri- ods of anatomy. Where, however, this individual's drawing is executed carefully and under the supervision of an expert anatomist, it becomes effective through its indi- vidual truth, its harmony with nature, not only for purposes of instruction, but also for the development of anatomic science; since this norm [*Mittelform*], which is no longer individual but has become ideal, can only be attained through an exact knowledge of the countless peculiarities of which it is the summation." *History and Bibliography of Anatomic Illustration in its Relation to Anatomic Science and the Graphic Arts*, ed. and trans. Mortimer Frank (1852; Chicago: University of Chicago Press, 1920), p. 23.

28. William Hunter, preface to *The Anatomy of the Human Gravid Uterus, Exhibited in Figures* (Birmingham: Baskerville, 1774), n.p. The drawings were done by the Dutch-born artist Jan van Riemsdyk and engraved by several hands. The printer John Baskerville was renowned for his luxury editions of the classics: Lyle Massey, "Pregnancy and Pathology: Picturing Childbirth in Eighteenth-Century Obstetric Atlases," *Art Bulletin* 87 (2005), pp. 73-91.

29. See L.J. Jordanova, "Gender, Generation and Science: William Hunter's Obstetrical Atlas," in W.F. Bynum and Roy Porter (eds.), *William Hunter and the Eighteenth-Century Medical World* (Cambridge: Cambridge University Press, 1985), pp. 385-412.

30. On preparing the cadaver for illustration, see Marielene Putscher, *Geschichte der medizinischen Abbildung*, vol. 2, *Von 1600 bis zur Gegenwart* (Munich: Moos, 1972), p. 49.

31. William Hunter, *The Anatomy of the Human Gravid Uterus, Exhibited in Figures* (Birmingham: John Baskerville, 1774), pl. 2, n.p.

32. On artistic anticipations of "photographic" vision, see Peter Galassi, *Before Photography: Painting and the Invention of Photography* (New York: Museum of Modern Art, 1981). On the history of the terms "naturalism" and "realism" in art history, see Boris Röhrl, *Kunsttheorie des Naturalismus und Realismus: Historische Entwicklung, Terminologie und Definitionen* (Hildesheim: Olms, 2003), esp. pp. 17-24, on the early history of the older term "naturalism."

33. William Cheselden, "To the Reader," *Osteographia, or, The Anatomy of the Bones* (London: Bowyer, 1733), n.p.

34. George Edwards, *A Natural History of Uncommon Birds, and of Some Other Rare and Undescribed Animals* (London: College of Physicians, 1743-51), vol. 1, pp. xv and xix.

35. Ann Shelby Blum, *Picturing Nature: American Nineteenth-Century Zoological Illustration* (Princeton, NJ: Princeton University Press, 1993), pp. 92-106. See also Amy R.W. Meyers, "Observations of an American Woodsman: John James Audubon as a Field Naturalist," in Annette Blaugrund and Theodore E. Stebbins Jr. (eds.), *John James Audubon: The Watercolours for the Birds of America*, Catalog entries by Carole Anne Slatkin (New York: Villard, 1993), pp. 43-54, for the moralized content of Audubon's depictions of birds; also Duff Hart-Davis, *Audubon's Elephant: America's Greatest Naturalist and the Making of the Birds of America* (New York: Henry Holt, 2004), pp. 133-36. On the importance of symmetry in eighteenth-century natural-history illustration, see E.C. Spary, "Scientific Symmetries," *History of Science* 42 (2004), pp. 1-46.

36. David Hume, *A Treatise of Human Nature*, 2nd ed., ed. L.A. Selby-Bigge, rev. P.H. Nidditch (1739-40; Oxford: Oxford University Press, 1978), 2.2.8, pp. 374-75. On the analogy among aesthetic, moral, and epistemological perceptions in Hume, see Dario Perinetti, *Hume, History and the Science of Human Nature*, forthcoming, ch. 4.

37. Sachiko Kusukawa, "From Counterfeit to Canon: Picturing the Human Body, especially by Andreas Vesalius," Max Planck Institute for the History of Science, preprint no. 281.

38. Albrecht von Haller, *Icones anatomicae*, vol. 2, fasc. 5 (Göttingen: B. Abram Vandenhoek, 1752), sig. A2.r–v.

39. Hans-Konrad Schmutz, "Barocke und klassizistische Elemente in der anatomischen Abbildung," *Gesnerus* 35 (1978), p. 61.

40. Joshua Reynolds, *Sir Joshua Reynolds's Discourses*, ed. Helen Zimmern (London: Scott, 1887), pp. 29 and 280.

41. Chevalier de Jaucourt, "La belle nature," in Jean Le Rond d'Alembert and Denis Diderot, *Encyclopédie, ou, Dictionnaire raisonné des sciences, des arts et des métiers* (Paris: Briasson, 1751–65), vol. 11, pp. 42–44.

42. Atlases of characteristic images presented individual cases as exemplary and illustrative of broader classes and causal processes. For example, Bénédict Auguste Morel's *Traité des dégénérescences physiques, intellectuelles, et morales de 1'espèce humaine et des causes qui produisent ces variétés maladives* (Paris: Baillière, 1857) insisted that constant causes "tend to create types of a determinate form" and that these pathologies would prove as "distinctive, fixed, and invariable" as normal types: pp. 5 and 9.

43. See, for example, Gottlieb Gluge, *Atlas of Pathological Histology*, trans. Joseph Leidy (Philadelphia: Blanchard and Lea, 1853), p. 6. Gluge was a disciple of Belgian statistician Adolphe Quetelet and subscribed to Quetelet's brand of statistical essentialism: see Theodore M. Porter, *The Rise of Statistical Thinking, 1820–1900* (Princeton, NJ: Princeton University Press, 1986), pp. 54–55.

44. Maurice Caullery, *Les papiers laissés par de Réaumur et le tome VII des Mémoires pour servir à l'histoire des insectes* (Paris: Lechevalier, 1929), pp. 8–9. The will was dated Paris, April 1, 1735.

45. René-Antoine Ferchault de Réaumur, *Mémoires pour servir à l'histoire des insectes* (Paris: Imprimerie Royale, 1734–42), vol. 1, pp. 53–56. Philippe Simonneau, who also did many of the drawings for the *Histoire et mémoires de l'Académie Royale des Sciences*, did most of the drawings and engravings for vol. 1; illustrations for subsequent volumes are mostly unsigned (and were presumably done by Dumoustier de Marsilly, who, according to Réaumur, wished to remain anonymous: vol. 1, p. 55) but through vol. 5 continue to be engraved by Simonneau.

46. Wolfgang Kemp has documented the ascent of drawing as part of genteel breeding in the eighteenth century: "... *Einen wahrhaft bildenden Zeichenunterricht überall einzuführen*": *Zeichnen und Zeichenunterricht der Laien 1500–1870* (Frankfurt: Syndikat, 1979), pp. 57–120. Martin Rudwick has traced the numerous sketches in the field notebooks of geologists in the late eighteenth and early nineteenth centuries back to youthful instruction in drawing among the upper classes: Martin J.S. Rudwick, "The Emergence of a Visual Language for Geological Science, 1760–1840," *History of Science* 14 (1976), p. 153.

47. William M. Ivins Jr., *Prints and Visual Communication* (1953; Cambridge, MA: MIT Press, 1969), p. 67; Ann Shelby Blum, *Picturing Nature: American Nineteenth-Century Zoological Illustration* (Princeton, NJ: Princeton University Press, 1993), p. 122.

48. Antoine-Laurent Lavoisier, whose notebooks are full of sketches, understood this kind of scientific drawing as an extension and refinement of textual intelligence, not only as a supplement to language but as a language in itself: Madeleine Pinault-Sørensen, "Dessins et archives," in Béatrice Didier and Jacques Neefs (eds.), *Editer des manuscrits: Archives, complétude, lisibilité* (Saint-Denis, France: Presses Universitaires de Vincennes, 1996), pp. 40 and 48.

49. René Descartes to Constantijn Huygens, July 30, 1636, *Oeuvres de Descartes*, Charles Adam and Paul Tannery, ed. (Paris: Vrin, 1982), vol. 1, p. 611.

50. Sachiko Kusukawa, "Leonhart Fuchs on the Importance of Pictures," *Journal of the History of Ideas* 58 (1997), pp. 403–27.

51. Walter Koschatzky, *Die Kunst der Graphik: Technik, Geschichte, Meisterwerke*, 13th ed. (Munich: Deutscher Taschenbuch Verlag, 2003), pp. 18–27.

52. Some of Sophie Cuvier's drawings, annotated by Georges Cuvier, are preserved at the Muséum National d'Histoire Naturelle, Paris, MS 412; on the British women botanical artists, see Ann B. Shteir, *Cultivating Women, Cultivating Science: Flora's Daughters and Botany in England, 1760–1860* (Baltimore: Johns Hopkins University Press, 1996); Abigail Jane Lustig, "The Creation and Uses of Horticulture in Britain and France in the Nineteenth Century," Ph.D. diss., University of California at Berkeley, 1997; Barbara T. Gates, *Kindred Nature: Victorian and Edwardian Women Embrace the Living World* (Chicago: University of Chicago Press, 1998); and the appendix to Martyn Rix (ed.), *Art in Nature: Over 5000 Plants Illustrated from Curtis's Magazine* (London: Studio Editions, 1991). On women helpmeets in science, see Pnina G. Abir-Am and Dorinda Outram (eds.), *Uneasy Careers and Intimate Lives: Women in Science, 1789–1979* (New Brunswick, NJ: Rutgers University Press, 1987); Londa Schiebinger, *The Mind Has No Sex? Women in the Origins of Modern Science* (Cambridge, MA: Harvard University Press, 1989).

53. Madeleine Pinault, *Le peintre et l'histoire naturelle* (Paris: Flammarion, 1990), pp. 45–51. See also the appendix to Martyn Rix (ed.), *Art in Nature: Over 5000 Plants Illustrated from Curtis's Magazine* (London: Studio Editions, 1991), and Henry Jouin and Henri Stein, "Vélins signés des peintres en titre," *Histoire et description du Jardin des Plantes et du Muséum d'Histoire Naturelle* (Paris: Plon, 1887).

54. Dossier Réaumur, box 7, folder "1716: Different Simonneau-Reaumur," Archives de l'Académie des Sciences, Paris. Louis Simonneau was the uncle of Philippe Simonneau, who drew the sketches of the tortoise for the Académie's comparative anatomy project and later engraved the illustrations for Réaumur's *Histoire des insectes*. We are grateful to Madeleine Pinault-Sørensen for this information. On the altercation with Réaumur, see also Madeleine Pinault-Sørensen, "Dessins et

archives," in Beatrice Didier and Jacques Neefs (eds.), *Editer des manuscrits: Archives, complétude, lisibilité* (Saint-Denis, France: Presses Universitaires de Vincennes, 1996), p. 46.

55. E.T. Hamy, "Les derniers jours du Jardin du Roi et la fondation du Muséum d'histoire naturelle," in *Centenaire de la fondation du Muséum d'histoire naturelle 10 juin 1793–10 juin 1893* (Paris: Imprimerie nationale, 1893), pp. 146–60. Spaendonck had been Basseporte's successor at the Jardin du Roi and "peintre ordinaire du roi pour les miniatures," but he became a committed revolutionary after 1789: Luc Vezin, *Les artistes au Jardin des plantes* (Paris: Herscher, 1990), pp. 41–42. In 1822, after the Bourbon restoration in France, the professorship of "iconographie naturelle" was abolished and replaced by "maîtres de dessin," who were clearly subordinated as adjuncts to the professors of botany and zoology: *Centenaire de la fondation du Muséum d'histoire naturelle 10 juin 1793–10 juin 1893* (Paris: Imprimerie nationale, 1893), p. v. More generally on the Muséum d'Histoire Naturelle during this period, see E.C. Spary, *Utopia's Garden: French Natural History from Old Regime to Revolution* (Chicago: University of Chicago Press, 2000).

56. On the conventions for identifying the authors of graphic art since the seventeenth century, see Walter Koschatzky, *Die Kunst der Graphik: Technik, Geschichte, Meisterwerke*, 13th ed. (Munich: Deutscher Taschenbuch Verlag, 2003), pp. 18–20.

57. On the contrasting situation of laboratory technicians, see Steven Shapin, "The Invisible Technician," *American Scientist* 77 (1989), pp. 554–63.

58. Pierre-Joseph Redouté, "Discours préliminaire," *Les liliacées* (Paris: Didot Jeune, 1802–16), vol. 1, pp. TK. The *Liliaceae* (the family, with some four thousand species, includes lilies, daffodils, tulips, and hyacinths; most members are perennials arising from a rhizome, bulb, or corm) are difficult to flatten and dry in herbaria because of their fleshy stalks. Some of Redouté's plates became the type specimens (iconotypes) — substituting for the plant itself or a herbarium sample as the plant of reference — for their species, a remarkable tribute to their perceived botanical accuracy. Candolle supplied the descriptions for vols. 1–4.

59. Abigail Jane Lustig, "The Creation and Uses of Horticulture in Britain and France in the Nineteenth Century," Ph.D. diss., University of California at Berkeley, 1997.

60. James Edward Smith, *Exotic Botany* (London: Taylor, 1804–1805), p. vii.

61. Sowerby Collection, Natural History Museum, London, folder B61, box 50, no. 22.

62. James Edward Smith, preface to *The English Flora*, 2nd ed. (London: Longman, Rees, Orme, Brown, and Green, 1828–30), quoted in Diana M. Simpkins, "Biographical Sketch of James Sowerby, Written by His Son, James de Carle Sowerby, 1825," *Journal of the Society for the Bibliography of Natural History* 6 (1974), p. 412.

63. On parallels in the relationships between travelers and stay-at-home naturalists, see E.C. Spary, *Utopia's Gardens: French Natural History from Old Regime to Revolution* (Chicago: University of Chicago Press, 2000), pp. 69–78.

64. Claude Bernard, *Introduction à l'étude de la médecine expérimentale*, ed. François Dagognet (1865; Paris: Garnier-Flammarion, 1966), pp. 53–54.

65. François Huber, *Nouvelles observations sur les abeilles: Adressées à M. Charles Bonnet* (Geneva: Barde, Manget, 1792), pp. 5–7; see also Patrick Singy, "Huber's Eyes: The Art of Observation Before the Emergence of Positivism," *Representations* 95 (2006), pp. 54–75.

66. Georges Friedmann, "*L'Encyclopédie* et le travail humain," *Annales: Economies, sociétés, civilisations* 8 (1953), pp. 53–61.

67. See, for example, the winning entry in the 1767 prize competition set by the Académie Française "sur l'utilité des écoles gratuites de dessin en faveur des métiers": J.B. Descamps, *Sur l'utilité des établissemens [sic] des écoles gratuites de dessin en faveur des métiers* (Paris: Regnard, 1768), and Yves Deforge, *Le graphisme technique* (Paris: Champion, 1976), pp. 149–51; Wolfgang Kemp, ". . . *Einen wahrhaft bildenden Zeichenunterricht überall einzuführen:*" *Zeichnen und Zeichenunterricht der Laien 1500–1870* (Frankfurt: Syndikat, 1979), pp. 150–51. A specialized but increasingly important part of the drawing reform movement emphasized technical drawing, which in the form of descriptive geometry dominated early nineteenth-century French and German efforts to rationalize handiwork: Antonius Lipsmeier, *Technik und Schule: Die Ausformung des Berufsschulcurriculums unter dem Einfluß der Technik als Geschichte des Unterrichts im technischen Zeichnen* (Wiesbaden: Steiner, 1971); Lorraine Daston, "The Physicalist Tradition in Early Nineteenth-Century French Geometry," *Studies in the History and Philosophy of Science* 17 (1986), pp. 269–95; Ken Alder, *Engineering the Revolution: Arms and Enlightenment in France, 1763–1815* (Princeton, NJ: Princeton University Press, 1997).

68. Wolfgang Kemp, ". . . *Einen wahrhaft bildenden Zeichenunterricht überall einzuführen*": *Zeichnen und Zeichenunterricht der Laien 1500–1870* (Frankfurt: Syndikat, 1979), pp. 175–88. On the Gobelins school, see E. Gerspach, *La Manufacture nationale des Gobelins* (Paris: Delagrave, 1892), pp. 151–76.

69. Wolfgang Kemp, "Disegno: Beiträge zur Geschichte des Begriffs zwischen 1547 und 1607," *Marburger Jahrbuch für Kunstwissenschaft* 19 (1974), pp. 219–40.

70. On Basseporte's career and work, see Madeleine Pinault, *Le peintre et l'histoire naturelle* (Paris: Flammarion, 1990), pp. 24, 26, 128, 154, 241. Basseporte worked for thirteen years painting birds and monkeys at Versailles for Louis XV, paying for her own materials and travel until Madame de Pompadour (for whom Basseporte also served as interior decorator) intervened to have her paid for her work and reimbursed for her expenses: AJ 15.510, p. 331 (Magdelaine Françoise Basseporte), Archives Nationales, Paris.

71. Quoted in Diana M. Simpkins, "Biographical Sketch of James Sowerby,

Written by His Son, James de Carle Sowerby, 1825," *Journal of the Society for the Bibliography of Natural History* 6 (1974), p. 409.

72. E.S. Barton, "A Memoir of Georg Dionysius Ehret," *Proceedings of the Linnean Society of London* (1894–95), p. 52.

73. Claudia Swan, "*Ad vivum, naer het leven*, from the life: Considerations on a Mode of Representation," *Word and Image* 11 (1995), pp. 353–72.

74. William Herbert was unusual in having the engravings of his *Amaryllidaceae* (London: Ridgeway and Sons, 1837) colored according to "the existing tints of the dry specimens," which, he warned, "are in many cases very fallacious": "Advertisement," n.p.

75. Madeleine Pinault, *Le peintre et l'histoire naturelle* (Paris: Flammarion, 1990), p. 72.

76. "Dessein," in Jean Le Rond d'Alembert and Denis Diderot, *Encyclopédie, ou, Dictionnaire raisonné des sciences, des arts et des métiers* (Paris: Briasson, 1751–65), vol. 4, pp. 889–92.

77. Antonius Lipsmeier, *Technik und Schule: Die Ausformung des Berufsschulcurriculums unter dem Einfluß der Technik als Geschichte des Unterrichts im technischen Zeichnen* (Wiesbaden: Steiner, 1971); Michel Foucault regarded the Gobelins drawing school as exemplary for a new way of ordering time and disciplining bodies: *Surveiller et punir: Naissance de la prison* (Paris: Gallimard, 1975), pp. 184–90. On the persistence of drawing as a component of French technical education, see Ken Alder, *Engineering the Revolution: Arms and Enlightenment in France, 1763–1815* (Princeton, NJ: Princeton University Press, 1997), pp. 138–46.

78. Wolfgang Kemp, "...*Einen wahrhaft bildenden Zeichenunterricht überall einzuführen*": *Zeichnen und Zeichenunterricht der Laien 1500–1870* (Frankfurt: Syndikat, 1979), pp. 100–101 and 134–46.

79. Yves Laissus and Anne-Marie Monseigny, "*Les Plantes du Roi*: Note sur un grand ouvrage de botanique préparé au XVIIe siècle par l'Académie Royale des Sciences," *Revue d'histoire des sciences et de leurs applications* 22 (1969), pp. 193–236; Madeleine Pinault, *Le peintre et l'histoire naturelle* (Paris: Flammarion, 1990), pp. 22–26.

80. Basseporte advised Madame de Pompadour on matters of interior decoration; Spaendonck came from the Flemish tradition of flower still lifes; Redouté was painter first to Marie-Antoinette and then to the empress Joséphine. On the close connection between eighteenth-century natural-history illustration and the luxury trade, see E.C. Spary, "Scientific Symmetries," *History of Science* 42 (2004), pp. 1–46, especially p. 14; see also the articles in Winfried Baer, *Das Flora Danica-Service 1790–1802: Höhepunkt der Botanischen Porzellanmalerei; Schloss Charlottenburg Berlin, 21 Oktober 1999–9 Januar 2000* (Copenhagen: Kongelige Udstillingsfond Køpenhavn, 1999), and Sam Segal and Michiel Roding, *De tulp en de kunst* (Zwolle, the Netherlands: Waanders, 1994).

81. By 1743, apprentices at the porcelain manufacture at Meissen received drawing instruction. The draftsmen trained at Glasgow Academy of the Fine Arts (established in 1753) created new patterns for local textile manufacturers; Josiah Wedgwood planned to open a private drawing school near his pottery factories in Etruria and Soho: Wolfgang Kemp, "... *Einen wahrhaft bildenden Zeichenunterricht überall einzuführen*": *Zeichnen und Zeichenunterricht der Laien 1500–1870* (Frankfurt: Syndikat, 1979), pp. 175–81.

82. For example, Marie-Thérèse Vien, who worked for the botanist Michel Adanson, exhibited her still lifes at the Paris salon of 1757 (Diderot praised her "patience et exactitude"): Madeleine Pinault, *Le peintre et l'histoire naturelle* (Paris: Flammarion, 1990), pp. 48–49. Even in the Netherlands, where the still life found its most avid cultivators and admirers, such paintings were priced in notarial records as craftworks, cheaper than historical paintings or even fine linen and lace: Simon Schama, *The Embarrassment of Riches: An Interpretation of Dutch Culture in the Golden Age* (Berkeley: University of California Press, 1988), pp. 318–19.

83. "Dessein," in Jean Le Rond d'Alembert and Denis Diderot, *Encyclopédie, ou, Dictionnaire raisonné des sciences, des arts et des métiers* (Paris, Briasson: 1751–65), vol. 4, pp. 890–91.

84. For example, Jean Joseph Sue, *Elémens d'anatomie, à l'usage des peintres, des sculpteurs et des amateurs* (Paris: Chez l'auteur, et chez Méquignon, Royer, Barrois, 1788), and Charles Bell, *Essays on the Anatomy of Expression in Painting* (London: Longman, Hurst, Rees, and Orme, 1806).

85. Ludwig Choulant, *History and Bibliography of Anatomical Illustration in its Relation to Anatomic Science and the Graphic Arts*, ed. and trans. Mortimer Frank (1852; Chicago: University of Chicago Press, 1920), p. 30.

86. Samuel Thomas von Soemmerring, *Abbildungen des menschlichen Auges* (Frankfurt am Main: Varrentrapp und Wenner, 1801), p. 2.

87. For a comprehensive overview of these techniques and their history, see Walter Koschatzky, *Die Kunst der Graphik: Technick, Geschichte, Meisterwerke*, 13th ed. (Munich: Deutscher Taschenbuch Verlag, 2003).

88. William M. Ivins Jr., *Prints and Visual Communication* (1953; Cambridge, MA: MIT Press, 1969), p. 70.

89. Antony Griffiths, *Prints and Printmaking: An Introduction to the History and Techniques* (Berkeley: University of California Press, 1996), pp. 51–55.

90. Bernhard Siegfried Albinus, "Historia hujus operis," *Tabulae sceleti et musculorum corporis humani* (Leiden: J. & H. Verbeek, 1747), n.p.; translated as *Tables of the Skeleton and Muscles of the Human Body* (London: John and Paul Knapton, 1749), sigs. a–c.

91. Pierre-Joseph Redouté, "Discours préliminaire," *Les liliacées* (Paris: Didot Jeune, 1802–16), vol. 1, p. i.

92. But the printer remained crucial: "It is perhaps as appropriate to write a

history of lithography in terms of the major lithographic printing establishments as in terms of the artists." Antony Griffiths, *Prints and Printmaking: An Introduction to the History and Techniques* (Berkeley: University of California Press, 1996), p. 102. Dependence on the skill of the printer was still greater for mezzotints and etchings: the lines of etchings cannot be drawn too close together without risking puddles of ink that will ruin the plate, but the waxy surface is less resistant than metal, so the etcher can draw more freely than the engraver. Hence etching was often the preferred medium of painters and draftsmen who wished to transcribe their own work for reproduction. Mezzotints permit finely graduated tones of light and shadow (and lend themselves well to coloring), since they start from a wholly black background that is "scraped" smooth by burnishing. Both processes are highly dependent on the skill of the printer, and the quality of an etching may vary greatly from proof to proof. Mezzotint plates need to be refreshed ("regrounded") after a number of impressions have been drawn, since the printing process wears out the background burr.

93. Jean Cruveilhier, "Avant propos," *Anatomie pathologique du corps humain* (Paris: Baillière, 1829–42), p. vii.

94. Eugène Trutat, *La photographie appliquée à l'histoire naturelle* (Paris: Gauthier-Villars, 1884), pp. ix, xi, 94.

95. A. Naumann, "Botanik," in K.W. Wolf-Czapek (ed.), *Angewandte Photographie in Wissenschaft und Technik* (Berlin: Union Deutsche Verlagsgesellschaft Zweigniederlassung, 1911), pt. 2, pp. 15–16.

96. Alphonse de Candolle, *La phytographie* (Paris: Masson, 1880), pp. 321 and 363. On the *Naturselbstdruck* technique, see Alois Auer, "Die Entdeckung des Naturselbstdruckes," *Denkschriften der Kaiserlichen Akademie der Wissenschaften, Mathematisch-Naturwissenschaftliche Classe*, Bd. 5, 1.Abt., pp. 107–110 (Vienna: Kaiserlich-Königliche Hof- und Staatsdruckerei, 1853); also Armin Geus (ed.), *Natur im Druck: Eine Ausstellung zur Geschichte und Technik des Naturselbstdrucks* (Marburg: Basilisken-Presse, 1995).

97. Ludolph C. Treviranus, *Die Anwendung des Holzschnittes zur bildlichen Darstellungen von Pflanzen nach Entstehung, Blüthe, Verfall und Restauration* (1855; Utrecht: De Haan, 1949), pp. 1–2, 71.

98. Claus Nissen, *Die botanische Buchillustration: Ihre Geschichte und Bibliographie* (Stuttgart: Hiersemann, 1951/1966), pp. 7–8.

99. Alphonse de Candolle, *La phytographie* (Paris: Masson, 1880), pp. 51–52.

100. For the former: O.F. Cook, "The Method of Types in Botanical Nomenclature," *Science* n.s. 12 (1900), p. 478. For the latter: A.S. Hitchcock, "The Type Concept in Systematic Botany," *American Journal of Botany* 8 (1921), p. 253.

101. Ernst Mayr, *Principles of Systematic Zoology* (New York: McGraw-Hill, 1969), p. 367. On the International Code of Botanical Nomenclature, see Lorraine Daston, "Type Specimens and Scientific Memory," *Critical Inquiry* 31 (2004), pp. 153–82. For an electronic version of the most recent version of the *International*

Code of Botanical Nomenclature (the Saint Louis code of 1999), see http://www. bgbm.org/iapt/nomenclature/code/SaintLouis/0000St.Luistitle. htm.

102. "The E-Type Initiative@Harvard Entomology," http://insects.oeb.harvard.edu/etypes/.

CHAPTER THREE: MECHANICAL OBJECTIVITY

1. Camillo Golgi, "The Neuron Doctrine – Theory and Facts," Nobel Lecture, Dec. 11, 1906; available online at http://nobelprize.org/medicine/laureates/1906/golgi-lecture.html, p. 216.

2. Santiago Ramón y Cajal, *Recollections of My Life*, trans. E. Horne Craigie with Juan Cano (Cambridge, MA: MIT Press, 1989), p. 553.

3. Camillo Golgi, *Sulla fina anatomia degli organi centrali del sistema nervoso* (Reggio-Emilia, Italy: Calderini, 1885) – which would have been read by Cajal and most of the scientific community in German translation: *Untersuchungen über den feineren Bau des centralen und peripherischen Nervensystems*, trans. R. Teuscher (Jena: Fischer, 1894), "Genau der Natur nach angefertigt"; compare fig. 25, "Die Besonderheiten des Baues erscheinen hier viel weniger compliciert, als in der Natur."

4. On Golgi and his debate with Cajal, excellent sources are Paolo Mazzarello, *The Hidden Structure: A Scientific Biography of Camillo Golgi*, ed. and trans. Henry A. Buchtel and Aldo Badiani (Oxford: Oxford University Press, 1999), pp. 89-90, and Nicholas J. Wade and Marco Piccolino, "Nobel Stains," *Perception* 35 (2006), pp. 1–8; also helpful is a book addressed to a wide audience, Richard Rapport, *Nerve Endings: The Discovery of the Synapse* (New York: Norton, 2005). We are extremely grateful to Elio Raviola and Paolo Mazzarello for many detailed comments and suggestions about our discussion of the black method, Golgi, and Cajal. Mazzarello persuasively suggests that Golgi's first Nobel image (figure 3.2) was taken from a previous publication, a letter from Golgi to the physiologist Luigi Luciani: Luciani, *Fisiologia dell'uomo*, 2nd ed. (Milan: Società Editrice Libreria, 1905), vol. 2, pp. 212–15, esp. p. 215. Differences between the two versions of the image (the Nobel image is considerably simpler, although it is of the same basic form) suggest that the Nobel version is a hand-drawn, simplified copy of the earlier one, which very probably was made with a camera lucida. Paolo Mazzarello, private communication to Peter Galison, April 4, 2006.

5. Camillo Golgi, "The Neuron Doctrine – Theory and Facts," Nobel Lecture, Dec. 11, 1906, available online at http://nobelprize.org/medicine/laureates/1906/golgi-lecture.html, p. 192.

6. Santiago Ramón y Cajal, *Recollections of My Life*, trans. E. Horne Craigie with Juan Cano (Cambridge, MA: MIT Press, 1989), p. 553.

7. Santiago Ramon y Cajal, "¿Neuronismo o reticularismo? Las pruebas objetivas de la unidad anatómica de las células nerviosas," *Archivos de neurobiología* 13 (1933), pp. 1–144, translated by M. Ubeda Purkiss and Clement A. Fox as *Neuron Theory or*

Reticular Theory? Objective Evidence of the Anatomical Unity of Nerve Cells (Madrid: Consejo Superior de Investigaciones Científicas, Instituto Ramón y Cajal, 1954).

8. Richard Greeff, *Atlas der äusseren Augenkrankheiten für Ärzte und Studierende* (Berlin: Urban & Schwarzenberg, 1909), p. v.

9. See, for example, Josef M. Eder and Eduard Valenta, *Atlas typischer Spektren*, 3rd ed. (Vienna: Holder, 1928), and Wilhelm His, *Anatomie menschlicher Embryonen*, vol. 1, *Atlas, Embryonen des ersten Monats; Tafel 1–8* (Leipzig: Vogel, 1880). One publisher alone, Lehmann Verlag, issued a series of seventeen atlases in medicine, and a much larger run of smaller hand-atlases (those that can be held in one hand).

10. Beaumont Newhall, *The History of Photography: From 1839 to the Present*, rev. ed. (London: Secker & Warburg, 1982), pp. 27–42; Monique Sicard, *La fabrique du regard: Images de science et appareils de vision (XVe–XXe siècle)* (Paris: Jacob, 1998), pp. 95–100.

11. The camera lucida was a portable optical instrument for drawing in perspective invented by William Wollaston: William Hyde Wollaston, "Description of the Camera Lucida," *Journal of Natural Philosophy, Chemistry and the Arts* 17 (1807), pp. 1–5. On its scientific and artistic uses in the early nineteenth century, see Erna Fiorentini, "Subjective Objective: The Camera Lucida and Protomodern Observers," *Bildwelten des Wissens: Kunsthistorisches Jahrbuch für Bildkritik* 2 (2004), pp. 58–66, and "Nuovi punti di vista: Giacinto Gigante e la camera lucida a Napoli," in Martina Hansmann and Max Seidel (eds.), *Pittura italiana nell'Ottocento* (Venice: Marsilio, 2005), pp. 535–57.

12. Larry J. Schaaf, *Out of the Shadows: Herschel, Talbot and the Invention of Photography* (New Haven: Yale University Press, 1992), pp. 35–44 and 82–83.

13. Jennifer Tucker, *Nature Exposed: Photography as Eyewitness in Victorian Science* (Baltimore: Johns Hopkins University Press, 2005).

14. On photography as an instrument of discovery, see Joel Snyder, "Visualization and Visibility," in Caroline A. Jones and Peter Galison (eds.), *Picturing Science, Producing Art* (New York: Routledge, 1998), pp. 379–97; Theresa Levitt, "Biot's Paper and Arago's Plates: Photographic Practice and the Transparency of Representation," *Isis* 94 (2003), pp. 456–76; and Peter Geimer, "Picturing the Black Box: On Blanks in Nineteenth-Century Paintings and Photographs," *Science in Context* 17 (2004), pp. 467–501.

15. Ann Shelby Blum, *Picturing Nature: American Nineteenth-Century Zoological Illustration* (Princeton, NJ: Princeton University Press, 1993), pp. 181–209 and 275–78.

16. Alexander Agassiz, "Application of Photography to Illustrations of Natural History: With Two Figures Printed by the Albert and Woodbury Processes," *Bulletin of the Museum of Comparative Zoology at Harvard College* 3 (1871), pp. 47–48.

17. François Arago, "Le Daguerréotype," *Annales de chimie et de physique* 71 (1839), pp. 327–28. Scientists who used photography to capture detail developed their own aesthetics of sharp contrast or "snap": see Alex Pang, "Technology, Aes-

thetics, and the Development of Astrophotography at the Lick Observatory," in Timothy Lenoir (ed.), *Inscribing Science: Scientific Texts and the Materiality of Communication* (Stanford, CA: Stanford University Press, 1998), pp. 223–48. For more on early assessment of the aesthetic virtues of photography as seen within the art world, see e.g. Andre Jammes and Eugenia Parry Jannis, *The Art of French Calotype* (Princeton, NJ: Princeton University Press, 1983); and Henri Zerner, *Gustave Le Gray, Heliographer-Artist* in Sylvie Aubenas, et al., *Gustave Le Gray, 1820–1884* (Paris: Gallimard, 2002), pp. 209–32. Our thanks to Robin Kelsey for discussion on these points.

18. William Henry Fox Talbot, quoted in Larry J. Schaaf, *Out of the Shadows: Herschel, Talbot and the Invention of Photography* (New Haven: Yale University Press, 1992), p. 52.

19. William Henry Fox Talbot, "Introduction," *The Pencil of Nature* (1844–1846; Introduction by Beaumont Newhall, New York: Da Capo Press, 1969), n.p. The calotype sometimes mistaken for an engraving was "The Open Door," which was compared to a painting by Philips Wouwerman.

20. Stephen Bann, "Photography, Printmaking, and the Visual Economy in Nineteenth-Century France," *History of Photography* 26 (2002), pp. 16–25.

21. Photography was not the only process that aimed to produce images automatically. Photograms did so by impressing objects on light-sensitive paper and other surfaces: Mike Ware, *Cyanotype: The History, Science and Art of Photographic Printing in Prussian Blue* (London: Science Museum, 1999); Caroline Armstrong and Catherine de Zegher (eds.), *Ocean Flowers: Impressions from Nature* (Princeton, NJ: Princeton University Press, 2004). The *Naturselbstdruck* pressed an object into soft lead, leaving an imprint from which copies could be printed: Alois Auer, "Die Entdeckung des Naturselbstdruckes," *Denkschriften der Kaiserlichen Akademie der Wissenschaften. Mathematisch-Naturwissenschaftliche* Classe 5 (1853), pp. 107–10.

22. So large did nature's agency loom that the terms "invention" and "discovery" were often applied interchangeably to photography, as if it were itself a part of nature, like oxygen or the moons of Jupiter: Mary Warner Marien, *Photography: A Cultural History* (London: Laurence King, 2002), p. 23.

23. Alfred Donné and Léon Foucault, *Cours de microscopie complémentaire des études médicales: Anatomie microscopique et physiologie des fluides de l'économie* (Paris: Baillère, 1844–45), pp. 36–37; Renata Taureck, *Die Bedeutung der Photographie für die medizinische Abbildung im 19. Jahrhundert* (Cologne: Forschungsstelle des Instituts für Geschichte der Medizin der Universität zu Köln, 1980).

24. Charles Baudelaire, "Salon de 1859," *Curiosités esthétiques: L'art romantique, et autres oeuvres critiques*, ed. Henri Lemaître (Paris: Garnier, 1962), pp. 319–21; emphasis in the original.

25. Louis Figuier, *La photographie au Salon de 1859* [1860] and *La photographie & le stéréoscope* (New York: Arno, 1979), p. 6.

26. Beaumont Newhall, *The History of Photography: From 1839 to the Present*, rev. ed. (London: Secker & Warburg, 1982), p. 105.

27. Monique Sicard, *La fabrique du regard: Images de science et appareils de vision (XVe–XXe siècle)* (Paris: Jacob, 1998), pp. 139–45.

28. See, for example, Ludwig Schrank, "Negative-Retouche," *Photographische Correspondenz* 3 March 1866, pp. 152–54, and Anton Martin, *Handbuch der gesammten Photographie*, 6th ed. (Vienna: Gerold, 1865), pp. 443–68. Writing in 1893, the French astronomer Crépaux thought the principal advantage of new techniques of color photography for science, as opposed to portraiture, was that they made retouching impossible: "Science, which seeks the truth, will not complain, on the contrary; but coquetry will find this less to its liking." Crépaux, "La photographie en couleurs," *L'Astronomie* 12 (1893), p. 340.

29. Joel Snyder, "Res Ipsa Loquitur," in Lorraine Daston (ed.), *Things That Talk: Object Lessons from Art and Science* (New York: Zone Books, 2004), pp. 194–221.

30. Martin Kemp, "'A Perfect and Faithful Record': Mind and Body in Medical Photography Before 1900," in Ann Thomas (ed.), *Beauty of Another Order: Photography in Science* (New Haven: Yale University Press, 1997), pp. 120–49.

31. Jennifer Tucker, "Photography as Witness, Detective, and Impostor: Visual Representation in Victorian Science," in Bernard Lightman (ed.), *Victorian Science in Context* (Chicago: University of Chicago Press, 1997), pp. 378–408. Despite all evidence to the contrary, the mythology of the image as evidence untouched by human hands remained powerful. Even a critic as sophisticated as Roland Barthes, who was so quick to spot culture masquerading as nature in his acute analyses of modern myths, was in its thrall. Writing of a photograph of a former slave that he had seen as a child, he explains its effect as one not of "exactitude but of reality: the historian was no longer the mediator, slavery was presented [*donné*] without mediation, the fact was established *without method*." See Roland Barthes, *Mythologies* (Paris: Seuil, 1957); quote from Barthes, *La chambre claire: Note sur la photographie* (Paris: Gallimard, 1980), p. 125.

32. On spirit photography, see Andreas Fischer and Veit Loers (eds.), *Im Reich der Phantome: Fotografie des Unsichtbaren* (Ostfildern-Ruit, Germany: Cantz, 1997), and Clément Chéroux, *The Perfect Medium: Photography and the Occult* (New Haven: Yale University Press, 2005).

33. Eugène Trutat, *La photographie appliquée à l'histoire naturelle* (Paris: Gauthier-Villars, 1884), p. vii.

34. Anton Martin, *Handbuch der gesammten Photographie*, 6th ed. (Vienna: Gerold, 1865), p. 429.

35. The point of departure for discussions of "mechanical reproduction" and its impact on modern art is Walter Benjamin, *Das Kunstwerk im Zeitalter seiner technischen Reproduzierbarkeit: Drei Studien zur Kunstsoziologie* (Frankfurt-am-Main: Suhrkamp, 1963).

36. Eugene Ostroff, "Etching, Engraving and Photography: History of Photo-mechanical Reproduction," *Photographic Journal*, 109 (1969), pp. 560-77 and "Photography and Photogravure," *Journal of Photographic Science* 17 (1969), pp. 101-15; Ann Thomas, "The Search for Pattern," in Ann Thomas (ed.), *Beauty of Another Order: Photography in Science* (New Haven: Yale University Press, 1997), p. 79.

37. Ann Shelby Blum, *Picturing Nature: American Nineteenth-Century Zoological Illustration* (Princeton, NJ: Princeton University Press, 1993), pp. 279-81.

38. Gaston Tissandier, "Les progrès et les applications de l'héliogravure," *La nature*, no. 65 (1874), pp. 199-202.

39. On the semantic field of "mechanical" in English, see Christopher Hill, *Change and Continuity in Seventeenth-Century England* (Cambridge, MA: Harvard University Press, 1975), pp. 251-60; E.P. Thompson, *The Making of the English Working Class* (Harmondsworth: Penguin, 1968), pp. 259-62; in French, Georges Friedmann, "*L'Encyclopédie* et le travail humain," *Annales: Economies, Sociétés, Civilisations* 8 (1953), pp. 53-61; and in German, Otto Mayr, *Authority, Liberty, and Automatic Machinery in Early Modern Europe* (Baltimore: Johns Hopkins University Press, 1986), pp. 54-121. On Romantic attitudes toward machines, especially scientific instruments, see John Tresch, "Humboldt's Romantic Technologies," in David Aubin, Charlotte Bigg, and H. Otto Sibum (eds.), *The Heavens on Earth: Observatory Techniques in the Nineteenth-Century* (Durham, NC: Duke University Press, 2007).

40. Charles Babbage, "A Letter to Sir Humphry Davy, Bart., President of the Royal Society, on the application of machinery to the purpose of calculating and printing mathematical tables" [1822], *The Works of Charles Babbage*, ed. Martin Campbell-Kelly, vol. 2, *The Difference Engine and Table Making* (New York: New York University Press, 1989), p. 6.

41. Simon Schaffer, "Babbage's Intelligence: Calculating Engines and the Factory System," *Critical Inquiry* 21 (1994), pp. 203-27.

42. James Clerk Maxwell, "Atom," *The Scientific Papers of James Clerk Maxwell*, ed. W.D. Niven (New York: Dover, 1965), pp. 445-84. As David Cahan has shown, the Physikalisch-Technische Reichsanstalt in the then-recently unified Germany sought to impose the same level of standardization on scientific wares that the customs agency (Zollverein) had set for commercial wares, and international commissions all over Europe and North America convened to establish standard units of electricity and other physical quantities: David Cahan, *An Institute for an Empire: The Physikalisch-Technische Reichsanstalt, 1871-1918* (Cambridge: Cambridge University Press, 1989). Rudolf Virchow caught some of the cultural luster associated with standardization when he extolled "geistige Einheit" to the 1871 meeting of the Gesellschaft Deutscher Naturforscher und Ärzte shortly after German unification: "The task of the future, now that external unity of the Reich has been established, is to establish the inner unity ... the true unification of minds, putting the many members of the nation on a common intellectual footing." *Tageblatt der Versammlung*

Deutscher Naturforscher und Ärzte 44 (1871), p. 77. See also Simon Schaffer, "Late Victorian Metrology and Its Instrumentation: A Manufactory of Ohms," in Robert Bud and Susan E. Cozzens (eds.), *Invisible Connections: Instruments, Institutions, and Science* (Bellingham, WA: SPIE Optical Engineering Press, 1992), pp. 23–56.

43. Charles Babbage, *On the Economy of Machinery and Manufactures*, 4th ed. (London: Knight, 1835), p. 54.

44. Otto Funke, *Atlas of Physiological Chemistry* (London: Cavendish Society, 1853), pp. viii–ix.

45. Otto Funke, *Atlas of Physiological Chemistry* (London: Cavendish Society, 1853), quotations from pp. iv–v and 17.

46. Otto Funke, *Atlas of Physiological Chemistry* (London: Cavendish Society, 1853), p. vi.

47. Otto Funke, *Atlas of Physiological Chemistry* (London: Cavendish Society, 1853): "not a single line," p. x; "extraordinary fidelity," p. x–xi; "delicate and uniform," p. xi; "subjective condition," p. xi; "optical part" (emphasis added), p. xi.

48. William Anderson, "An Outline of the History of Art in Its Relation to Medical Science," *St. Thomas's Hospital Reports* 15 (1886), p. 170.

49. William Anderson, "An Outline of the History of Art in Its Relation to Medical Science," *St. Thomas's Hospital Reports* 15 (1886), p. 172.

50. William Anderson, "An Outline of the History of Art in Its Relation to Medical Science," *St. Thomas's Hospital Reports* 15 (1886), p. 175.

51. Quoted in Linda Nochlin, *Realism* (Harmondsworth: Penguin, 1971), p. 36.

52. Charles W. Cathcart and F.M. Caird, preface to *Johnston's Students' Atlas of Bones and Ligaments* (Edinburgh: Johnston, 1885), n.p.

53. Emil Ponfick, "Methode" (i.e., methodological preface), *Topographischer Atlas der medizinisch-chirurgischen Diagnostik* (Jena: Fischer, 1901), n.p.

54. Robert Hooke, *Micrographia, or, Some Physiological Descriptions of Minute Bodies Made by Magnifying Glasses, with Observations and Inquiries Thereupon* (London: Martyn and Allestry, 1665).

55. John Nettis, "An Account of a Method of Observing the Wonderful Configurations of the Smallest Shining Particles of Snow, with Several Figures of Them," *Philosophical Transactions* 49 (1755), pp. 646 and 648.

56. Edward Belcher, *The Last of the Arctic Voyages* (London: Reeve, 1855), pp. 300–301.

57. James Glaisher, "On the Severe Weather at the Beginning of the Year, and on Snow and Snow-Crystals," *Report of the Council of the British Meteorological Society: Read at the Fifth Annual General Meeting*, May 22, 1855 (London: n.p., 1855), pp. 16–30.

58. Gustav Hellmann, with microphotographs by Richard Neuhauss, *Schneekrystalle: Beobachtungen und Studien* (Berlin: Mückenberger, 1893), pp. 23–24.

59. Richard Neuhauss, *Lehrbuch der Mikrophotographie*, 3rd rev. ed. (Leipzig: Hirzel, 1907), pp. 200-201.

60. Gustav Hellmann, with microphotographs by Richard Neuhauss, *Schneekrystalle: Beobachtungen und Studien* (Berlin: Mückenberger, 1893), pp. 23-24.

61. On the history of soap bubbles, their physics, and their wider sociocultural place, see Simon Schaffer, "A Science Whose Business Is Bursting: Soap Bubbles as Commodities in Classical Physics," in Lorraine Daston (ed.), *Things That Talk: Object Lessons from Art and Science* (New York: Zone Books, 2004), pp. 147-94; on bullet photography and droplets, see Peter Galison, "Bullets, Splashes, Objectivity," Paper presented at Freie Universität, Berlin, 2006; Lord Rayleigh, "Some Applications of Photography," *Nature* 44 (1891), pp. 249-54; and R.S. Cole, "The Photography of the Splash of a Drop," *Nature* 50 (1894), p. 222. On Boys, see Graeme J.N. Gooday, "Sir Charles Vernon Boys," *Oxford Dictionary of National Biography* (Oxford: Oxford University Press, 2004), http://www.oxforddnb.com/view/article/32016.

62. Worthington said he had taken the term "objective view" from the British physicist, engineer, and deacon Frederick J. Smith. Though the work in question is surely Smith's "Photography of an Image by Reflection," *Nature* 47 (1892-1893), p. 10, Smith does not use the term there. But he does insist on the value of taking pictures by reflection and not merely shadow photographs — his method was to use convex mirrors rather than lenses, to gather more light than would otherwise have been possible. For the spark discharge, R.S. Cole cites Lord Rayleigh; for the technique of shadow photography, he invokes Sir Charles Verson Boys's bullets; see R.S. Cole, "The Photography of the Splash of a Drop," *Nature* 50 (1894), pp. 222-23; "objective views," p. 223.

63. All quotations from Arthur Worthington, *The Splash of a Drop* (London: Society for Promoting Christian Knowledge, 1895), pp. 74-75.

64. Karl von Bardeleben and Heinrich Haeckel, *Atlas der topographischen Anatomie des Menschen für Studierende und Ärzte* (Jena: Fischer, 1894), p. iv.

65. Karl von Bardeleben and Heinrich Haeckel, *Atlas der topographischen Anatomie des Menschen für Studierende und Ärzte* (Jena: Fischer, 1894), p. iv.

66. Wilhelm His, *Anatomie menschlicher Embryonen* (Leipzig: Vogel, 1880), p. 6.

67. Robert Koch, "Zur Untersuchung von pathogenen Organismen," *Mittheilungen aus dem Kaiserlichen Gesundheitsamte* 1 (1881), p. 10; Thomas Schlich, "'Wichtiger als der Gegenstand selbst' — Die Bedeutung des fotografischen Bildes in der Begründung der bakteriologischen Krankheitsauffassung durch Robert Koch," in Martin Dinges and Thomas Schlich (eds.), *Neue Wege in der Seuchengeschichte* (Stuttgart: Steiner, 1995), pp. 143-74.

68. Robert Koch, "Zur Untersuchung von pathogenen Organismen," *Mittheilungen aus dem Kaiserlichen Gesundheitsamte* 1 (1881), pp. 11-12.

69. Johannes Sobotta, *Atlas and Textbook of Human Anatomy*, ed. J. Playfair McMurrich (Philadelphia: Saunders, 1909), p. 13, emphasis added.

70. Johannes Sobotta, *Atlas und Grundriss der Histologie und mikroskopischen Anatomie des Menschen* (Munich: Lehmann, 1902), pp. vi–vii.

71. Francis Galton, "Composite Portraits," *Nature* 18 (1878), p. 97.

72. Francis Galton, "Composite Portraits," *Nature* 18 (1878); "mechanical precision," p. 97, "resemblance to all, . . . not more like to one of them than to another," p. 98; weights and family composition, p. 100.

73. Francis Galton, "Composite Portraits," *Nature* 18 (1878), p. 98. Galton was by no means alone in his hunt for automatic (objective) composites. When Wilhelm Weygandt aimed in 1902 to depict the facial expressions of the psychiatric patient, he too combined images in his atlas to achieve an effect that would "eliminate individual factors" of judgment. And he made it clear that his goal was to render his depictions in a manner that would be "as objective as possible." Wilhelm Weygandt, *Atlas und Grundriss der Psychiatrie* (Munich: Lehmann, 1902), pp. iv–v.

74. Erwin Christeller, *Atlas der Histotopographie gesunder und erkrankter Organe* (Leipzig: Georg Thieme, 1927).

75. Erwin Christeller, *Atlas der Histotopographie gesunder und erkrankter Organe* (Leipzig: Georg Thieme, 1927), p. 18.

76. Erwin Christeller, *Atlas der Histotopographie gesunder und erkrankter Organe* (Leipzig: Georg Thieme, 1927), p. 18.

77. Erwin Christeller, *Atlas der Histotopographie gesunder und erkrankter Organe* (Leipzig: Georg Thieme, 1927), p. 18.

78. Erwin Christeller, *Atlas der Histotopographie gesunder und erkrankter Organe* (Leipzig: Georg Thieme, 1927), p. 19.

79. E. Walter Maunder, *The Royal Observatory, Greenwich: A Glance at Its History and Work* (London: Religious Tract Society, 1900), pp. 176–77. We would like to thank Simon Schaffer for bringing this quotation to our attention; see his article, "Astronomers Mark Time: Discipline and the Personal Equation," *Science in Context* 2 (1988), pp. 115–45.

80. Hermann Pagenstecher and Carl Genth, *Atlas der pathologischen Anatomie des Augapfels* (Wiesbaden: Kreidel, 1875), p. vii, emphasis added.

81. Hermann Pagenstecher and Carl Genth, *Atlas der pathologischen Anatomie des Augapfels* (Wiesbaden: Kreidel, 1875), pp. vii–viii, emphasis added.

82. Eduard Jaeger, *Ophthalmoskopsicher Hand-Atlas*, rev. Maximilian Salzmann (Leipzig: Deuticke, 1894), pp. vi–viii.

83. Eduard Jaeger, *Ophthalmoskopsicher Hand-Atlas*, rev. Maximilian Salzmann (Leipzig: Deuticke, 1894), p. vii.

84. Eduard Jaeger, *Ophthalmoskopsicher Hand-Atlas*, rev. Maximilian Salzmann (Leipzig: Deuticke, 1894), p. viii. Moral-epistemic probity meant not succumbing to schematization or aesthetization. This theme was repeated over and over in atlases by many authors across a myriad of fields of inquiry. Jena's "physiological chemist" Carl Gotthelf Lehmann painstakingly analyzed animal fluids, judging that micro-

444

scopic images of crystals could yield the chemical content and provide a "purely objective handling" of specific topics. He had put every effort into presenting the morphotic objects in visual form (woodcuts); his aim was to show the crystals precisely as "armed eyes" (*bewaffnete Augen*) would grasp them. Such perceptual armament was needed, Lehmann cautioned: "Drawings of microscopic crystal forms have their own difficulties; for only too easily can one fall into idealization even in an otherwise correct representation." Temptation loomed — to darken a line, to make the mathematical form clearer or the three-dimensionality more manifest. C.G. Lehmann, *Handbuch der physiologischen Chemie*, 2nd ed. (Leipzig: Engelmann, 1859), pp. ix and x. The author of a 1900 atlas of pathological anatomy captured this same need for self-control when he promised to find "an absolute truth to nature" free of any "schematizing," avoiding the siren song that lured illustrators to combine different objects in one depiction. He wanted "visual fields that were really seen and were represented with the utmost precision." Hermann Dürck, *Atlas und Grundriss der speziellen pathologischen Histologie* (Munich: Lehmann, 1900–1901), vol. 1, p. viii.

85. M. Allen Starr, *Atlas of Nerve Cells* (New York: Macmillan, 1896), pp. v–vi.

86. Carl Fraenkel and Richard Pfeiffer, *Mikrophotographischer Atlas der Bakterienkunde* (Berlin: Hirschwald, 1887), p. 1.

87. Carl Fraenkel and Richard Pfeiffer, *Mikrophotographischer Atlas der Bakterienkunde* (Berlin: Hirschwald, 1887), pp. 2–3.

88. Carl Fraenkel and Richard Pfeiffer, *Mikrophotographischer Atlas der Bakterienkunde* (Berlin: Hirschwald, 1887), pp. 4–5.

89. K.B. Lehmann, *Atlas und Grundriss der Bakteriologie und Lehrbuch der speziellen bakteriologischen Diagnostik* (Munich: Lehmann, 1896), p. 2.

90. Percival Lowell, foreword to *Drawings of Mars, 1905* (n.p.: Lowell Observatory, 1906), n.p. For a brilliant exploration of the ambiguities of astronomical drawings in the nineteenth century, see Simon Schaffer, "On Astronomical Drawing," in Caroline A. Jones and Peter Galison (eds.), *Picturing Science, Producing Art* (New York: Routledge, 1998), pp. 441–74.

91. Percival Lowell, *Mars and Its Canals* (New York: Macmillan, 1906), cited in William Graves Hoyt, *Lowell and Mars* (Tucson: University of Arizona Press, 1976), pp. 179 and 182–85. As Hoyt notes, the pictures that were reproduced (in the *New York Times*, *Scientific American*, *Popular Astronomy*, and *Knowledge and Illustrated Scientific News*) all failed to show the lines of the canals; the pictures in one journal did: *The Scottish Review*. Lowell was sufficiently concerned by this fiasco that he personally brought the original photographs to show some of the more prominent astronomers.

92. William Graves Hoyt, *Lowell and Mars* (Tucson: University of Arizona Press, 1976), pp. 185 and 195–96.

93. Santiago Ramón y Cajal, *Cajal on the Cerebral Cortex: An Annotated Translation of the Complete Writings*, ed. Javier DeFelipe and Edward G. Jones (New York: Oxford University Press, 1988), p. 3.

94. Cited in Laura Otis's insightful reflection on Ramón y Cajal: *Membranes: Metaphors of Invasion in Nineteenth-Century Literature, Science, and Politics* (Baltimore: Johns Hopkins University Press, 1999), p. 77.

95. Santiago Ramón y Cajal, *Recollections of My Life*, trans. E. Horne Craigie with Juan Cano (Cambridge, MA: MIT Press, 1996), p. 335.

96. Santiago Ramón y Cajal, *Recollections of My Life*, trans. E. Horne Craigie with Juan Cano (Cambridge, MA: MIT Press, 1996), p. 337.

97. Santiago Ramón y Cajal, *Recollections of My Life*, trans. E. Horne Craigie with Juan Cano (Cambridge, MA: MIT Press, 1996), p. 338.

98. Santiago Ramón y Cajal, *Recollections of My Life*, trans. E. Horne Craigie with Juan Cano (Cambridge, MA: MIT Press, 1996), p. 338, emphasis added.

99. Johann Wolfgang von Goethe, "Der Versuch als Vermittler von Objekt und Subjekt" [1792, pub. 1823], *Goethes Werke*, vol. 13, *Naturwissenschaftliche Schriften*, ed. Dorothea Kuhn and Rike Wankmüller (Munich: Beck, 1975–1976), pp. 14–15; translated by Douglas Miller in *Scientific Studies* (New York: Suhrkamp, 1988), p. 14. On these rules, see Zeno G. Swijtink, "The Objectification of Observation: Measurement and Statistical Methods in the Nineteenth Century," in Lorenz Krüger, Lorraine J. Daston, and Michael Heidelberger (eds.), *The Probabilistic Revolution*, vol. 1, *Ideas in History* (Cambridge, MA: MIT Press, 1990), pp. 261–85; Simon Schaffer, "Astronomers Mark Time: Discipline and the Personal Equation," *Science in Context* 2 (1988), pp. 115–46; Richard R. Yeo, "Scientific Method and the Rhetoric of Science in Britain, 1830–1917," in John A. Schuster and Richard R. Yeo (eds.), *The Politics and Rhetoric of Scientific Method: Historical Studies* (Dordrecht, The Netherlands: Reidel, 1986), pp. 259–97.

100. Charles Rosen and Henri Zerner, *Romanticism and Realism: The Mythology of Nineteenth-Century Art* (New York: Viking, 1984), p. 108.

101. Charles Baudelaire, "Salon de 1859," *Curiosités esthetiques: L'art romantique, et autres oeuvres critiques*, ed. Henri Lemaître (Paris: Garnier, 1962), p. 329.

102. All quotations from Richard Neuhauss, *Lehrbuch der Mikrophotographie*, 2nd ed. (Brunswick, Germany: Bruhn, 1898), pp. 234–36.

103. Rudolf Virchow, "Die Freiheit der Wissenschaften im modernen Staatsleben," *Amtlicher Bericht über die Versammlung Deutscher Naturforscher und Ärtzte* 50 (1877), p. 74.

CHAPTER FOUR: THE SCIENTIFIC SELF

1. Wilhelm His, *Unsere Körperform und das physiologische Problem ihrer Entstehung* (Leipzig: Vogel, 1874), p. 171. On the evolutionary context of the embryological debate, see Robert J. Richards, *The Meaning of Evolution: The Morphological Construction and Ideological Reconstruction of Darwin's Theory* (Chicago: University of Chicago Press, 1992), pp. 91–166 and 171–80.

2. Ernst Haeckel, *Anthropogenie, oder Entwicklungsgeschichte des Menschen*, 4th

ed. (Leipzig: Engelmann, 1891), pp. 858–60. On Haeckel's illustrations, see Rein-
hard Gursch, *Die Illustrationen Ernst Haeckels zur Abstammungs- und Entwicklungsgeschichte: Diskussion im wissenschaftlichen und nichtwissenschaftlichen Schrifttum*
(Frankfurt am Main: Lang, 1981), and on Haeckel's ontogenetic principles more
generally, see Bernhard Kleeberg, *Theophysis: Ernst Haeckels Philosophie des Natur-
ganzen* (Cologne: Böhlau, 2005), pp. 130–69. For a detailed account of the protracted controversy over Haeckel's embryological illustrations, see Robert J.
Richards, *The Tragic Sense of Life: Ernst Haeckel and the Struggle over Evolutionary
Thought in Germany* (Chicago: University of Chicago Press, in press), ch. 8. We are
grateful to Professor Richards for allowing us to read this chapter in manuscript.

    3. Wilhelm His, *Anatomie menschlicher Embryonen* (Leipzig: Vogel, 1880), pp.
6–12. On His's techniques, see Nick Hopwood, "'Giving Body' to Embryos: Modeling, Mechanism, and the Microtome in Late Nineteenth–Century Anatomy," *Isis*
90 (1999), pp. 462–96, and "Producing Development: The Anatomy of Human
Embryos and the Norms of Wilhelm His," *Bulletin of the History of Medicine* 74
(2000), pp. 29–79. As Hopwood points out in the latter article, His himself was
striving for types in his *Normentafel*.

    4. Ernst Haeckel, *Freie Wissenschaft und freie Lehre: Eine Entgegnung auf Rudolf
Virchows Münchener Rede über "Die Freiheit der Lehre im modernen Staat,"* 2nd ed.
(Leipzig: Kröner, 1908), p. 18.

    5. Zeno G. Swijtink, "The Objectification of Observation: Measurement and
Statistical Methods in the Nineteenth Century," in Lorenz Krüger, Lorraine J. Das-
ton, and Michael Heidelberger (eds.), *The Probabilistic Revolution*, vol. 1, *Ideas in
History* (Cambridge, MA: MIT Press, 1990), pp. 261–86; see also Gerd Gigerenzer,
"Probabilistic Thinking and the Fight Against Subjectivity," in Lorenz Krüger, Gerd
Gigerenzer, and Mary S. Morgan (eds.), *The Probabilistic Revolution*, vol. 2, *Ideas in
the Sciences* (Cambridge, MA: MIT Press, 1990), pp. 11–34, for similar developments
in twentieth-century psychology.

    6. Frederic L. Holmes and Kathryn M. Olesko, "The Images of Precision:
Helmholtz and the Graphical Method in Physiology," in M. Norton Wise (ed.), *The
Values of Precision* (Princeton, NJ: Princeton University Press, 1995), pp. 198–221.
But for a contrasting view, see M. Norton Wise, *Bourgeois Berlin and Laboratory Sci-
ence* (in preparation), ch. 8.

    7. Karl Pearson, *The Grammar of Science* (London: Scott, 1892), pp. 6–8; see
also the acute analysis of Pearson's moral understanding of objectivity as renunciation in Theodore M. Porter, *Karl Pearson: The Scientific Life in a Statistical Age*
(Princeton, NJ: Princeton University Press, 2004), pp. 67–68, 267, 309–10.

    8. Adolf Trendelenburg, "Zur Geschichte des Wortes Person," *Kant-Studien* 13
(1908), pp. 1–17; Bruno Snell, *Die Entdeckung des Geistes: Studien zur Entstehung des
europäischen Denkens bei den Griechen* (Hamburg: Claaszen & Goverts, 1946); Marcel
Mauss, "Une catégorie de l'esprit humain: La notion de personne, celle de 'moi,'"

Journal of the Royal Anthropological Institute 68 (1938), pp. 236–81; Charles Taylor, *Sources of Self: The Making of the Modern Identity* (Cambridge, MA: Harvard University Press, 1989); Pierre Hadot, *Exercices spirituels et philosophie antique*, 2nd rev. ed. (Paris: Etudes augustiniennes, 1987); Michel Foucault, *Histoire de la sexualité*, vol. 3, *Le souci de soi* (Paris: Gallimard, 1984), and "L'herméneutique du sujet," *Résumé des cours, 1970–1982* (Paris: Julliard, 1989), pp. 145–66; Jerrold Siegel, *The Idea of the Self: Thought and Experience in Western Europe since the Seventeenth Century* (Cambridge: Cambridge University Press, 2005).

9. On these and other methodological problems in writing the history of the self, see the lucid discussion in Jan Goldstein, "Introduction," *The Post-Revolutionary Self: Politics and Psyche in France, 1750–1850* (Cambridge, MA: Harvard University Press, 2005), pp. 1–17.

10. Michel Foucault, "Subjectivité et vérité," *Résumé des cours, 1970–1982* (Paris: Julliard, 1989), p. 134.

11. Michel Foucault, "Writing the Self," in Arnold I. Davidson (ed.), *Foucault and His Interlocutors* (Chicago: University of Chicago Press, 1997), pp. 234–48; Arnold I. Davidson, "Ethics as Ascetics: Foucault, the History of Ethics, and Ancient Thought," in Jan Goldstein (ed.), *Foucault and the Writing of History* (Oxford: Blackwell, 1994), pp. 63–80.

12. Denis Diderot, *Le rêve de d'Alembert* [1769, pub. 1830], *Oeuvres complètes de Diderot*, ed. J Assézat (Paris: Garnier frères, 1875–77), vol. 2, pp. 163ff; translated by Francis Birrell as "D'Alembert's Dream," *Dialogues* (New York: Capricorn, 1969), pp. 71, 82–83, 88–89.

13. William James, *The Principles of Psychology* (1890; New York: Dover, 1950), vol. 1, pp. 297–98.

14. Although these two visions represent dominant views about the self, neither was without alternatives: for the Enlightenment, see Fernando Vidal, *Les sciences de l'âme: XVIe–XVIIIe siècle* (Paris: Honoré Champion, 2005), and for the late nineteenth century, see Katherine Arens, *Structures of Knowing: Psychologies of the Nineteenth Century* (Dordrecht, The Netherlands: Kluwer, 1989), and Jacqueline Carroy, *Hypnose, suggestion et psychologie: l'Invention de sujets* (Paris: Presses Universitaires de France, 1991).

15. Jan Goldstein, "Foucault and the Post-Revolutionary Self: The Uses of Cousinian Pedagogy in Nineteenth-Century France," in Jan Goldstein (ed.), *Foucault and the Writing of History* (Oxford: Blackwell, 1994), pp. 109–10.

16. Jean Senebier, *L'Art d'observer* (Geneva: Chez Philibert et Chirol, 1775), vol. 1, pp. 15–16.

17. See George Levine, *Dying to Know: Scientific Epistemology and Narrative in Victorian England* (Chicago: University of Chicago Press, 2002), for a perceptive analysis of how literature took up the scientific theme of self-elimination.

18. Arthur Schopenhauer, *Die Welt als Wille und Vorstellung*, ed. Arthur

Hübscher (1819, 1844, 1859; Stuttgart: Reclam, 1987), bk. IV, sec. 68, vol. 1, pp. 545–46.

19. Friedrich Nietzsche, "Vom Nutzen und Nachteil der Historie für das Leben" [1874], *Unzeitgemässe Betrachtungen*, 2nd ed., ed. Peter Pütz (Munich: Goldmann, 1992), p. 106; translated by R.J. Hollingdale as "On the Uses and Disadvantages of History for Life," *Untimely Meditations* (Cambridge: Cambridge University Press, 1983), pp. 86–87.

20. Lorraine Daston and H. Otto Sibum, "Introduction: Scientific Personae and Their Histories," *Science in Context* 16 (2003), pp. 1–8.

21. Peter Galison, "History from the Outside," *Cahiers parisiens/Parisian Notebooks* 1 (2004–5), University of Chicago in Paris Center.

22. See Chapter One for Kant's reformulation of the objective and the subjective.

23. See, for example, René Wellek, *Immanuel Kant in England, 1793–1838* (Princeton, NJ: Princeton University Press, 1931); Joachim Kopper, "La signification de Kant pour la philosophie française," *Archives de philosophie* 44 (1981), pp. 63–83; Frederick C. Beiser, *The Fate of Reason: German Philosophy from Kant to Fichte* (Cambridge, MA: Harvard University Press, 1987); François Azouvi and Dominique Bourel (eds.), *De Königsberg à Paris: La réception de Kant en France (1788–1804)* (Paris: Vrin, 1991); Rolf-Peter Horstmann, *Die Grenzen der Vernunft: Eine Untersuchung zu Zielen und Motiven des Deutschen Idealismus* (Frankfurt: Hain, 1991); and Sally Sedgwick (ed.), *The Reception of Kant's Critical Philosophy: Fichte, Schelling, and Hegel* (Cambridge: Cambridge University Press, 2000).

24. John Cottingham, *A Descartes Dictionary* (Oxford: Blackwell, 1993), s.v. "Objective Reality"; Louis N. Bescherelle, *Dictionnaire national, ou, Dictionnaire universel de la langue française* (Paris: Garnier frères, 1847–1848), s.v. "Objectif." On the changing definitions, see Michael Karskens, "The Development of the Opposition Subjective Versus Objective in the 18th Century," *Archiv für Begriffsgeschichte* 35 (1992), pp. 214–56.

25. See, for example, Theodor Heinsius, *Volksthümliches Wörterbuch der deutschen Sprache* (Hanover: Hahn, 1820), s.v. "Objektivität"; Louis N. Bescherelle, *Dictionnaire national, ou, Dictionnaire universel de la langue française* (Paris: Garnier frères, 1847–48), s.v. "Objectif"; Emile Littré, *Dictionnaire de la langue française* (Paris: Hachette, 1878), s.v. "Objectif, ive"; John Ogilvie, *The Imperial Dictionary of the English Language*, rev. Charles Annandale (London: Blackie & Son, 1871), s.v. "Objective." Cf. the eighteenth-century definition in Ephraim Chambers, *Cyclopaedia, or, An Universal Dictionary of Arts and Sciences* (London: James and John Knapton, 1728), s.v. "Objective": "Hence a thing is said to *exist* OBJECTIVELY, Objective when it exists no otherwise than in being known; or in being an Object of the Mind."

26. Johann Wolfgang von Goethe, *Theory of Colours* [1810], trans. Charles Lock Eastlake (1840; Cambridge, MA: MIT Press, 1970), p. 1n.

27. G.W.F. Hegel, *Enzyklopädie der philosophischen Wissenschaften im Grund-*

risse [1830], *Werke*, ed. Eva Moldenhauer and Karl Markus Michel (Frankfurt: Suhr-kamp, 1986), pt. 1, sec. 41, vol. 8, p. 115. For pre-Kantian philosophical usages, see Johann Nicolas Tetens, *Philosophische Versuche über die menschliche Natur und ihre Entwickelung* (Leipzig: Weidmanns Erben und Reich, 1777), pp. 363–67, and Johann Georg Heinrich Feder and Christoph Meiners (eds.), *Philosophische Biblio-thek* (Göttingen, Germany: Dietrich, 1788–1791), vol. 1, pp. 1–42.

28. Immanuel Kant, *Kritik der reinen Vernunft* [1781, 1787], ed. Raymund Schmidt (Hamburg: Meiner, 1926), A96, p.138a; A22–30/B37–45, pp. 66–73. On Kant's notion of *objektive Gültigkeit* more generally, see Henry E. Allison, *Kant's Transcendental Idealism: An Interpretation and Defense* (New Haven: Yale University Press, 1983), pp. 134–55, and Günter Zöller, *Theoretische Gegenstandsbeziehung bei Kant: Zur systematischen Bedeutung der Termini "objektive Realität" und "objektive Gültigkeit" in der "Kritik der reinen Vernunft"* (Berlin: Walter de Gruyter, 1984).

29. Immanuel Kant, *Kritik der reinen Vernunft* [1781, 1787], ed. Raymund Schmidt (Hamburg: Meiner, 1926), A106–14, pp. 153a–167a; B140–42, pp. 151b–155b.

30. Immanuel Kant, *Prolegomena to Any Future Metaphysics that Will Be Able to Come Forward as a Science* [1783], trans. Paul Carus, rev. James W. Ellington (Indi-anapolis: Hackett, 1977), sec. 19, p. 42.

31. Immanuel Kant, *Kritik der reinen Vernunft* [1781, 1787], ed. Raymund Schmidt (Hamburg: Meiner, 1926), B139–140; pp. 151b–152b.

32. Immanuel Kant, *Kritik der praktischen Vernunft* [1788], *Kants Werke* (1908; Berlin: Walter de Gruyter, 1968), vol. 5, p. 19.

33. Immanuel Kant, *Grundlegung zur Metaphysik der Sitten* [1785], ed. Theodor Valentiner (Stuttgart: Reclam, 2000), pp. 56–58. For an overview of the subsequent philosophical history of the Kantian subject, see Manfred Frank, "Subjectivity and Individuality: Survey of a Problem," in David E. Klemm and Günter Zöller (eds.), *Figuring the Self: Subject, Absolute, and Others in Classical German Philosophy* (Albany: State University of New York Press, 1997), pp. 3–30.

34. Adam Smith, *The Principles which Lead and Direct Philosophical Enquiries: Illustrated by the History of Astronomy* [1795], *The Works of Adam Smith, with an Account of his Life and Writings*, ed. Dugald Stewart (London: Cadell and Davies, 1811), vol. 5, pp. 188–89.

35. Rachel Laudan, "Histories of Science and Their Uses: A Review to 1913," *History of Science* 31 (1993), pp. 5–12.

36. John F.W. Herschel, *A Preliminary Discourse on the Study of Natural Philoso-phy* [1830] (Chicago: University of Chicago Press, 1987), pp. 360–61.

37. Alexander von Humboldt, *Kosmos* [1845–1862], ed. Bernhard von Cotta (Stuttgart: J.G. Cotta, 1874), vol. 1, p. xxiv. William James also dated this realiza-tion to *circa* 1850: "Up to about 1850 almost everyone believed that sciences expressed truths that were exact copies of a definite code of non-human realities. But the enormously rapid multiplication of theories in these latter days has well-

nigh upset the notion of any one of them being a more literally objective kind of thing than another." William James, "Humanism and Truth" [1904], *The Meaning of Truth*, ed. Fredson Bowers (Cambridge, MA: Harvard University Press, 1975), p. 40.

38. Charles Delaunay, *Rapport sur le progrès de l'astronomie* (Paris: Imprimerie impériale, 1867), p. 14.

39. Henri Poincaré, *Les méthodes nouvelles de la mécanique céleste* (Paris: Gauthier-Villars, 1892–1899), vol. 1, pp. 3–4.

40. J.C. Poggendorff, *Geschichte der Physik* (Leipzig: Barth, 1879), p. 643; Jed Z. Buchwald, *The Rise of the Wave Theory of Light: Optical Theory and Experiment in the Early Nineteenth Century* (Chicago: University of Chicago Press, 1989).

41. Henry Adams, *The Education of Henry Adams: An Autobiography* (1907; Boston: Houghton Mifflin, 1961), pp. 495 and 497.

42. Lorraine Daston, "The Historicity of Science," in Glenn W. Most (ed.), *Historicization-Historisierung* (Göttingen, Germany: Vandenhoeck & Ruprecht, 2001), pp. 201–21.

43. Thomas Henry Huxley, "The Progress of Science" [1887], *Methods and Results: Essays* (London: Macmillan, 1893), p. 65.

44. Hermann von Helmholtz, "Die Thatsachen in der Wahrnehmung" [1878], *Vorträge und Reden*, 4th ed. (Brunswick, Germany: Vieweg und Sohn, 1896), vol. 2, pp. 224–25.

45. Claude Bernard, *Cahier de notes, 1850–1860: Edition intégrale du "Cahier rouge*," ed. Mirko Drazen Grmek (Paris: Gallimard, 1965), p. 184, and Bernard, "Notes on Comte's *Cours de philosophie positive*," [ca. 1865–1866], *Philosophie: Manuscript inédit*, ed. Jacques Chevalier (Paris: Boivin, 1937), p. 43.

46. Thomas Henry Huxley, "Notebook: 'Thoughts and Doings'" [1842], in Charles Darwin and Thomas Henry Huxley, *Autobiographies*, ed. Gavin De Beer (Oxford: Oxford University Press, 1983), pp. 95–96.

47. Claude Bernard, "Notes on Tennemann's *Manuel de l'histoire de la philosophie*," *Philosophie: Manuscript inédit*, ed. Jacques Chevalier (Paris: Boivin, 1937), p. 22.

48. Michael Heidelberger, "Force, Law, and Experiment: The Evolution of Helmholtz's Philosophy of Science," in David Cahan (ed.), *Hermann von Helmholtz and the Foundations of Nineteenth-Century Science* (Berkeley: University of California Press, 1993), pp. 461–97.

49. Hermann von Helmholtz, "Ueber das Ziel und die Fortschritte der Naturwissenschaft" [1869], *Vorträge und Reden*, 4th ed. (Brunswick, Germany: Vieweg und sohn, 1896), vol. 1, p. 376.

50. Hermann von Helmholtz, *Einleitung zu den Vorlesungen über theoretische Physik*, ed. Arthur König and Carl Runge (Leipzig: Barth, 1903), p. 14.

51. Thomas Henry Huxley, "A Liberal Education and Where to Find It" [1868], *The Major Prose of Thomas Henry Huxley*, ed. Alan P. Barr (Athens: University of Georgia Press, 1997), p. 209.

52. Santiago Ramón y Cajal, *Advice for a Young Investigator*, trans. Neely Swanson and Larry W. Swanson (1897; Cambridge, MA: MIT Press, 1999), p. 86.

53. On the changing portrayals of Isaac Newton, see Patricia Fara, *Newton: The Making of Genius* (London: Macmillan, 2002); on portraits of scientists more generally, see Ludmilla Jordanova, *Defining Features: Scientific and Medical Portraits, 1660–2000* (London: Reaktion, 2000); Régine Pietra, *Sage comme une image: Figures de la philosophie dans les arts* (Paris: Félin, 1992); and Claudia Valter, "Gelehrte Gesellschaft: Wissenschaftler und Erfinder im Porträt," in Hans Holländer (ed.), *Erkenntnis, Erfindung, Konstruktion: Studien zur Bildgeschichte von Naturwissenschaften und Technik vom 16. bis zum 19. Jahrhundert* (Berlin: Mann, 2000), pp. 833–59.

54. Paolo Frisi, "Elogio" [1778], in A. Rupert Hall, *Isaac Newton: Eighteenth-Century Perspectives* (Oxford: Oxford University Press, 1999), pp. 121, 146, 156, 166, 171.

55. Henry C. Ewart, *Heroes and Martyrs of Science* (London: Ibister, 1886), pp. 151–52, 154, 169.

56. Frank E. Manuel, *A Portrait of Isaac Newton* (Washington, DC: New Republic Books, 1979), pp. 193–95; Richard S. Westfall, *The Life of Isaac Newton* (Cambridge: Cambridge University Press, 1993), pp. 105, 110, 161.

57. Diogenes Laertius, *Lives of Eminent Philosophers*, trans. R.D. Hicks (London: Heinemann, 1925).

58. Charles Bazerman, *Shaping Written Knowledge: The Genre and Activity of the Experimental Article in Science* (Madison: University of Wisconsin Press, 1988).

59. Dorinda Outram, "The Language of Natural Power: The 'Eloges' of Georges Cuvier and the Public Language of Nineteenth-Century Science," *History of Science* 16 (1978), pp. 153–78; Charles B. Paul, *Science and Immortality: The Eloges of the Paris Academy of Sciences (1699–1791)* (Berkeley: University of California Press, 1980); Michael Shortland and Richard Yeo (eds.), *Telling Lives in Science: Essays in Scientific Biography* (Cambridge: Cambridge University Press, 1996); Christopher Lawrence and Steven Shapin (eds.), *Science Incarnate: Historical Embodiments of Natural Knowledge* (Chicago: University of Chicago Press, 1998).

60. John Locke, *An Essay Concerning Human Understanding* [1690], ed. Peter H. Nidditch (Oxford: Oxford University Press, 1979), II.xxvii.9, p. 335. This chapter was added to the second edition, published in 1694.

61. David Hume, *A Treatise of Human Nature* [1739], ed. L.A. Selby-Bigge (1888; Oxford: Clarendon, 1975), I.iv.6, p. 252.

62. John Locke, *An Essay Concerning Human Understanding* [1690], ed. Peter H. Nidditch (Oxford: Oxford University Press, 1979), II.xxvii.17, p. 341; II.xxvii.20, p. 342; see Christopher Fox, *Locke and the Scriblerians: Identity and Consciousness in Early Eighteenth-Century Britain* (Berkeley: University of California Press, 1988), on early eighteenth-century British debates over Locke's theory of personal identity.

63. Ute Mohr, *Melancholie und Melancholiekritik im England des 18. Jahrhunderts* (Frankfurt am Main/Bern: Lang, 1990), pp. 22–26; T.H. Jobe, "Medical Theories of

Melancholia in the Seventeenth and Early Eighteenth Centuries," *Clio Medica* 2 (1976), pp. 7–31; Esther Fischer-Homberger, "Hypochondriasis of the Eighteenth Century — Neurosis of the Present Century," *Bulletin of the History of Medicine* 46 (1972), pp. 391–401.

64. Etienne Bonnot de Condillac, *Essai sur l'origine des connaissances humaines* [1746], *Oeuvres philosophiques*, ed. Georges Le Roy (Paris: Presses Universitaires de France, 1947), I.ii.ix. 86–88, vol. 1, p. 31. On Enlightenment ambivalence toward the faculty of the imagination, which was seen as at once essential for synthesizing sensations and dangerous as a source of seductive fantasies, see Lorraine Daston, "Fear and Loathing of the Imagination in Science," *Daedalus* 127 (1998), pp. 73–85; Jessica Riskin, *Science in the Age of Sensibility: The Sentimental Empiricists of the French Enlightenment* (Chicago: University of Chicago Press, 2002), pp. 212–14; and Jan Goldstein, *The Post-Revolutionary Self: Politics and Psyche in France, 1750–1850* (Cambridge, MA: Harvard University Press, 2005), pp. 21–59.

65. Raymond Martin and John Barresi, *Naturalization of the Soul: Self and Personal Identity in the Eighteenth Century* (London: Routledge, 2000).

66. Jean A. Perkins, *The Concept of the Self in the French Enlightenment* (Geneva: Droz, 1969), p. 67.

67. Georg Christoph Lichtenberg, *Sudelbücher*, Heft K, no. 162, quoted in Helmut Pfotenhauer, "Sich selber schreiben: Lichtenbergs fragmentarisches Ich," *Um 1800: Konfigurationen der Literatur, Kunstliteratur und Ästhetik* (Tübingen: Niemeyer, 1991), p. 13.

68. Etienne Bonnot de Condillac, *Essai sur l'origine des connaissances humaines* [1746], *Oeuvres philosophiques*, ed. Georges Le Roy (Paris: Presses Universitaires de France, 1947), I.ii.10.89, vol. 1, p. 32.

69. Georges Cuvier, *Recueil des éloges historiques lus dans les séances publiques de l'Institut de France* (Paris: Levrault, 1819–1827), vol. 3, p. 180. See also Richard W. Burkhardt Jr., *The Spirit of System: Lamarck and Evolutionary Biology* (Cambridge, MA: Harvard University Press, 1995), pp. 61–62 and 196–97. Cuvier made similar objections to Georges Leclerc de Buffon's work in natural history: impelled by "les efforts de l'imagination, sans démonstration et sans analyse," Buffon had produced "un ouvrage, dont presque partout le fond et la forme sont également admirables, d'une foule de ces hypothèses vagues, de ces systèmes fantastiques qui n'a servent qu'à le déparer" (vol. 3, p. 297).

70. Samuel Johnson, *The History of Rasselas, Prince of Abissinia* [1759], ed. J.P. Hardy (Oxford: Oxford University Press, 1968), p. 114.

71. Etienne Bonnot de Condillac, *Traité des systèmes* [1749], ed. Francine Markovits and Michel Authier (Paris: Fayard, 1991), p. 27.

72. Samuel-Auguste Tissot, *De la santé des gens de lettres* [1768], (Paris: Editions de la différence, 1991), pp. 45 and 55.

73. Voltaire, "Imagination, Imaginer," in Jean Le Rond d'Alembert and Denis

Diderot, *Encyclopédie, ou, Dictionnaire raisonné des sciences, des arts et des métiers* (1751–1780) (Stuttgart: Frommann, 1988), vol. 8, p. 561.

74. Lorraine Daston, "Strange Facts, Plain Facts, and the Texture of Scientific Experience in the Enlightenment," in Suzanne Marchand and Elizabeth Lunbeck (eds.), *Proof and Persuasion: Essays on Authority, Objectivity, and Evidence* (Turnhout, Belgium: Brepols, 1996), pp. 42–59.

75. Marie-Jean-Antoine de Caritat Condorcet, "Eloge de Mariotte," *Eloges des académiciens de l'Académie royale des sciences, morts depuis 1666, jusqu'en 1699* (Paris: Hôtel de Thou, 1773), pp. 62–63.

76. Charles Du Fay, "Mémoire sur un grand nombre de phosphores nouveaux," *Mémoires de l'Académie Royale des Sciences,* Année 1730, p. 527.

77. Denis Diderot, "Le rêve de d'Alembert" [1769, pub. 1830], *Oeuvres complètes de Diderot,* ed. J Assézat (Paris: Garnier frères, 1875–77), vol. 2, pp. 163–64; translated by Jacques Barzun and Ralph H. Bowen as "D'Alembert's Dream," in *Rameau's Nephew and Other Works* (New York: Macmillan, 1956), p. 148.

78. Wilhelm von Humboldt to Friedrich Schiller, June 23, 1798, in Siegfried Seidel (ed.), *Der Briefwechsel zwischen Friedrich Schiller und Wilhelm von Humboldt* (Berlin: Aufbau, 1962), pp. 153–58; quoted in François Azouvi and Dominique Bourel (eds.), *De Königsberg à Paris: La réception de Kant en France (1788–1804)* (Paris: Vrin, 1991), p. 110. Among those present at the seminar were Antoine Louis Claude Destutt de Tracy, Pierre Jean George Cabanis, and Emmanuel Joseph Sieyès.

79. René Wellek, *Immanuel Kant in England, 1793–1838* (Princeton, NJ: Princeton University Press, 1931), p. 82.

80. For the French case, see Jan Goldstein, *The Post-Revolutionary Self: Politics and Psyche in France, 1750–1850* (Cambridge, MA: Harvard University Press, 2005).

81. George Gore, *The Art of Scientific Discovery, or, The General Conditions and Methods of Research in Physics and Chemistry* (London: Longmans, Green, and Co., 1878), pp. 60–61.

82. Philip Gilbert Hamerton, *The Intellectual Life* (London: Macmillan and Co., 1873), p. 57.

83. George L. Craik, *The Pursuit of Knowledge under Difficulties* [1845], rev. ed. (London: Murray, 1858), p. 10.

84. Samuel Smiles, *Self-Help* (London: Murray, 1869), p. 95.

85. See also the character traits (diligence, industry) singled out by Alexander von Humboldt in his letters of recommendation for various academic posts to the Prussian Ministry of Culture: Alexander von Humboldt, *Vier Jahrzehnte Wissenschaftsförderung: Briefe an das preußische Kultusministerium 1818–1859,* ed. Kurt-R. Biermann (Berlin: Akademie, 1985), pp. 40, 49, 93, 96, 120, 138, 164.

86. Gaston Tissandier, *Les martyrs de la science* (Paris: Dreyfous, 1882), p. 2; compare David Brewster, *The Martyrs of Science, or, The Lives of Galileo, Tycho Brahe, and Kepler* (New York: Harper & Bros., 1841).

87. Samuel Smiles, *Self-Help* (London: IEA Health and Welfare Unit, 1997), p. 78.

88. Dorothea Goetz (ed.), *Hermann von Helmholtz über sich selbst: Rede zu seinem 70. Geburtstag* (Leipzig: Teubner, 1966), p. 13.

89. Charles Darwin, *The Autobiography of Charles Darwin, 1809–1882: With Original Omissions Restored,* ed. Nora Barlow (New York: Norton, 1969), pp. 140–41.

90. Thomas Henry Huxley, "A Liberal Education and Where to Find It" [1868], *The Major Prose of Thomas Henry Huxley*, ed. Alan P. Barr (Athens: University of Georgia Press, 1997), p. 209.

91. Simon Schaffer, "Astronomers Mark Time: Discipline and the Personal Equation," *Science in Context* 2 (1988), pp. 115–45, and "Babbage's Intelligence: Calculating Engines and the Factory System," *Critical Inquiry* 21 (1994), pp. 203–27; Crosbie Smith and M. Norton Wise, *Energy and Empire: A Biographical Study of Lord Kelvin* (Cambridge: Cambridge University Press, 1989).

92. Paul White, *Thomas Henry Huxley: Making the "Man of Science"* (Cambridge: Cambridge University Press, 2003); Andreas W. Daum, *Wissenschaftspopularisierung im 19. Jahrhundert: Bürgerliche Kultur, naturwissenschaftliche Bildung und die deutsche Öffentlichkeit, 1848–1914* (Munich: Oldenbourg, 1998).

93. Christopher Lawrence, "Incommunicable Knowledge: Science, Technology and the Clinical Art in Britain, 1850–1914," *Journal of Contemporary History* 20 (1985), pp. 503–20; Theodore M. Porter, *Trust in Numbers: The Pursuit of Objectivity in Science and Public Life* (Princeton, NJ: Princeton University Press, 1995), pp. 89–113.

94. Quoted in Samuel Smiles, *Character* (London: Murray, 1871), p. 168.

95. George L. Craik, *The Pursuit of Knowledge under Difficulties* [1845], rev. ed. (London: Murray, 1858), p. 150.

96. Karl Pearson, *The Grammar of Science* (London: Walter Scott, 1892), p. 38. On Pearson himself, see Theodore M. Porter, *Karl Pearson: The Scientific Life in a Statistical Age* (Princeton, NJ: Princeton University Press, 2004), p. 308.

97. Ernest Renan, *L'Avenir de la science* (comp. 1848; Paris: Calmann-Levy, 1890), p. 235.

98. Friedrich Nietzsche, "On the Uses and Disadvantages of History for Life" [1874], *Untimely Meditations*, trans. R.J. Hollingdale (Cambridge: Cambridge University Press, 1983), p. 105.

99. Dorothea Goetz (ed.), *Hermann von Helmholtz über sich selbst: Rede zu seinem 70. Geburtstag* (Leipzig: Teubner, 1966), p. 13; Walter B. Cannon, *The Way of an Investigator: A Scientist's Experiences in Medical Research* (New York: Norton, 1945), p. 61.

100. Etienne Bonnot de Condillac, *Traité des sensations* [1754], *Oeuvres philosophiques,* ed. Georges Le Roy (Paris: Presses Universitaires de France, 1947), I.iv.6.2–3, vol. 1, p. 238.

101. Charles Bonnet to Spallanzani, Sept. 14, 1765, *Oeuvres d'histoire naturelle et de philosophie* (Neuchâtel: Fauche, 1779–1783), vol. 5 (1781), p. 10.

102. Benjamin-Samuel-Georges Carrard, *Essai qui a remporté le prix de la Société Hollandoise des Sciences de Haarlem en 1770: Sur cette question, Qu'est-ce qui est requis dans l'art d'observer; & jusques-où cet art contribue-t-il à perfectionner l'entendement?* (Amsterdam: Marc-Michel Rey, 1777), pp. 10–14. On the *coup d'oeil*: Valeria Pansini, "L'oeil du topographe et la science de la guerre: Travail scientifique et perception militaire (1760–1820)," Thèse, Ecole des Hautes Etudes en Sciences Sociales, Paris, 2002.

103. Robert A. Fothergill, *Private Chronicles: A Study of English Diaries* (London: Oxford University Press, 1974).

104. David Hume, *A Treatise of Human Nature* [1739], ed. L.A. Selby-Bigge (1888; Oxford: Clarendon, 1975), I.iv.6, p. 262.

105. John Locke, "A Register of the Weather for the Year 1692, Kept at Oates in Essex," *Philosophical Transactions* 24 (1704), pp. 1917–37; Gustav Hellmann, "Die Entwicklung der meteorologischen Beobachtungen in Deutschland von den ersten Anfängen bis zur Einrichtung staatlicher Beobachtungsnetze," *Abhandlungen der Preussischen Akademie der Wissenschaften, Physikalisch-Mathematische Klasse* 1 (1926), pp. 1–25; Vladimir Jankovic, *Reading the Skies: A Cultural History of the English Weather, 1650–1820* (Manchester: University of Manchester Press, 2000); Arthur Sherbo, "The English Weather, *The Gentleman's Magazine*, and the Brothers White," *Archives of Natural History* 12 (1985), pp. 23–29.

106. Sibylle Schönborn, *Das Buch der Seele: Tagebuchliteratur zwischen Aufklärung und Kunstperiode* (Tübingen: Niemeyer, 1999), pp. 59 and 276.

107. Georg Christoph Lichtenberg, "Tagebuch 1770–1772," *Schriften und Briefe*, ed. Wolfgang Promies (Munich: Hanser, 1971), vol. 2, p. 611.

108. Albrecht von Haller to Charles Bonnet, May 18, 1777, in Raymond Savioz (ed.), *Mémoires autobiographiques de Charles Bonnet de Genève* (Paris: Vrin, 1948), p. 108.

109. On early psychological theories of attention, see Gary Hatfield, "Attention in Early Scientific Psychology," in Richard D. Wright (ed.), *Visual Attention* (New York: Oxford University Press, 1998), pp. 3–25.

110. Hans Poser, "Die Kunst der Beobachtung: Zur Preisfrage der Holländischen Akademie von 1768," in Hans Poser (ed.), *Erfahrung und Beobachtung: Erkenntnistheoretische und Wissenschaftshistorische Untersuchungen zur Erkenntnisbegründung* (Berlin: Technische Universität Berlin, 1992), pp. 99–119; J.G. de Bruijn, *Inventaris van de prijsvragen uitgeschreven door de Hollandsche Maatschappij der Wetenschappen 1753–1917* (Haarlem: Hollandsche Maatschappij der Wetenschappen, 1977).

111. Jean Senebier, *L'Art d'observer* (Geneva: Chez Philibert et Chirol, 1775). Senebier's treatise is dedicated to the Dutch Society of Sciences, which he thanks for its "approbation." The Society's prize was won by Benjamin-Samuel-Georges Carrard, *Essai qui a remporté le prix de la Société Hollandoise des Sciences de Haarlem en 1770: Sur cette question, Qu'est-ce qui est requis dans l'art d'observer; & jusques-où cet*

art contribue-t-il à perfectionner l'entendement? (Amsterdam: Marc-Michel Rey, 1777).

112. Jean Senebier, *L'Art d'observer* (Geneva: Chez Philibert et Chirol, 1775), vol. 1, pp. 15–16.

113. Charles Bonnet, *Essai analytique sur les facultés de l'âme* [1759], *Oeuvres d'histoire naturelle et de philosophie* (Neuchâtel: Fauche, 1779–1783), vol. 6, p. vii; cf. p. 134.

114. Charles Bonnet, "Observations diverses sur les insectes," *Oeuvres d'histoire naturelle et de philosophie* (Neuchâtel: Fauche, 1779–1783), vol. 2, pp. 212–13.

115. Charles Bonnet, *Traité d'insectologie, ou, Observations sur les pucerons* [1745], *Oeuvres d'histoire naturelle et de philosophie* (Neuchâtel: Fauche, 1779–1783), vol. 1, p. 21.

116. Jean Wüest, "Bonnet face aux insectes," in Marino Buscaglia (ed.), *Charles Bonnet: Savant et philosophe, 1720–1793* (Geneva: Passé Présent, 1994), p. 153.

117. Jean Senebier, *L'Art d'observer* (Geneva: Chez Philibert et Chirol, 1775), vol. 2, p. 15.

118. Jean Senebier, *L'Art d'observer* (Geneva: Chez Philibert et Chirol, 1775), vol. 1, pp. 145–49.

119. Charles Bonnet, *Essai analytique sur les facultés de l'âme* [1759], *Oeuvres d'histoire naturelle et de philosophie* (Neuchâtel: Fauche, 1779–1783), vol. 6, p. 119.

120. Charles Bonnet, *Essai analytique sur les facultés de l'âme* [1759], *Oeuvres d'histoire naturelle et de philosophie* (Neuchâtel: Fauche, 1779–1783), vol. 6, p. 124.

121. René-Antoine Ferchault de Réaumur, *Mémoires pour servir à l'histoire des insectes* (Paris: Imprimerie royale, 1734–1742), vol. 1, p. 10.

122. Charles Bonnet, *Traité d'insectologie, ou, Observations sur les pucerons* [1745], *Oeuvres de l'histoire naturelle et de philosophie* (Neuchâtel: Fauche, 1779–1783), vol. 1, p. 36.

123. Jean Senebier, *L'Art d'observer* (Geneva: Chez Philibert et Chirol, 1775), vol. 1, p. 5.

124. See, for example, Jean de La Bruyère, *Les Caractères de Théophraste: Traduits du Grec avec Les caractères ou Les moeurs de ce siècle*, ed. Robert Pignarre (Paris: Garnier-Flammarion, 1965), pp. 337–38.

125. See, for example, Théodule Ribot, *Psychologie de l'attention* (Paris: Alcan, 1889), pp. 95–113; Josef Clemens Kreibig, *Die Aufmerksamkeit als Willenserscheinung* (Vienna: Hölder, 1897), pp. 1–12 and 49–66; and Jean-Paul Nayrac, *Physiologie et psychologie de l'attention: Evolution, dissolution, rééducation, éducation* (Paris: Alcan, 1906), pp. 73–97, for surveys of the contemporary literature on the relationship between attention and will; see also the perspicuous discussion in Jonathan Crary, *Suspensions of Perception: Attention, Spectacle, and Modern Culture* (Cambridge, MA: MIT Press, 1999), pp. 42–48.

126. William James, *The Principles of Psychology* (1890; New York: Dover, 1950), vol. 2, p. 564.

127. David Braunschweiger, *Die Lehre von der Aufmerksamkeit in der Psychologie des 18. Jahrhunderts* (Leipzig: Haacke, 1899), pp. 50–96.

128. Théodule Ribot, *Psychologie de l'attention* (Paris: Alcan, 1889), pp. 60–63.

129. Jean-Paul Nayrac, *Physiologie et psychologie de l'attention: Evolution, dissolution, reéducation, éducation* (Paris: Alcan, 1906), pp. x–xi.

130. Hermann von Helmholtz, "Über das Verhältnis der Naturwissenschaften zur Gesammtheit der Wissenschaft" [1862], *Vorträge und Reden*, 5th ed. (Brunswick, Germany: Vieweg und Sohn, 1903), vol. 1, p. 178.

131. Benjamin-Samuel-Georges Carrard, *Essai qui a remporté le prix de la Société Hollandoise des Sciences de Haarlem en 1770: Sur cette question, Qu'est-ce qui est requis dans l'art d'observer; & jusques-où cet art contribue-t-il à perfectionner l'entendement?* (Amsterdam: Marc-Michel Rey, 1777), p. 245.

132. Claude Bernard, *Introduction à l'étude de la médecine expérimentale* [1865], ed. François Dagnognet (Paris: Garnier-Flammarion, 1966), p. 53.

133. H. Otto Sibum, "Narrating by Numbers: Keeping an Account of Early Nineteenth-Century Laboratory Experiences," in Frederic L. Holmes, Jürgen Renn, and Hans-Jörg Rheinberger (eds.), *Reworking the Bench: Research Notebooks in the History of Science* (Dordrecht, The Netherlands: Kluwer, 2003), pp. 141–58.

134. Christoph Hoffmann, "The Pocket Schedule: Note-taking as a Research Technique; Ernst Mach's Ballistic-Photographic Experiments," in Frederic L. Holmes, Jürgen Renn, and Hans-Jörg Rheinberger (eds.), *Reworking the Bench: Research Notebooks in the History of Science* (Dordrecht, The Netherlands: Kluwer, 2003), pp. 183–202.

135. Michael Faraday, *Chemical Manipulation* (1827; New York: Wiley, 1974), p. 546.

136. Friedrich Steinle, "The Practice of Studying Practice: Analyzing Research Records of Ampère and Faraday," in Frederic L. Holmes, Jürgen Renn, and Hans-Jörg Rheinberger (eds.), *Reworking the Bench: Research Notebooks in the History of Science* (Dordrecht, The Netherlands: Kluwer, 2003), pp. 93–118.

137. Michael Faraday, *Faraday's Diary*, ed. Thomas Martin (London: Bell and Sons, 1934), vol. 5, p. 162.

138. Charles Baudelaire, "Salon de 1859," *Curiosités esthétiques: L'art romantique, et autres oeuvres critiques*, ed. Henri Lemaître (Paris: Garnier, 1962), pp. 320–21; emphasis in the original.

139. Félix Bracquemond, *Du dessin et de la couleur* (Paris: Charpentier, 1885), p. 221.

140. George Sand, *Valvèdre* (1861; Paris: Michel Levy frères, 1875), p. 334.

141. Santiago Ramón y Cajal, *Vacation Stories: Five Science Fiction Tales*, trans. Laura Otis (Urbana: University of Illinois Press, 2001), p. 199.

142. Ernst Haeckel, *Monographie der Medusen* (Jena: Fischer, 1879–1881), and *Kunstformen der Natur* (Leipzig: Verlag des Bibliographischen Instituts, 1899–1904).

143. Hermann von Helmholtz, "Ueber Goethe's naturwissenschaftliche Arbeiten" [1853], and "Goethe's Vorahnungen kommender naturwissenschaftlichen

Ideen" [1892], *Vorträge und Reden*, 5th ed. (Brunswick, Germany: Vieweg und Sohn, 1903), vol. 1, pp. 23–45 and 335–61. Different as these two lectures are in many respects, they concur in their portrayal of Goethe's science as the work of an artist.

144. Friedrich Nietzsche, "On the Uses and Disadvantages of History for Life" [1874], *Untimely Meditations*, trans. R.J. Hollingdale (Cambridge: Cambridge University Press, 1983), p. 93.

145. E.g. the algebraic atlas of Christoph Jansen, Klaus Lux, Richard Parker, and Robert Wilson, *An Atlas of Brauer Characteristics* (Oxford: Clarendon Press, 1995).

CHAPTER FIVE: STRUCTURAL OBJECTIVITY

1. Hermann von Helmholtz, "Ueber das Ziel und die Fortschritte der Naturwissenschaft" [1869], *Vorträge und Reden*, 4th ed. (Brunswick, Germany: Vieweg und Sohn, 1896), vol. 1, pp. 367–98, on p. 393. On the context of Helmholtz's sign theory, see Timothy Lenoir, "The Politics of Vision: Optics, Painting, and Ideology in Germany, 1845–95," *Instituting Science: The Cultural Production of Scientific Disciplines* (Stanford, CA: Stanford University Press, 1997), pp. 131–78.

2. Hermann von Helmholtz, "Die Thatsachen in der Wahrnehmung" [1878], *Vorträge und Reden*, 4th ed. (Brunswick, Germany: Vieweg und Sohn, 1896), vol. 2, pp. 224–25.

3. We here follow Michael Friedman's coinage in his account of Carnap's position on objectivity: Michael Friedman, *Reconsidering Logical Positivism* (Cambridge: Cambridge University Press, 1999), p. 96n.

4. Max Planck, *Acht Vorlesungen über theoretische Physik: Gehalten an der Columbia University in the City of New York im Frühjahr 1909* (Leipzig: Hirzel, 1910), p. 6.

5. Rudolf Carnap, *Der logische Aufbau der Welt: Scheinprobleme in der Philosophie*, 2nd ed. [1928] (Hamburg: Meiner, 1961), sec. 3, p. 3. Carnap studied with Frege at the University of Jena; his student notes reveal how deeply he was steeped in Fregean logic: on Frege's influence, see Rudolf Carnap, *Frege's Lectures on Logic: Carnap's Student Notes, 1910–1914*, ed. Gottfried Gabriel, ed. and trans. Eric H. Reck and Steve Awodey (Chicago: Open Court, 2004), pp. 17–44.

6. Rudolf Carnap, *Der logische Aufbau der Welt: Scheinprobleme in der Philosophie*, 2nd ed. [1928] (Hamburg: Meiner, 1961), sec. 3, p. 3; sec. 2, p. 3; also sec. 12, p. 15 (where Carnap cites Russell as his main source for the definition of structure), and sec. 16, p. 16 (where Carnap draws attention to similarities between his own view of objectivity and that of Poincaré, although he credits Russell as the first to show how structures related to objectivity). On the connection between objectivity and "logical form or structure" in Carnap's *Aufbau*, see Michael Friedman, *Reconsidering Logical Positivism* (Cambridge: Cambridge University Press, 1999), pp. 95–101, esp. p. 99, concerning Carnap's revision of Kantian notions of form in light of Frege's and Russell's work on formal logic.

7. See the cognate entries in the *Oxford English Dictionary*, *Grimms Wörterbuch der deutschen Sprache*, and *Le Robert: Dictionnaire historique de la langue française*, which trace a largely parallel etymology in English, German, and French.

8. Herbert Mehrtens, *Moderne Sprache Mathematik: Eine Geschichte des Streits um die Grundlagen der Disziplin und des Subjekts formaler Systeme* (Frankfurt: Suhrkamp, 1990), pp. 315–26; Leo Corry, *Modern Algebra and the Rise of Mathematical Structures* (Basel: Birkhäuser, 1996), pp. 21–65 and 221–53, 293–342. In linguistics, "structuralism" became associated in the 1920s and 1930s with the antihistorical approaches of Ferdinand de Saussure and Roman Jakobson. On the post-Second World War fortunes of structuralism as a primarily French intellectual movement in the human sciences, see François Dosse, *Histoire du structuralisme* (Paris: La Découverte, 1991–1992).

9. Zeno G. Swijtink, "The Objectification of Observation: Measurement and Statistical Methods in the Nineteenth Century," in Lorenz Krüger, Lorraine J. Daston, and Michael Heidelberger (eds.), *The Probabilistic Revolution*, vol. 1, *Ideas in History* (Cambridge, MA: MIT Press, 1990), pp. 261–85; Gerd Gigerenzer, "Probabilistic Thinking and the Fight Against Subjectivity," in Lorenz Krüger, Gerd Gigerenzer, and Mary S. Morgan (eds.), *The Probabilistic Revolution*, vol. 2: *Ideas in the Sciences* (Cambridge, MA: MIT Press, 1990), pp. 11–33; Gerd Gigerenzer, *The Empire of Chance: How Probability Changed Science and Everyday Life* (Cambridge: Cambridge University Press, 1989), pp. 83–84, 107–108, 233–34, 267–68; Theodore M. Porter, "Objectivity as Standardization: The Rhetoric of Impersonality in Measurement, Statistics, and Cost-Benefit Analysis," in Allan Megill (ed.), *Rethinking Objectivity* (Durham, NC: Duke University Press, 1994), pp. 197–237; Theodore M. Porter, *Trust in Numbers: The Pursuit of Objectivity in Science and Public Life* (Princeton, NJ: Princeton University Press, 1995).

10. Charles Sanders Peirce, "Three Logical Sentiments," *Collected Papers*, ed. Charles Hartshorne and Paul Weiss (Cambridge, MA: Harvard University Press, 1960–1966), vol. 2, p. 398.

11. There is a specialist historical literature on each of these topics, but for general orientation, see Edwin G. Boring, *A History of Experimental Psychology*, 2nd ed. (Englewood Cliffs, NJ: Prentice Hall, 1957); Hans Hiebsch, *Wilhelm Wundt und die Anfänge der experimentellen Psychologie* (Berlin: Akademie, 1977); Robert W. Rieber and David K. Robinson (eds.), *Wilhelm Wundt in History: The Making of a Scientific Psychology* (New York: Kluwer Academic/Plenum, 1980); Kurt Danziger, *Constructing the Subject: Historical Origins of Psychological Research* (Cambridge: Cambridge University Press, 1990).

12. Jimena Canales, "Sensational Differences: Individuality in Observation, Experimentation, and Representation (France, 1853–1895)," Ph.D. diss., Harvard University, 2003.

13. François Gonnessiat, *Recherches sur l'équation personnelle dans les observations astronomiques de passage* (Paris: Masson, 1892).

14. Wilhelm Wundt, *Grundzüge der physiologischen Psychologie* (Leipzig: Engelmann, 1874), p. 709.

15. On the force of late nineteenth-century historicism and its echoes in anthropology, especially in the German-speaking tradition, see Otto Gerhard Oexle, *Geschichtswissenschaft im Zeichen des Historismus: Studien zur Problemgeschichten der Moderne* (Göttingen: Vandenhoeck & Ruprecht, 1996); Theodore Ziolkowski, *Clio the Romantic Muse: Historicizing the Faculties in Germany* (Ithaca, NY: Cornell University Press, 2004); George W. Stocking Jr. (ed.), *Volksgeist as Method and Ethic: Essays on Boasian Ethnography and the German Anthropological Tradition* (Madison: University of Wisconsin Press, 1996).

16. Ernst Cassirer, *Substanzbegriff und Funktionsbegriff* [1910], ed. Reinold Schmücker (Hamburg: Meiner, 2000), pp. 297, 302, 334; cf. Bertrand Russell, *Human Knowledge: Its Scope and Limits* (1948; London: Allen and Unwin, 1966), pp. 18–19, on sensation as "the source of privacy."

17. Henri Poincaré, *La valeur de la science* (1905; Paris: Flammarion, 1970), p. 184.

18. John Worrall, "Structural Realism: The Best of Both Worlds?" *Dialectica* 43 (1989), pp. 99–124; Elie Zahar, "Poincaré's Structural Realism and His Logic of Discovery," in Jean-Louis Greffe, Gerhard Heinzmann, and Kuno Lorenz (eds.), *Henri Poincaré: Science and Philosophy* (Berlin: Akademie, 1996), pp. 45–68; Dan McArthur, "Reconsidering Structural Realism," *Canadian Journal of Philosophy* 33 (2003), pp. 517–36; for a thorough survey of the status of structural realism in current philosophy of science, see Ioannis Votsis, "The Epistemological Status of Scientific Theories: An Investigation of the Structural Realist Account," Ph.D. diss., London School of Economics, 2004, pp. 8–67.

19. Henri Poincaré, *La valeur de la science* (1905; Paris: Flammarion, 1970), p. 184.

20. Immanuel Kant, *Critique of Pure Reason* [1781, 1787], trans. Norman Kemp Smith (New York: Saint Martin's Press, 1965), p. 645, A820/B848.

21. Hermann von Helmholtz, "Über den Ursprung und die Bedeutung der geometrischen Axiome" [1878], *Schriften zur Erkenntnistheorie*, ed. Paul Hertz and Moritz Schlick (1921; Vienna: Springer, 1998), pp. 31 and 39.

22. Hermann von Helmholtz, "Zählen und Messen, erkenntnistheoretisch betrachtet" [1887], *Schriften zur Erkenntnistheorie*, ed. Paul Hertz and Moritz Schlick (1921; Vienna: Springer, 1998), p. 101.

23. Hermann von Helmholtz, "Messungen über den zeitlichen Verlauf der Zuckung animalischer Muskeln und die Fortpflanzunsgeschwindigkeit der Reizung in den Nerven" [1850] and "Messungen über Fortpflanzungsgeschwindigkeit der Reizung in den Nerven" [1850], *Wissenschaftliche Abhandlungen* (Leipzig: Barth, 1882–1895), vol. 3, pp. 764–843 and 844–61; Timothy Lenoir, "Farbensehen, Tonempfindung und der Telegraph: Helmholtz und die Materialität der Kommu-

nikation," in Hans-Jörg Rheinberger and Michael Hagner (eds.), *Die Experimental-isierung des Lebens: Experimentalsysteme in den biologischen Wissenschaften 1850/1950* (Berlin: Akademie, 1993), pp. 50–73, and "Helmholtz and the Materialities of Communication," *Osiris* 9 (1994), pp. 184–207; Kathryn Olesko and Frederic L. Holmes, "Experiment, Quantification, and Discovery: Helmholtz's Early Physiological Researches, 1843–50," in David Cahan (ed.), *Hermann von Helmholtz and the Foundations of Nineteenth-Century Science* (Berkeley: University of California Press, 1993), pp. 50–108; Frederic L. Holmes and Kathryn M. Olesko, "The Images of Precision: Helmholtz and the Graphical Method in Physiology," in M. Norton Wise (ed.), *The Values of Precision* (Princeton, NJ: Princeton University Press, 1995), pp. 198–221.

24. Wilhelm Wundt, "Über die mathematische Induction," *Philosophische Studien* 1 (1881), p. 121.

25. Wilhelm Wundt, "Logische Streitfragen," *Vierteljahrsschrift für wissenschaftliche Philosophie* 6 (1882), pp. 342 and 345.

26. Wilhelm Wundt, "Die Aufgaben der experimentellen Psychologie," *Unsere Zeit: Deutsche Revue der Gegenwart* 1 (1882), pp. 399 and 405–406.

27. Henning Schmidgen, "Of Frogs and Men: The Origins of Psychophysiological Time Experiments, 1850–1865," *Endeavour* 26 (2004), pp. 142–48; "Time and Noise: The Stable Surroundings of Reaction Experiments, 1860–1890," *Studies in History and Philosophy of Biological and Biomedical Sciences* 34 (2003), pp. 237–75; and "Physics, Ballistics, and Psychology: A History of the Chronoscope in/as Context," *History of Psychology* 8 (2005), pp. 46–78.

28. Wilhelm Wundt, *Grundzüge der physiologischen Psychologie* (Leipzig: Engelmann, 1874), p. 685.

29. Wilhelm Wundt, *System der Philosophie* [1889], 3rd ed. (Leipzig: Engelmann, 1907), vol. 1, pp. 142–47.

30. Wilhelm Wundt, *Logik der exakten Wissenschaften* [1880–83], 4th ed. (Stuttgart: Enke, 1920), p. 131.

31. Wilhelm Wundt, *Logik der exakten Wissenschaften* [1880–83], 4th ed. (Stuttgart: Enke, 1920), p. 126.

32. Hermann von Helmholtz, "Zählen und Messen, erkenntnistheoretisch betrachtet" [1887], *Schriften zur Erkenntnistheorie*, ed. Paul Hertz and Moritz Schlick (1921; Vienna: Springer, 1998), p. 101. From his earliest work in mathematics until late in life, Frege seems to have consistently held the Kantian position that while geometry required synthetic *a priori* intuitions, arithmetic was purely analytic. In his dissertation, he attempted to provide a geometric and therefore "intuitive" representation of imaginary numbers; in his Habilitationsschrift, he contrasted intuitive geometry with abstract arithmetic: Gottlob Frege, "On the Geometrical Representation of Imaginary Forms in the Plane" [1873] and "Methods of Calculation Based on an Extension of the Concept of Quantity" [1874], *Collected Papers on Mathematics, Logic, and Philosophy*, ed. Brian McGuiness, trans. Max Black (Oxford: Black-

well, 1984), pp. 1–3 and 56–57. In what was probably a response (*circa* 1899–1906) to David Hilbert's *Grundlagen der Geometrie* (Leipzig: Teubner, 1899), Frege argued that either Euclidean or non-Euclidean geometry, but not both, was true, apparently because of geometry's synthetic component: Gottlob Frege, "Über Euklidische Geometrie," *Nachgelassene Schriften*, ed. Gottfried Gabriel, 2nd rev. ed. (Hamburg: Meiner, 1983), pp. 182–84.

33. Gottlob Frege, *Grundgesetze der Arithmetik* (1893; Hildesheim: Olms, 1998), p. 140n.

34. Michael Dummett, *Frege: Philosophy of Language* (New York: Harper and Row, 1973); Hans D. Sluga, *Gottlob Frege* (London: Routledge and Kegan Paul, 1980); Leila Haaparanta and Jaakko Hintikka (eds.), *Frege Synthesized: Essays on the Philosophical and Foundational Work of Gottlob Frege* (Dordrecht, The Netherlands: Reidel, 1986); Wolfgang Carl, *Frege's Theory of Sense and Reference: Its Orgins and Scope* (Cambridge: Cambridge University Press, 1994); Martin Kusch, *Psychologism: A Case Study of the Sociology of Philosophical Knowledge* (London: Routledge, 1995); Lothar Kreiser, *Gottlob Frege: Leben, Werk, Zeit* (Hamburg: Meiner, 2001); Albert Newen, Ulrich Nortmann, and Rainer Stuhlmann-Laeisz (eds.), *Building on Frege: New Essays on Sense, Content, and Concept* (Stanford, CA: CSLI Publications, 2001).

35. Hermann Lotze, *Logik*, 2nd ed., ed. Georg Misch (Leipzig: Meiner, 1928), sec. 3, p. 16. On Frege's debt to Lotze, see Hans D. Sluga, *Gottlob Frege* (London: Routledge and Kegan Paul, 1980), p. 118.

36. Gottlob Frege, *Die Grundlagen der Arithmetik: Eine logisch mathematische Untersuchung über den Begriff der Zahl* (1884; Wroclaw, Poland: Marcus, 1934), p. 34.

37. Gottlob Frege, *Die Grundlagen der Arithmetik: Eine logisch mathematische Untersuchung über den Begriff der Zahl* (1884; Wroclaw, Poland: Marcus, 1934), p. 35.

38. Michael Dummett, *Frege: Philosophy of Language* (New York: Harper & Row, 1973); Hans D. Sluga, *Gottlob Frege* (London: Routledge and Kegan Paul, 1980), ch. 1.

39. Gottlob Frege, *Die Grundlagen der Arithmetik: Eine logisch mathematische Untersuchung über den Begriff der Zahl* (1884; Wroclaw, Poland: Marcus, 1934), p. 11 and passim.

40. Salomon Stricker, *Studien über die Association der Vorstellungen* (Vienna: Braumüller, 1883), p. 77; cf. Gottlob Frege, *Die Grundlagen der Arithmetik: Eine logisch mathematische Untersuchung über den Begriff der Zahl* (1884; Wroclaw, Poland: Marcus, 1934), p. xvii.

41. Gottlob Frege, "Logik [1897]," in Gottlob Frege, *Nachgelassene Schriften*, ed. Hans Hermes, Friedrich Kambartel, and Friedrich Kaulback, 2nd ed. (Hamburg: Felix Meiner, 1983), pp. 137–63, on p. 158; Thomas Achelis, "Völkerkunde und Philosophie," *Beilage zur Allgemeinen Zeitung* no. 26 (26 Februar 1897), p. 4.

42. Gottlob Frege, *Die Grundlagen der Arithmetik: Eine logisch mathematische*

Untersuchung über den Begriff der Zahl (1884; Wroclaw, Poland: Marcus, 1934), pp. xx–xxi.

43. Gottlob Frege, *Die Grundlagen der Arithmetik: Eine logisch mathematische Untersuchung über den Begriff der Zahl* (1884; Wroclaw, Poland: Marcus, 1934), pp. 104–106 and 108; Hermann Hankel, *Theorie der Complexen Zahlensysteme* (Leipzig: Voss, 1867), pp. 48, 124; Georg Cantor, *Grundlagen einer allgemeinen Mannich-faltigkeitslehre: Ein mathematisch-philosophisch Versuch in der Lehre des Unendlichen* (Leipzig: Teubner, 1883).

44. Gottlob Frege, *Grundgesetze der Arithmetik* (1893; Hildesheim: Olms, 1998), p. xv; Benno Erdmann, *Logik* (Halle a.S.: Max Niemeyer, 1892), pp. 272–75.

45. Gottlob Frege, *Die Grundlagen der Arithmetik: Eine logisch mathematische Untersuchung über den Begriff der Zahl* (1884; Wroclaw, Poland: Marcus, 1934), p. 63; cf. Ernst Schröder, *Lehrbuch der Arithmetik und Algebra für Lehrer und Studirende* (Leipzig: Teubner, 1873), p. 6.

46. Paul Du Bois-Reymond, *Die allgemeine Functionentheorie* (Tübingen: Laupp, 1882), pp. 86–87.

47. F. Kaulbach, "Anschauung," and E.-O. Onnasch and O.R. Schulz, "Vorstellung," in Joachim Ritter (ed.), *Historisches Wörterbuch der Philosophie* (Darmstadt: Wissenschaftliche Buchgesellschaft, 1971), vol. 1, cols. 339–47, and vol. 11, cols. 1227–46.

48. Gottlob Frege, "Der Gedanke" [1918], *Logische Untersuchungen*, ed. Günther Patzig, 2nd ed. (Göttingen, Germany: Vandenhoeck und Ruprecht, 1976), p. 41.

49. Wilhelm Wundt, *Grundzüge der physiologischen Psychologie* (Leipzig: Engelmann, 1874), pp. 464–65.

50. Hermann von Helmholtz, *Handbuch der physiologischen Optik* (Leipzig: Voss, 1867), p. 432.

51. Gottlob Frege, *Die Grundlagen der Arithmetik: Eine logisch mathematische Untersuchung über den Begriff der Zahl* (1884; Wroclaw, Poland: Marcus, 1934), p. 37n.

52. Gottlob Frege, "Thoughts," *Logical Investigations*, ed. and trans. P.T. Geach (New Haven, CT: Yale University Press, 1977), p. 24; Gottlob Frege, "Der Gedanke" [1918], *Logische Untersuchungen*, 2nd ed., ed. Günther Patzig (Göttingen, Germany: Vandenhoeck und Ruprecht, 1976), p. 49.

53. Gottlob Frege, *The Foundations of Arithmetic: A Logico-Mathematical Enquiry into the Concept of Number*, trans. J.L. Austin (Evanston, IL: Northwestern University, 1980), p. 115.

54. Gottlob Frege, *Begriffsschrift* [1879], *Begriffsschrift und andere Aufsätze*, ed. Ignacio Angelelli, 2nd ed. (Hildesheim: Olms, 1998), p. iv.

55. Gottlob Frege, *Begriffsschrift* [1879], *Begriffsschrift und andere Aufsätze*, ed. Ignacio Angelelli, 2nd ed. (Hildesheim: Olms, 1998), p. xi.

56. On Abbe, see Friedrich Stier, "Ernst Abbes akademische Tätigkeit an der Universität Jena," *Jenaer Reden und Schriften* 3 (1955), pp. 26–28; see also Lothar Kreiser, *Gottlob Frege: Leben, Werk, Zeit* (Hamburg: Meiner, 2001), pp. 356–460, on Frege's career at the University of Jena. For Russell, see Bertrand Russell, *The Autobiography of Bertrand Russell*, vol. 1, 1872–1914, (Boston: Little, Brown and Co., 1967), p. 91.

57. Gottlob Frege, *Begriffsschrift* [1879], *Begriffsschrift und andere Aufsätze*, ed. Ignacio Angelelli, 2nd ed. (Hildesheim: Olms, 1993), p. xii.

58. Gottlob Frege, "Der Gedanke" [1918], *Logische Untersuchungen*, ed. Günther Patzig, 2nd ed. (Göttingen, Germany: Vandenhoeck und Ruprecht, 1976), p. 50.

59. Gottlob Frege, "Ueber den Zweck der Begriffsschrift" [1883], *Begriffsschrift und andere Aufsätze*, ed. Ignacio Angelelli, 2nd ed. (Hildesheim: Olms, 1998), p. 97. On Frege's views on the inadequacy of most, if not all, mathematical proofs, see also Thomas G. Ricketts, "Objectivity and Objecthood: Frege's Metaphysics of Judgment," in Leila Haaparanta and Jaakko Hintikka (eds.), *Frege Synthesized: Essays on the Philosophical and Foundational Work of Gottlob Frege* (Dordrecht, The Netherlands: Reidel, 1976), pp. 65–95.

60. Gottlob Frege, *Begriffsschrift* [1879], *Begriffsschrift und andere Aufsätze*, ed. Ignacio Angelelli, 2nd ed. (Hildesheim: Olms, 1998), pp. 2–3 and 9.

61. Gottlob Frege, "Ueber die wissenschatliche Berechtigung einer Begriffsschrift" [1882], *Begriffsschrift und andere Aufsätze*, ed. Ignacio Angelelli, 2nd ed. (Hildesheim: Olms, 1998), pp. 110–11.

62. Gottlob Frege, *Die Grundlagen der Arithmetik: Eine logisch mathematische Untersuchung über den Begriff der Zahl* (1884; Wroclaw, Poland: Marcus, 1934), p. 103.

63. Color has been a philosophical problem since Antiquity, but not always the same problem: see Charles A. Riley II, *Color Codes: Modern Theories of Color in Philosophy, Painting and Architecture, Literature, Music, and Psychology* (Hanover, NH: University Press of New England, 1995), pp. 20–69; J.B. Maund, "The Nature of Color," *History of Philosophy Quarterly* 8 (1991), pp. 253–63.

64. René Descartes, *Optics* (1637), "Sixth Discourse," *Discourse on Method, Optics, Geometry, and Meteorology*, trans. Paul J. Olscamp (Indianapolis: Bobbs-Merrill, 1965), p. 101; cf. John Locke, *An Essay Concerning Human Understanding* [1689], ed. Peter H. Nidditch (Oxford: Oxford University Press, 1979), IV.iii.11–16, pp. 544–48.

65. Henri Poincaré, *La valeur de la science* (1905; Paris: Flammarion, 1970), p. 179.

66. Descartes, *Méditations*, in *Oeuvres de Descartes*, ed. Charles Adam and Paul Tannery (Paris: Cerf, 1910), vol. 9, pp. 31–33 (Meditation III). We cite the French translation, in which Descartes expands on the meaning of this scholastic terminology, probably for the benefit of nonacademic readers. He added the same explicative phrase on participation by representation in degrees of being or perfection after

alluding to "réalité objective" in the French translation of the Synopsis, p. 11. On Descartes's use of this scholastic terminology, see Calvin Normore, "Meaning and Objective Being: Descartes and His Sources," in Amélie Oksenberg Rorty (ed.), *Essays on Descartes' Meditations* (Berkeley: University of California Press, 1986), pp. 223–41.

67. Arthur König and Conrad Dieterici, "Die Grundempfindungen in normalen und anomalen Farbensystemen und ihre Intensitätsverteilung im Spektrum," *Zeitschrift für Psychologie und Physiologie der Sinnesorgane* 4 (1892), pp. 241–347.

68. Henri Poincaré, *La valeur de la science* (1905; Paris: Flammarion, 1917), p. 271.

69. Immanuel Kant, *Critique of Pure Reason* [1781, 1787], trans. Norman Kemp Smith (New York: Saint Martin's Press, 1965), pp. 645–50, A822–829/ B850–857.

70. Michael Heidelberger, "Beziehungen zwischen Sinnesphysiologie und Philosophie im 19. Jahrhundert," in Hans Jörg Sandkühler (ed.), *Philosophie und Wissenschaften: Formen und Prozesse ihrer Interaktion* (Frankfurt: Lang, 1997), pp. 37–58.

71. Johann Wolfgang von Goethe, *Zur Farbenlehre* [1810], *Goethes Werke*, rev. ed., ed. Erich Trunz, vol. 13, *Naturwissenschaftliche Schriften*, ed. Dorothea Kuhn and Rike Wankmüller (Munich: Beck, 1981), arts. 136–38 and 188, pp. 359–60 and 371. Goethe explicitly pairs the "subjective" experiments on refraction (arts. 195–302) with the "objective" ones (arts. 303–49).

72. Edwin G. Boring, *Sensation and Perception in the History of Experimental Psychology* (1942; New York: Irvington, 1970), pp. 115–16.

73. Jan E. Purkinje, "Beobachtungen und Versuche zur Physiologie der Sinne" [1819], *Opera omnia* (Prague: Purkyňova spoleçnost, 1918–1941), vol. 1, p. 89.

74. Hermann von Helmholtz, *Handbuch der physiologischen Optik* (Leipzig: Voss, 1867), p. 426.

75. Hermann von Helmholtz, *Handbuch der physiologischen Optik* (Leipzig: Voss, 1867), p. 422.

76. Ewald Hering, "Über individuelle Verschiedenheiten des Farbensinnes," *Lotos* 6 (1885), p. 156. On the remarkable efforts of the Austrian scientists Siegmund and Franz Serafin Exner to establish a universal theory of human color perception in response to the experimental evidence of individual differences, see Deborah R. Coen, *Vienna in the Age of Uncertainty: Science, Liberalism, and Private Life*, (Chicago: University of Chicago Press, 2007).

77. R. Steven Turner, *In the Eye's Mind: Vision and the Helmholtz-Hering Controversy* (Princeton, NJ: Princeton University Press, 1994), p. 177.

78. Hugo Magnus, *Die geschichtliche Entwickelung des Farbensinnes* (Leipzig: Veit, 1877), pp. 11 and 29–41.

79. Richard Andree, "Ueber den Farbensinn der Naturvölker," *Zeitschrift für Ethnologie* 10 (1878), pp. 323–34; Hermann Rabl-Rückhard, "Zur historischen Entwickelung des Farbensinnes," *Zeitschrift für Ethonologie* 12 (1880), pp. 210–21.

80. Gottlob Frege, *Die Grundlagen der Arithmetik: Eine logisch mathematische Untersuchung über den Begriff der Zahl* (1884; Wroclaw, Poland: Marcus, 1934), p. 105.

81. Gottlob Frege, *Die Grundlagen der Arithmetik: Eine logisch mathematische Untersuchung über den Begriff der Zahl* (1884; Wroclaw, Poland: Marcus, 1934), p. 36.

82. Ludwig Wittgenstein, *Remarks on Colour/Bemerkungen über die Farben*, ed. G.E.M. Anscombe, trans. Linda L. McAlister and Margarete Schättle (Oxford: Blackwell, 1977), pp. 4–14.

83. Henri Poincaré, *The Value of Science: Essential Writings of Henri Poincaré* (New York: Modern Library, 2001), p. 345.

84. Henri Poincaré, *The Value of Science: Essential Writings of Henri Poincaré* (New York: Modern Library, 2001), p. 345.

85. Peter Galison, *Einstein's Clocks, Poincaré's Maps: Empires of Time* (New York: Norton, 2003), pp. 63–66 and 298–99.

86. Henri Poincaré, *The Value of Science: Essential Writings of Henri Poincaré* (New York: Modern Library, 2001), p. 350.

87. Henri Poincaré, *The Value of Science: Essential Writings of Henri Poincaré* (New York: Modern Library, 2001), pp. 347–48.

88. Theodore M. Porter, "The Death of the Object: Fin-de-siècle Philosophy of Physics," in Dorothy Ross (ed.), *Modernist Impulses in the Human Sciences, 1870–1930* (Baltimore: Johns Hopkins University Press, 1994), pp. 128–51.

89. Ernst Mach, *The Analysis of Sensations, and the Relation of the Physical to the Psychical* [1897], trans. C.M. Williams, rev. ed. (New York: Dover, 1959), pp. 29–31.

90. Ernst Mach, *The Analysis of Sensations, and the Relation of the Physical to the Psychical* [1897], trans. C.M. Williams, rev. ed. (New York: Dover, 1959), p. 364; cf. p. 346, on mathematics.

91. J.L. Heilbron, *The Dilemmas of an Upright Man: Max Planck and the Fortunes of German Science* (Cambridge, MA: Harvard University Press, 2000), pp. 47–60.

92. Edouard Le Roy, "Science et philosophie," *Revue de métaphysique et de morale* 7 (1899), pp. 375–425, 503–62, 708–31.

93. Henri Poincaré, *La valeur de la science* (1905; Paris: Flammarion, 1970), p. 152.

94. Peter Galison, *Einstein's Clocks, Poincaré's Maps: Empires of Time* (New York: Norton, 2003), pp. 203–11.

95. Henri Poincaré, *La valeur de la science* (1905; Paris: Flammarion, 1970), pp. 180–82.

96. Henri Poincaré, *La valeur de la science* (1905; Paris: Flammarion, 1970), p. 184.

97. June Barrow-Green, *Poincaré and the Three Body Problem* (Providence, RI: American Mathematical Society, 1997), ch. 3.

98. Henri Poincaré, *Electricité et optique: La lumière et les théories électrodynamiques*, eds. Jules Blondin and Eugène Néculcéa, 2nd ed. (Paris: Gauthier-Villars, 1901), p. viii.

99. Henri Poincaré, *La science et l'hypothèse* (Paris: Flammarion, 1903), p. 281.

100. Henri Poincaré, *La science et l'hypothèse* (Paris: Flammarion, 1903), p. 163.

101. Henri Poincaré, *The Value of Science: Essential Writings of Henri Poincaré* (New York: Modern Library, 2001), pp. 348–49.

102. Henri Poincaré, "La morale et la science," *Dernières pensées* (Paris: Flammarion, 1913), pp. 232–33.

103. Bertrand Russell, "The Place of Science in a Liberal Education" [1913], *On the Philosophy of Science*, ed. Charles A. Fritz Jr. (Indianapolis: Bobbs-Merrill, 1965), p. 214.

104. Henri Poincaré, *The Value of Science: Essential Writings of Henri Poincaré* (New York: Modern Library, 2001), p. 353.

105. Rudolf Carnap, *Der logische Aufbau der Welt*, 2nd ed. (Hamburg: Meiner, 1961), sec. 3, p. 2.

106. Rudolf Carnap, "Intellectual Autobiography," *The Philosophy of Rudolf Carnap*, ed. Paul Arthur Schilpp (Cambridge: Cambridge University Press, 1963), pp. 12 and 17–20, quotation on p. 17.

107. Rudolf Carnap, *Der logische Aufbau der Welt: Scheinprobleme in der Philosophie*, 2nd ed. (Hamburg: Meiner, 1961), sec. 2, p. 2.

108. Rudolf Carnap, *Der logische Aufbau der Welt: Scheinprobleme in der Philosophie*, 2nd ed. (Hamburg: Meiner, 1961), sec. 14, pp. 17–18.

109. Rudolf Carnap, *Der logische Aufbau der Welt: Scheinprobleme in der Philosophie*, 2nd ed. (Hamburg: Meiner, 1961), sec. 5, pp. 4–5.

110. Rudolf Carnap, *Der logische Aufbau der Welt: Scheinprobleme in der Philosophie*, 2nd ed. (Hamburg: Meiner, 1961), sec. 4, p. 4; sec. 15, pp. 19–20. See also Michael Friedman, "Carnap's Aufbau Reconsidered," *Noûs* 21 (1987), pp. 521–45, esp. pp. 526–29.

111. Rudolf Carnap, *Der logische Aufbau der Welt: Scheinprobleme in der Philosophie*, 2nd ed. (Hamburg: Meiner, 1961), sec. 16, p. 20. A fascinating borderline case between structural-algebraic representation and the visualizable-mimetic is the Feynman diagram. A remarkable book that tracks the flow and mutation of these diagrams as different groups of theorists take them up is David Kaiser's, *Drawing Theories Apart: The Dispersion of Feynman Diagrams in Postwar Physics* (Chicago: University of Chicago Press, 2005). On the electrodynamic and quantum work that led up to these diagrams see S. S. Schweber, *QED and the Men Who Made It* (Princeton, NJ: Princeton University Press, 1994); and on Feynman's wartime research behind those diagrams, Peter Galison, "Feynman's War: Modelling Weapons, Modelling Nature," *Studies in the History* and *Philosophy of Modern Physics* 29 (1998), pp. 391–434.

112. Rudolf Carnap, *The Logical Structure of the World: Pseudoproblems in Philosophy*, trans. Rolf A. George (Berkeley: University of California Press, 1969), pp. xvi–xvii, translation emended. See Rudolf Carnap, *Der logische Aufbau der Welt: Scheinprobleme in der Philosophie*, 2nd ed. (Hamburg: Meiner, 1961), p. xix.

113. Rudolf Carnap, *The Logical Structure of the World: Pseudoproblems in Philosophy*, trans. Rolf A. George (Berkeley: University of California Press, 1969), p. xviii.

114. Nancy Cartwright, Jordi Cat, Lola Fleck, and Thomas E. Uebel, *Otto Neurath: Philosophy Between Science and Politics* (New York: Cambridge University Press, 1996); more specifically, on the views of Carnap, Cassirer, and Schlick on objectivity, see Michael Friedman, *A Parting of the Ways: Carnap, Cassirer, and Heidegger* (Chicago: Open Court, 2000), pp. 111-27.

115. Rudolf Carnap, *The Logical Syntax of Language*, trans. Amethe Smeaton (London: Routledge & Kegan Paul, 1937), p. 52.

116. Rudolf Carnap, *The Logical Structure of the World: Pseudoproblems in Philosophy*, trans. Rolf A. George (Berkeley: University of California Press, 1969), p. 30.

117. Bertrand Russell to William James, Nov. 6, 1908, in William James, *The Meaning of Truth*, ed. Fredson Bowers (Cambridge, MA: Harvard University Press, 1975), p. 300.

118. Bertrand Russell, *Introduction to Mathematical Philosophy* (New York: Macmillan, 1919), p. 61.

119. Bertrand Russell, *Introduction to Mathematical Philosophy* (New York: Macmillan, 1919), p. 61.

120. Moritz Schlick, "On the Relation Between Psychological and Physical Concepts" [1935], *Philosophical Papers*, ed. Henk L. Mulder and Barbara F.B. van de Velde-Schlick, trans. Peter Heath (Dordrecht, The Netherlands: Reidel, 1979), vol. 2, p. 424.

121. Moritz Schlick, "On the Relation Between Psychological and Physical Concepts" [1935], *Philosophical Papers*, ed. Henk L. Mulder and Barbara F.B. van de Velde-Schlick, trans. Peter Heath (Dordrecht, The Netherlands: Reidel, 1979), vol. 2, p. 425.

122. Moritz Schlick, "The Universe and the Human Mind" [1936], *Philosophical Papers*, ed. Henk L. Mulder and Barbara F.B. van de Velde-Schlick, trans. Peter Heath (Dordrecht, The Netherlands Reidel, 1979), vol. 2, pp. 510-11.

123. Elmar Schenkel, "Wie die Menschen außerirdisch wurden: Aliens in der frühen Science Fiction, 1880-1940," in Thomas P. Weber (ed.), *Science & Fiction II: Leben auf anderen Sternen* (Frankfurt: Fischer, 2004), pp. 137-62; Michael J. Crowe, *The Extraterrestrial Life Debate, 1750-1900: The Idea of a Plurality of Worlds from Kant to Lowell* (Cambridge: Cambridge University Press, 1986), pp. 359-546.

124. H.G. Wells, "Intelligence on Mars," *Saturday Review* 81 (1896), p. 346.

125. H.G. Wells, *The First Men in the Moon* (London: Newnes, 1901), p. 329.

126. J.H. Rosny, *Les Xipéhuz* (Paris: Société du Mercure de France, 1896), p. 61.

127. Camille Flammarion, *La fin du monde* (Paris: Flammarion, 1894), p. 132.

128. MS IV.A-B (Carte du Ciel), Observatoire de Paris.

129. For these and other nineteenth-century international scientific collaborations, see Eric Brian, "Transactions statistiques au XIXe siècle: Mouvements inter-

nationaux de capitaux symboliques," *Actes de la recherche en sciences sociales*, nr. 145 (décembre 2002), pp. 34–46; Aant Elzinga and Catharina Landström (eds.), *Internationalism and Science* (London: Taylor Graham, 1996); Charlotte Bigg, "Photography and Labour History of Astrometry: The Carte du Ciel," in Klaus Hentschel and Axel D. Wittman (eds.), *The Role of Visual Representations in Astronomy: History and Research Practice* (Thun, Switzerland: Deutsch, 2000), pp. 90–106; Ulrich Völter, *Geschichte und Bedeutung der internationalen Erdmessung* (Munich: Verlag der Bayerischen Akademie der Wissenschaften, 1963); John Lankford, "Photography and the Nineteenth-Century Transits of Venus," *Technology and Culture* 28 (1987), pp. 648–57; Jimena Canales, "Photogenic Venus: The 'Cinematographic Turn' and Its Alternatives in Nineteenth-Century France," *Isis* 93 (2002), pp. 585–613; Gustav Hellmann, "Die Entwicklung der meteorologischen Beobachtungen in Deutschland von den ersten Anfängen bis zur Einrichtung staatlicher Beobachtungsnetze," *Abhandlungen der Preussischen Akademie der Wissenschaften, Physikalisch-Mathematische Klasse* 1 (1926), pp. 1–25; Robert Marc Friedman, *Appropriating the Weather: Vilhelm Bjerknes and the Construction of a Modern Meteorology* (Ithaca: Cornell University Press, 1989); James Rodger Fleming, *Meteorology in America, 1800–1870* (Baltimore: Johns Hopkins University Press, 1990); Peter Galison, *Einstein's Clocks, Poincaré's Maps: Empires of Time* (New York: Norton, 2003), pp. 84–220.

130. See, for example, Janet Browne, *Charles Darwin: Vol. 2, The Power of Place* (London: Jonathan Cape, 2002), pp. 181–82, concerning Darwin's dependence on his children's governess, Miss Ludwig, to help him with German scientific literature.

131. C. Lloyd Morgan, *An Introduction to Comparative Psychology* (London: Scott, 1894), pp. 41–42.

132. Gottlob Frege, "Der Gedanke" [1918], *Logische Untersuchungen*, ed. Günther Patzig, 2nd ed. (Göttingen, Germany: Vandenhoeck und Ruprecht, 1976), p. 36.

133. Bertrand Russell to Lucy Martin Donnelly, April 22, 1906, in Bertrand Russell, *The Autobiography of Bertrand Russell* vol. 1, 1872–1914 (Boston: Little, Brown and Co., 1967), pp. 280–81.

134. Albert Einstein, "Autobiographical Notes," in Paul Arthur Schilpp (ed.), *Albert Einstein: Philosopher-Scientist* (1949; La Salle, IL: Open Court, 1970), vol. 1, pp. 4–7.

135. Bertrand Russell, "The Place of Science in a Liberal Education" [1913], *On the Philosophy of Science*, ed. Charles A. Fritz Jr. (Indianapolis: Bobbs-Merrill, 1965), p. 219.

136. Charles Sanders Peirce, "Three Logical Sentiments," *Collected Papers*, ed. Charles Hartshorne and Paul Weiss (Cambridge, MA: Harvard University Press, 1960–1966), vol. 2, pp. 395–400, esp. p. 398; Karl Pearson, *The Grammar of Science* (London: Scott, 1892), pp. 7–8.

137. Charles Sanders Peirce, "The Century's Great Men in Science," *Annual Report of the Board of Regents of the Smithsonian Institution* (Washington, DC: Gov-

ernment Printing Office, 1901), p. 696. On the contradictions in Pearson's position, see Theodore M. Porter, *Karl Pearson: The Scientific Life in a Statistical Age* (Princeton, NJ: Princeton University Press, 2004), pp. 9–10 and 308–10.

138. Karl Pearson, *The Grammar of Science* (London: Scott, 1892), pp. 7–8.

139. Hermann Weyl, "Erkenntnis und Besinnung (ein Lebensrückblick)" [1954], *Gesammelte Abhandlungen*, ed. K. Chadrasekharan (Berlin: Springer, 1968), vol. 4, p. 644; cf. Hermann Weyl, *Philosophy of Mathematics and Natural Science* (Princeton, NJ: Princeton University Press, 1949), p. 123.

140. Albert Einstein, "The Problem of Space, Ether, and the Field in Physics," *Ideas and Opinions: Based on Mein Weltbild* (New York: Bonanza, 1954), pp. 281–82.

141. Albert Einstein, "Physics and Reality," *Ideas and Opinions: Based on Mein Weltbild* (New York: Bonanza, 1954). On Einstein, Poincaré, and the synchronization of clocks, see Peter Galison, *Einstein's Clocks, Poincaré's Maps: Empires of Time* (New York: Norton, 2003).

142. Albert Einstein, "Relativity and the Problem of Space," *Ideas and Opinions: Based on Mein Weltbild* (New York: Bonanza, 1954), pp. 370–71.

143. Albert Einstein, "Relativity and the Problem of Space," *Ideas and Opinions: Based on Mein Weltbild* (New York: Bonanza, 1954), p. 374: "All this simply means that an objective metrical significance is attached to the quantity $ds^2 = dx_1^2 + dx_2^2 + dx_3^2 - dx_4^2$."

144. Henry Margenau in Paul Arthur Schilpp (ed.), *Albert Einstein: Philosopher-Scientist* (La Salle, IL: Open Court, 1970), p. 252.

145. Henry Margenau in Paul Arthur Schilpp (ed.), *Albert Einstein: Philosopher-Scientist* (La Salle, IL: Open Court, 1970), p. 253. Margenau takes "invariance" to be both numbers (like the speed of light) and the preservation of the form of laws (covariance).

146. Albert Einstein, "Reply to Criticisms," in Paul Arthur Schilpp (ed.), *Albert Einstein: Philosopher-Scientist* (La Salle, IL: Open Court, 1970), pp. 680–81, emphasis added.

147. Albert Einstein, "Reply to Criticisms," in Paul Arthur Schilpp (ed.), *Albert Einstein: Philosopher-Scientist* (La Salle, IL: Open Court, 1970), pp. 673–74 and 680.

148. Thomas Nagel, *The View from Nowhere* (New York: Oxford University Press, 1986), p. 5.

149. Robert Nozick, *Invariances: The Structure of the Objective World* (Cambridge, MA; Belknap Press, Harvard University Press, 2001), p. 96.

CHAPTER SIX: TRAINED JUDGMENT

1. Rudolf Grashey, *Atlas typischer Röntgenbilder vom normalen Menschen*, 6th ed. (Munich: Lehmann, 1939), p. v.

2. Henri F. Ellenberger, *The Discovery of the Unconscious: The History and Evolution of Dynamic Psychiatry* (New York: Basic Books, 1970).

3. P.B. Medawar, *Advice to a Young Scientist* (New York: Harper & Row, 1979), p. 40.

4. Walter B. Cannon, *The Way of an Investigator: A Scientist's Experiences in Medical Research* (New York: Norton, 1945), pp. 58–64.

5. Hans Selye, *From Dream to Discovery: On Being a Scientist* (New York: McGraw-Hill, 1964), p. 47.

6. Hans Selye, *From Dream to Discovery: On Being a Scientist* (New York: McGraw-Hill, 1964), p. 61.

7. Charles Richet, *Le savant* (Paris: Hachette, 1923), p. 14.

8. P.B. Medawar, *Advice to a Young Scientist* (New York: Harper & Row, 1979), p. 7.

9. Charles Richet, *Le savant* (Paris: Hachette, 1923), p. 115.

10. Walter B. Cannon, *The Way of an Investigator: A Scientist's Experiences in Medical Research* (New York: Norton, 1945), p. 30.

11. On Wolfgang Gentner, Heinz Maier-Leibnitz, and Walther Bothe, see "Walther Bothe and the Physics Institute: The Early Years of Nuclear Physics," http://nobelprize.org/physics/articles/states/walther-bothe.html; their volume was first published in 1940 by Julius Springer Verlag, and the revised edition is *An Atlas of Typical Expansion Chamber Photographs*, 2nd ed. (London: Pergamon Press, 1954).

12. Rudolf Grashey, *Atlas typischer Röntgenbilder vom normalen Menschen*, 6th ed. (Munich: Lehmann, 1939), p. v.

13. On mimetic representation and its enemies in the realm of the epistemology of experimentation, see Peter Galison, *Image and Logic: A Material Culture of Microphysic* (Chicago: University of Chicago Press, 1997).

14. Henry Alsop Riley, *An Atlas of the Basal Ganglia, Brain Stem, and Spinal Cord*, rev. ed. (New York: Hafner, 1960), p. viii.

15. Henry Alsop Riley, *An Atlas of the Basal Ganglia, Brain Stem, and Spinal Cord*, rev. ed. (New York: Hafner, 1960), p. viii.

16. Moulton K. Johnson and Myles J. Cohen, *The Hand Atlas* (Springfield, IL: Thomas, 1975), p. vii.

17. Frederick A. Gibbs and Erna L. Gibbs, preface to *Atlas of Electroencephalography* (Cambridge, MA: Cummings, 1941), n.p.

18. Frederick A. Gibbs and Erna L. Gibbs, preface to *Atlas of Electroencephalography* (Cambridge, MA: Cummings, 1941), n.p.

19. Frederick A. Gibbs and Erna L. Gibbs, *Atlas of Electroencephalography*, vol. 1, *Methodology and Controls*, 2nd ed. (Reading, MA: Addison-Wesley Press, 1951), pp. 112–13.

20. Erwin Christeller, *Atlas der Histotopographie: Gesunder und erkrankter Organe* (Leipzig: Georg Thieme, 1927), p. 18.

21. Walther von Dyck, "Die mathematische, naturwissenschaftliche und technische Hochschulbildung," in Paul Hinneberg (ed.), *Die Kultur der Gegenwart: Ihre Entwickelung und ihre Ziele*, vol. 1, Allgemeine Geschichte der Philosophie, ed. Wilhelm Wundt (Berlin: Teubner, 1909), pt. 1, pp. 335–36; David Cahan, "The Institu-

NOTES TO PAGES 326-328

tional Revolution in German Physics, 1865–1914," *Historical Studies in the Physical Sciences* 15 (1985), pp. 1–65, and *An Institute for Empire: The Physikalisch-Technische Reichsanstalt, 1871–1918* (Cambridge: Cambridge University Press, 1989), pp. 20–24. On the growth of academic physics internationally *circa* 1900, see Paul Forman, John L. Heilbron, and Spencer Weart, *Physics circa 1900: Personnel, Funding, and Productivity of the Academic Establishments* (Princeton, NJ: Princeton University Press, 1975), pp. 3–128.

22. Robert Fox and George Weisz, "Introduction: The Institutional Basis of French Science in the Nineteenth Century," in Robert Fox and George Weisz (eds.), *The Organization of Science and Technology in France, 1808–1914* (Cambridge: Cambridge University Press, 1980), pp. 1–28, esp. table 2, on p. 12.

23. Nicole Hulin-Jung, *L'organisation de l'enseignement des sciences: La voie ouverte par le Second Empire* (Paris: Editions du Comité des Travaux Historiques et Scientifiques, 1989), pp. 297–303.

24. *Sixth Report from the Royal Commission on Scientific Instruction and the Advancement of Science* (London: H.M.S.O., 1875), p. 10, quoted in D.M. Turner, *History of Science Teaching in England* (London: Chapman & Hall, 1927), p. 99; *Seventh Report from the Royal Commission on Scientific Instruction and the Advancement of Science* (London: H.M.S.O., 1875), p. 3.

25. *A History of the Cavendish Laboratory 1871–1910* (London: Longmans, Green, and Co., 1910); W.B. Stephens, *Education in Britain, 1750–1914* (New York: St. Martin's Press, 1998), p. 137.

26. R. Steven Turner, "The Prussian Universities and the Concept of Research," *Internationales Archiv für Sozialgeschichte der deutschen Literatur* 5 (1980), pp. 68–93; William Clark, "On the Dialectical Origins of the Research Seminar," *History of Science* 27 (1989), pp. 111–54.

27. On Neumann's seminar: Kathryn M. Olesko, *Physics as a Calling: Discipline and Practice in the Königsberg Seminar for Physics* (Ithaca: Cornell University Press, 1991), pp. 367–86, and "The Meaning of Precision: The Exact Sensibility in Early Nineteenth-Century Germany," in M. Norton Wise (ed.), *The Values of Precision* (Princeton, NJ: Princeton University Press, 1995), pp. 103–34. On the Edinburgh medical students: L.S. Jacyna, "'A Host of Experienced Microscopists': The Establishment of Histology in Nineteenth-Century Edinburgh," *Bulletin for the History of Medicine* 75 (2001), p. 238.

28. Nick Hopwood, *Embryos in Wax: Models from the Ziegler Studio* (Cambridge: Whipple Museum of the History of Science, 2002), p. 35; on models more generally, see Soraya de Chadarevian and Nick Hopwood (eds.), *Models: The Third Dimension of Science* (Stanford, CA: Stanford University Press, 2004).

29. Lorraine Daston, "The Glass Flowers," in Lorraine Daston (ed.), *Things That Talk: Object Lessons from Art and Science* (New York: Zone Books, 2004), p. 248.

30. Finally, most difficult of all, comes paragenesis, the unraveling of the order

of growth — and it is to this aim that O. Oelsner, *Atlas of the Most Important Ore Mineral Parageneses Under the Microscope* (1961; Oxford: Pergamon, 1966) directs his work; see pp. v–vi. The atlas trains the eye by depicting microscopic samples and providing worksheets that the reader superposes on the picture to provide keys to interpretation.

31. Frederick A. Gibbs and Erna L. Gibbs, *Atlas of Electroencephalography*, vol. 1: *Methodology and Controls*, 2nd ed. (Reading, MA: Addison-Wesley Press, 1951), p. 113.

32. Hallowell Davis, introduction to William F. Caveness, *Atlas of Electroencephalography in the Developing Monkey Macaca mulatta* (Reading, MA: Addison-Wesley Publishing Company, 1962), p. 2.

33. The elites' struggle to maintain their earlier, bedside way of life is beautifully documented by Chistopher Lawrence in "A Tale of Two Sciences: Bedside and Bench in Twentieth-Century Britain," *Medical History* 43 (1999), pp. 421–49; see also his "Still Incommunicable: Clinical Holists and Medical Knowledge in Interwar Britain," in Christopher Lawrence and George Weisz (eds.), *Greater than the Parts: Holism in Biomedicine, 1920–1950* (New York: Oxford University Press, 1998), pp. 94–111. In another study of clinical strategies to avoid procedural-scientific counter-arguments, David S. Jones shows that deep into the 1960s and 1970s, among cardiac surgeons, the visualization techniques of surgeons were often deployed as the bottom-line argument in favor of new procedures. In doing so, surgeons directly opposed increasing pressure to evaluate their procedures by means of random clinical trials. See David S. Jones "Visions of a Cure: Visualization, Clinical Trials, and Controversies in Cardiac Therapeutics, 1968–1998," *Isis* 91 (2000), pp. 504–41.

34. Theodore M. Porter has written a superb comparative study of administrative-bureaucratic decision making in the public sphere: *Trust in Numbers: The Pursuit of Objectivity in Science and Public Life* (Princeton, NJ: Princeton University Press, 1995); on the French case of the *polytechniciens* and the Corps des Ponts et Chaussées, see esp. ch. 6.

35. Alvarez Group scanning training film, 1968, cited in Peter Galison, *Image and Logic: A Material Culture of Microphysics* (Chicago: University of Chicago Press, 1997), p. 382.

36. The struggle over judgment and rule-governed image assessment in bubble chamber physics is treated much more extensively in Peter Galison, *Image and Logic: A Material Culture of Microphysics* (Chicago: University of Chicago Press, 1997), p. 406.

37. C.F. Powell and G.P.S. Occhialini, *Nuclear Physics in Photographs: Tracks of Charged Particles in Photographic Emulsions* (Oxford: Clarendon, 1947); C.F. Powell, P.H. Fowler, and D.H. Perkins, *The Study of Elementary Particles by the Photographic Method: An Account of the Principal Techniques and Discoveries* (London: Pergamon, 1959).

38. On holism and politics before and during Nazism, see Anne Harrington, *Reenchanted Science: Holism in German Culture from Wilhelm II to Hitler* (Princeton,

NJ: Princeton University Press, 1996), and Mitchell G. Ash, *Gestalt Psychology in German Culture, 1890–1967: Holism and the Quest for Objectivity* (Cambridge: Cambridge University Press, 1995).

39. W.W. Morgan, Philip C. Keenan, and Edith Kellman, *An Atlas of Stellar Spectra, with an Outline of Spectral Classification* (Chicago: University of Chicago Press, 1943), p. 4.

40. W.W. Morgan, Philip C. Keenan, and Edith Kellman, *An Atlas of Stellar Spectra, with an Outline of Spectral Classification* (Chicago: University of Chicago Press, 1943), p. 5.

41. W.W. Morgan, Philip C. Keenan, and Edith Kellman, *An Atlas of Stellar Spectra, with an Outline of Spectral Classification* (Chicago: University of Chicago Press, 1943). In twentieth-century medicine, one often sees "clinical judgment" opposed to the view that nature can speak for itself. For example, the author of an atlas on electrocardiograms argues that traces cannot replace clinical judgment. See Joseph E.F. Riseman, P-Q-R-S-T, *A Guide to Electrocardiogram Interpretation*, 5th ed. (New York: Macmillan, 1968).

42. W.W. Morgan, Philip C. Keenan, and Edith Kellman, *An Atlas of Stellar Spectra, with an Outline of Spectral Classification* (Chicago: University of Chicago Press, 1943), p. 6.

43. We draw here on the rich and compelling article by Carlo Ginzburg, "Family Resemblances and Family Trees: Two Cognitive Metaphors," *Critical Inquiry* 30 (2004), pp. 537–56; and we follow Ginzburg in citing Ludwig Wittgenstein, aphorism 67, "A Lecture on Ethics," *Philosophical Review* 74 (1965), pp. 4–5. For the information about Wittgenstein's composite photograph we are indebted to James Conant, "Family Resemblance, Composite Photography, and Unity of Concept: Goethe, Galton, and Wittgenstein," available as a podcast at http://wab.aksis.uib.no/wab_contrib-audio-cj-photo05.page (accessed 7 August 2010).

44. As Wittgenstein made clear, he took the "intermediate terms" that bound objects together to be not genetic but conceptual. A circle could be linked to an ellipse by means of a form halfway between the two — but this did not mean that the ellipse grew out of the circle. Wittgenstein judged that the fallacy of believing these intermediate states to have actually occurred lay behind the worst speculative excesses of Frazer. Probably (according to the editors) these remarks come from around 1931. Ludwig Wittgenstein, *Bemerkungen über Frazers Golden Bough/Remarks on Frazer's Golden Bough*, ed. Rush Rhees, trans. A.C. Miles (Retford, England: Brynmill Press, 1979). James George Frazer's *Golden Bough: A Study in Magic and Religion* was first published in 1890.

45. Malcolm Nicolson, "Alexander von Humboldt and the Geography of Vegetation," in Andrew Cunningham and Nicholas Jardine (eds.), *Romanticism and the Sciences* (Cambridge: Cambridge University Press, 1990), pp. 169–85.

46. Gary D. Stark, "Der Verleger als Kulturunternehmer: Der J.F. Lehmanns

Verlag und Rassenkunde in der Weimarer Republik," *Archiv für Geschichte des Buchwesens* 16 (1976), pp. 291–318.

47. *75 Jahre J.F. Lehmanns Verlag München, 1890–1965* (Munich: Markus, 1965), on the number of atlases, p. 5.

48. Gary D. Stark, "Der Verleger als Kulturunternehmer: Der J.F. Lehmanns Verlag und Rassenkunde in der Weimarer Republik," *Archiv für Geschichte des Buchwesens* 16 (1976), p. 302, quoting letter from Julius Lehmann to Hans Günther, Oct. 11, 1920; on the portrait prize, p. 307.

49. Hans F.K. Günther, *Kleine Rassenkunde Europas* (Munich: Lehmann, 1925), p. 8; for another example of Lehmann's race volumes, see Ludwig Schemann, *Die Rasse in den Geisteswissenschaften: Studien zur Geschichte des Rassengedankens* (Munich: Lehmann, 1938), which also emphasized the nonquantifiable, indeed more than physical character of race: for example, p. 35.

50. Diane B. Paul, *Controlling Human Heredity: 1865 to the Present* (Amherst, NY: Humanity Books, 1998), and *The Politics of Heredity: Essays on Eugenics, Biomedicine, and the Nature-Nurture Debate* (Albany: State University of New York Press, 1998).

51. On laboratory, observatory, and factory, see, for example, Peter Galison, *Image and Logic: A Material Culture of Microphysics* (Chicago: University of Chicago Press, 1997); Peter Galison and Caroline A. Jones, "Laboratory, Factory, Studio: Dispersing Sites of Production," in Peter Galison and Emily Thompson (eds.), *The Architecture of Science* (Cambridge, MA: MIT Press, 1999); Simon Schaffer, "Astronomers Mark Time: Discipline and the Personal Equation," *Science in Context* 2 (1988), pp. 115–45.

52. Edward Pickering, preface to *Annals of the Astronomical Observatory of Harvard College* 91 (1918), pp. iii–iv.

53. On women astronomical workers and their often low-paid status within the observatory, see, for example, Margaret Rossiter, *Women Scientists in America: Struggles and Strategies to 1940* (Baltimore: Johns Hopkins University Press, 1982), and Londa Schiebinger, *The Mind Has No Sex? Women in the Origins of Modern Science* (Cambridge, MA: Harvard University Press, 1989); on Harvard College Observatory, see Pamela E. Mack, "Straying from Their Orbits: Women in Astronomy in America," G. Kass-Simon, P. Farnes, and D. Nash (eds.), *Women of Science: Righting the Record* (Bloomington: Indiana University Press, 1990), pp. 72–116. On women scanners in European and American high-energy physics laboratories, see Peter Galison, *Image and Logic: A Material Culture of Microphysics* (Chicago: University of Chicago Press, 1997), esp. chs. 3 and 5.

54. On the management and "correction" of workers by use of the personal equation, see Simon Schaffer, "Astronomers Mark Time: Discipline and the Personal Equation," *Science in Context* 2 (1988), pp. 115–45, and "Babbage's Intelligence: Calculating Engines and the Factory System," *Critical Inquiry* 21 (1994), pp. 201–327. On the notion of the "unskilled" generally and, in particular, about the mostly female workers who reduced the nuclear emulsion photographs in the 1940s and

1950s, see Peter Galison, *Image and Logic: A Material Culture of Microphysics* (Chicago: University of Chicago Press, 1997).

55. Owen Gingerich, "Cannon, Annie Jump," in Charles C. Gillispie (ed.), *Dictionary of Scientific Biography* (New York: Charles Scribner's Sons, 1970– 1980), vol. 3, p. 50. Cecilia Payne-Gaposchkin, *Cecilia Payne-Gaposchkin: An Autobiography and Other Recollections*, ed. Katherine Haramundanis (Cambridge: Cambridge University Press, 1984).

56. Theodore E. Keats, *An Atlas of Normal Roentgen Variants That May Simulate Disease* (Chicago: Year Book Medical Publishers, 1973), p. vii.

57. Rudolf Grashey, *Atlas typischer Röntgenbilder vom normalen Menschen* (Munich: Lehmann, 1939).

58. P.M.S. Blackett, foreword to G.D. Rochester and J.G. Wilson, *Cloud Chamber Photographs of the Cosmic Radiation* (New York: Academic Press, 1952), p. vii.

59. Ivan Baronofsky, *Atlas of Precautionary Measures in General Surgery* (Saint Louis: Mosby, 1968), p. x.

60. The problem of what we "see" raises a fundamental and exceedingly subtle point. Wittgenstein distinguishes between judgments that in some way lie so close to our perceptions that the insertion of a distance between what we perceive and what we judge is absurd. Can one really say "I *know* I am in pain" without making a joke? Ludwig Wittgenstein, *Philosophical Investigations,* trans. G.E.M. Anscomb (New York: Macmillan, 1958), p. 89e (proposition 246). This is to be contrasted with the expert judgment, as in algebraic extensions of a series, for example. This distinction between a kind of basic and an expert judgment lies deep in Wittgenstein, as Stanley Cavell has shown in *The Claim of Reason: Wittgenstein, Skepticism, Morality, and Tragedy* (Oxford: Clarendon, 1979), esp. chs. 3 and 4. But in the case of scientific figuration, the radical splitting between "seeing" and "seeing as" leads, we believe, to much confusion.

61. John L. Madden, *Atlas of Technics in Surgery* (New York: Appleton-Century-Crofts, 1958), p. xi.

62. Johannes Sobotta, *Atlas and Textbook of Human Anatomy*, ed. J. Playfair McMurrich (Philadelphia: Saunders, 1909), p. 13.

63. E.A. Zimmer, *Technique and Results of Fluoroscopy of the Chest* (Springfield, IL: Thomas, 1954), pp. v–vi.

64. V.A. Firsoff, "Introduction," *Moon Atlas* (London: Hutchinson, 1961), p. 7. On Firsoff, see the timeline at http://www.space.edu/moon/timeline/ timeline1970–89.html.

65. See Peter Galison, *Image and Logic: A Material Culture of Microphysics* (Chicago: University of Chicago Press, 1997), for the contrast between technologies that are homomorphic (retaining the *form* of that which is represented) and those that are homologous (retaining logical relations in that which is represented).

66. Robert Howard, Václav Bumba, and Sara F. Smith., *Atlas of Solar Magnetic*

Fields, August 1959 – June 1966 (Washington, D.C.: Carnegie Institute, 1967), pp. 1–2, emphasis added.

67. Gerhart S. Schwarz and Charles R. Golthamer, *Radiographic Atlas of the Human Skull: Normal Variants and Pseudo-Lesions* (New York: Hafner, 1965), no page number; on Golthamer see p. viii; on Schwarz and Grashey see p. ix.

68. Gerhart S. Schwarz and Charles R. Golthamer, *Radiographic Atlas of the Human Skull: Normal Variants and Pseudo-Lesions* (New York: Hafner, 1965), p. xi.

69. Gerhart S. Schwarz and Charles R. Golthamer, *Radiographic Atlas of the Human Skull: Normal Variants and Pseudo-Lesions* (New York: Hafner, 1965), n.p.

70. Gerhart S. Schwarz and Charles R. Golthamer, *Radiographic Atlas of the Human Skull: Normal Variants and Pseudo-Lesions* (New York: Hafner, 1965), p. x (citation); p. xi on medical artists.

71. Gerhart S. Schwarz and Charles R. Golthamer, *Radiographic Atlas of the Human Skull: Normal Variants and Pseudo-Lesions* (New York: Hafner, 1965), p. xi.

72. Gerhart S. Schwarz and Charles R. Golthamer, *Radiographic Atlas of the Human Skull: Normal Variants and Pseudo-Lesions* (New York: Hafner, 1965), p. x.

73. Henri Poincaré, *The Value of Science: Essential Writings of Henri Poincaré* (New York: Modern Library, 2001), p. 99.

74. Henri Poincaré, *The Value of Science: Essential Writings of Henri Poincaré* (New York: Modern Library, 2001), pp. 203 and 208–209.

75. Jacques Hadamard, *An Essay on the Psychology of Invention in the Mathematical Field* (New York: Dover, 1954), p. 40 and chs. 2 and 3.

76. Peter Galison, "Image of Self," in Lorraine Daston (ed.), *Things That Talk: Object Lessons from Art and Science* (New York: Zone Books, 2004), pp. 257–94.

77. Peter Galison, "Image of Self," in Lorraine Daston (ed.), *Things That Talk: Object Lessons from Art and Science* (New York: Zone Books, 2004), pp. 257–94.

CHAPTER SEVEN: REPRESENTATION TO PRESENTATION

1. For twentieth-century cases in which the exercise of scientific judgment was seen as trained judgment, see Gerald Holton, "Subelectrons, Presuppositions, and the Millikan-Ehrenhaft Dispute," *Historical Studies in the Physical Sciences* 9 (1978), pp. 161–224; for cases in which it was seen as tantamount to fraud, see Allan D. Franklin, "Millikan's Published and Unpublished Data on Oil Drops," *Historical Studies in the Physical Sciences* 11 (1981), pp. 185–201. On how fraud and real-world science cross in volatile ways, see Daniel J. Kevles, *The Baltimore Case: A Trial of Politics, Science, and Character* (New York: Norton, 1998), pp. 19–95 and 364–88.

2. Recent influential contributions and reviews of this debate include David Michael Levin (ed.), *Modernity and the Hegemony of Vision* (Berkeley: University of California Press, 1993); Martin Jay, *Downcast Eyes: The Denigration of Vision in Twentieth-Century French Thought* (Berkeley: University of California Press, 1993); and Teresa Brennan and Martin Jay (eds.), *Vision in Context: Historical and Contemporary*

Perspectives on Sight (New York: Routledge, 1996).

3. Art historians have pioneered this approach: see especially Michael Baxandall, *Painting and Experience in Fifteenth-Century Italy: A Primer in the Social History of Pictorial Style*, 2nd ed. (Oxford: Oxford University Press, 1988); and Jonathan Crary, *Techniques of the Observer: On Vision and Modernity in the Nineteenth Century* (Cambridge, MA: MIT Press, 1990), and *Suspensions of Perception: Attention, Spectacle, and Modern Culture* (Cambridge, MA: MIT Press, 1999).

4. Isaiah Berlin, *The Crooked Timber of Humanity: Chapters in the History of Ideas*, ed. Henry Hardy (London: John Murray, 1990); and Stuart Hampshire, *Justice Is Conflict* (London: Duckworth, 1999).

5. See, for example, Bernard Williams, *Descartes: The Project of Pure Enquiry* (Hassocks, England: Harvester Press, 1978), pp. 69–70; Thomas Nagel, *The View from Nowhere* (Oxford: Oxford University Press, 1986), p. 15; Karsten Harries, "Descartes, Perspective, and the Angelic Eye," *Yale French Studies* 49 (1973), pp. 28–42; and Susan R. Bordo, *The Flight to Objectivity: Essays on Cartesianism and Culture* (Albany: State University of New York Press, 1987).

6. Plato, *The Symposium*, trans. Walter Hamilton (Harmondsworth, England: Penguin, 1951), especially p. 108, 221a. The literature on this subject is vast, but see the remarkable essay by Pierre Hadot, "The Figure of Socrates," *Philosophy as a Way of Life*, ed. Arnold I. Davidson, trans. Michael Chase (Malden, MA: Blackwell, 1995), pp. 147–78.

7. Albert Einstein, "Autobiographical Notes," in Paul Arthur Schilpp (ed.), *Albert Einstein: Philosopher-Scientist* (La Salle, IL: Open Court, 1970), vol. 1, pp. 4–7.

8. Michael Polanyi, *Personal Knowledge: Towards a Post-Critical Philosophy* (Chicago: University of Chicago Press, 1958); David Bloor, *Wittgenstein: A Social Theory of Knowledge* (London: Macmillan, 1983); H.M. Collins, *Changing Order: Replication and Induction in Scientific Practice* (London: Sage Publications, 1985). For a more historicized view, see H. Otto Sibum, "Wissen aus erster Hand: Mikro-Dynamik wissenschaftlichen Wandels im frühviktorianischen England," *Historische Anthropologie* 13 (2005), pp. 301–24.

9. Cf. the discussion on "values" versus "rules" in Thomas S. Kuhn, "Objectivity, Value Judgment, and Theory Choice," *The Essential Tension: Selected Studies in Scientific Tradition and Change* (Chicago: University of Chicago Press, 1977), pp. 320–39, esp. pp. 330–33.

10. Raymond Williams, "Culture," *Keywords: A Vocabulary of Culture and Society*, rev. ed. (New York: Oxford University Press, 1985), pp. 87–93. On the several meanings of objectivity in philosophy, see Elisabeth A. Lloyd, "Objectivity and the Double Standard for Feminist Epistemologies," *Synthese* 104 (1995), pp. 351–81; Heather Douglas, "The Irreducible Complexity of Objectivity," *Synthese* 138 (2004), pp. 453–73; and Joseph F. Hanna, "The Scope and Limits of Scientific Objectivity," *Philosophy of Science* 71 (2004), pp. 339–61. Bruno Latour explores

how graphs and figures can serve to join and fix relations of human and nonhuman actors; he analyzes scientific images as crucial in the gradual constitution of objectivity as an assembly of "faithful allies." See Bruno Latour, "Visualization and Cognition: Thinking with Eyes and Hands," *Knowledge and Society* 6 (1986), pp. 1–40.

11. Peter Novick, *That Noble Dream: The "Objectivity Question" and the American Historical Profession* (Cambridge: Cambridge University Press, 1988); Julie Robin Solomon, *Objectivity in the Making: Francis Bacon and the Politics of Inquiry* (Baltimore: Johns Hopkins University Press, 1998).

12. Richard Rorty, "Solidarity or Objectivity?" *Objectivity, Relativism, and Truth* (Cambridge: Cambridge University Press, 1991), p. 21.

13. Nicholas Rescher, *Objectivity: The Obligations of Impersonal Reason* (Notre Dame, IN: University of Notre Dame Press, 1997), p. 6.

14. Helen E. Longino, *Science as Social Knowledge: Values and Objectivity in Scientific Inquiry* (Princeton, NJ: Princeton University Press, 1990), p. 74.

15. Robert Nozick, *Invariances: The Structure of the Objective World* (Cambridge, MA: Belknap Press, University Press, 2001), p. 96.

16. Crispin Wright, *Saving the Differences: Essay on Themes from Truth and Objectivity* (Cambridge, MA: Harvard University Press, 2003), p. 16. Wright calls this a "*mechanical* view" of how semantics functions in a given context: "To be sure, the semantics of the language depends on institution; it is we who built the machine. But, once built, it runs by itself."

17. R.W. Newell, *Objectivity, Empiricism, and Truth* (London: Routledge and Kegan Paul, 1986), pp. 16–17.

18. On nanotechnology viewed from a cultural perspective (that is, the relationship between nanoscience and art, literature, science fiction), see N. Katherine Hayles (ed.), *Nanoculture: Implications of the New Technoscience* (Portland, OR: Intellect Books, 2004); for historical essays, see Davis Baird, Alfred Nordmann, and Joachim Schummer (eds.), *Discovering the Nanoscale* (Amsterdam: IOS Press, 2004).

19. Heung Sik Kang *et al.*, "The Visible Man: Three-Dimensional Interactive Musculo-Skeletal Anatomic Atlas of the Lower Extremity," *Radiographics* 20 (2000), pp. 279–86; available online at http://radiographics.rsnajnls.org/ cgi/reprint/20/ 1/279.

20. "A New Galaxy Atlas — Sloan Digital Sky Survey Findings Comprise New Compendium of Galaxy Families," Sloan Digital Sky Survey, Jan. 11, 2005, http://www.sdss.org/news/releases/20050111.atlas.html.

21. "Meteorites," http://labs.sci.qut.edu.au/minerals/meteorites/meteorites1.htm.

22. "A New Galaxy Atlas — Sloan Digital Sky Survey Findings Comprise New Compendium of Galaxy Families," Sloan Digital Sky Survey, Jan. 11, 2005, http:// www.sdss.org/ news/releases/20050111.atlas.html.

23. Ian Hacking, *Representing and Intervening: Introducing Topics in the Philosophy of Natural Science* (Cambridge: Cambridge University Press, 1983), quotation

from p. 146; "So far as I'm concerned, if you can spray them then they are real," p. 23.

24. National Science and Technology Council, Committee on Technology, and the Interagency Working Group on NanoScience, Engineering and Technology, "Nanostructure Science and Technology: A Worldwide Study," 1999, www.wtec.org/loyola/nano/.

25. "UMass Amherst To Offer New Ph.D. Training in Nanotechnology," Nanotechwire, July 12, 2005, http://nanotechwire.com/news.asp?nid=2132.

26. Friedrich Kohlrausch, "Antrittsrede," *Sitzungsberichte der Preußischen Akademie der Wissenschaften* 33 (1896), p. 743.

27. http://www.asylumresearch.com/Applications/MicroAngelo/MicroAngelo.shtml.

28. Frank Fehrenbach, *Licht und Wasser: Zur Dynamik naturphilosophischer Leitbilder im Werk Leonardo da Vincis* (Tübingen: Wasmuth, 1997).

29. Milton Van Dyke, *An Album of Fluid Motion* (Stanford, CA: Parabolic Press, 1982), p. 6.

30. By the early 2000s, Van Dyke's book was very widely used in courses on topics from fluid dynamics to astronomy at, for example, the University of California at Berkeley, (http://astron.berkeley.edu/~jrg/ay202/lectures.html), Johns Hopkins University (http://pegasus.me.jhu.edu/~meneveau/courses/ gradfluids 1-04/Syllabus.pdf), and the University of Minnesota (http:// www.aem.umn.edu/teaching/curriculum/syllabi/UGrad/AEM_4201_ syllabus.shtml).

31. M. Samimy, K.S. Breuer, L.G. Leal, and P.H. Steen (eds.), *A Gallery of Fluid Motion* (Cambridge: Cambridge University Press, 2003), p. ix. The editors note that a "major contributory factor" in their selection was that earlier images were superseded by later ones with "color imaging or better visualization methods" (p. x).

32. Marie Farge, "Choix des palettes de couleur pour la visualisation de résultats d'expériences numériques," *Couleur, design et communication* (n.p.: Centre Français de la Couleur and IFEC, 1988), p. 33.

33. Marie Farge, "A Proposition of Normalization for High-Resolution Raster Display Applied to Turbulent Fields," unpublished paper (1985), p. 11.

34. Marie Farge, "Choix des palettes de couleur pour la visualisation de résultats d'expériences numériques," *Couleur, design et communication* (n.p.: Centre Français de la Couleur and IFEC, 1988), p. 27.

35. Jean-François Colonna and Marie Farge, *Science pour l'art* (Paris: Ecole Polytechnique, 1994).

36. Marie Farge, interview with Peter Galison, Sept. 15, 2005.

37. Peter Weiss, "An Electron Runs Through It: Surprising Rivulets and Ripples Complicate the Microchip Picture," *Science News*, Dec. 18, 2004, p. 394.

38. "About the Art," Resonance Fine Art, http://www.ericjhellergallery. com/index.pl?page=aboutart.

Index

Abbe, Ernst, 271,
Academia Naturae Curiosorum (later
Leopoldina), 67.
Académie des Beaux-Arts, 130.
Académie Royale de Peinture et de
Sculpture, 100.
Académie Royale des Sciences (later
Académie des Sciences), 37, 64, 67, 89,
91, 130, 327.
"Account of a Method of Observing the
Wonderful Configurations of the
Smallest Shining Particles of Snow, with
Several Figures of Them, An" (Nettis),
149.
*Account of the Arctic Regions with a History
and Description of the Northern Whale-
Fishery, An* (Scoresby), *152.*
Achelis, Thomas, 267.
Adams, Henry, 212, 213.
Advice to a Young Investigator (Ramón y
Cajal), 215.
Aesthetics, 70, 75, 77, 102, 120, 412;
identical objects and, 139; Kantian, 206,
265; in natural forms, 247, *249;*
photography and, 133; right depiction
and, 413; self-surveillance against, 174,
184, 187. *See also* Art/artists.
Africa, 57, 281.
Agassiz, Alexander, 126, 129.
Agassiz, George R., 180.
Agassiz, Louis, 129.

Alberti, Leon Battista, 73, 390.
Albertype, 129.
Albinus, Bernhard Siegfried, 89, 95, 125,
168, 247, 347, 367; art and science in
work of, 402; idealizing classicism of,
102; scientific drawings and, 194;
scientific self and, 196; *Tables of the
Skeleton*, 70, *72*, 73–75; on "true"
representation, 315.
Album of Fluid Motion, An (Van Dyke),
402–403, *404*, 481 n.30.
Alchemists, 39, 41.
Alembert, Jean d', 81, 97, 200, 211.
Algorithms, 330–35, 377, 390.
Alienation, 223.
"Allegorical Monument to Sir Isaac Newton,
An" (Pittoni the Younger), *218.*
*Allgemeine Functionentheorie, Die [General
Theory of Functions]* (Du Bois-Reymond),
268.
Alvarez, Luis, 330, 331.
American Physical Society, 403.
Ampère, André-Marie, 286, 288.
Anatomie menschlicher Embryonen (His), 193,
194.
*Anatomie pathologique du corps humain
[Pathological Anatomy of the Human Body]*
(Cruveilhier), 66, *83, 108.*
Anatomists, 43, 75, 77; artists and, 146;
"canonical" body and, 81; drawing
schools and, 12; mechanical objectivity

483

484

Zone Books series design by Bruce Mau
Typesetting by Archetype
Image placement and production by Julie Fry
Printed and bound by Thomson-Shore, Inc.